Research Topics in Wind Energy

Volume 9

The series Research Topics in Wind Energy publishes new developments and advances in the fields of Wind Energy Research and Technology, rapidly and informally but with a high quality. Wind Energy is a new emerging research field characterized by a high degree of interdisciplinarity. The intent is to cover all the technical contents, applications, and multidisciplinary aspects of Wind Energy, embedded in the fields of Mechanical and Electrical Engineering, Physics, Turbulence, Energy Technology, Control, Meteorology and Long-Term Wind Forecasts, Wind Turbine Technology, System Integration and Energy Economics, as well as the methodologies behind them. Within the scope of the series are monographs, lecture notes, selected contributions from specialized conferences and workshops, as well as selected PhD theses. Of particular value to both the contributors and the readership are the short publication time frame and the worldwide distribution, which enable both wide and rapid dissemination of research output. The series is promoted under the auspices of the European Academy of Wind Energy.

Indexed by SCOPUS, SCImago.

Daniele Rocchi · Giorgio Diana · Tommaso Argentini · Federico Zanelli · Stefano Giappino

Wind Engineering

Fundamentals and Applications

Springer

Daniele Rocchi
Department of Mechanical Engineering
Politecnico di Milano
Milan, Italy

Giorgio Diana
Department of Mechanical Engineering
Politecnico di Milano
Milan, Italy

Tommaso Argentini
Department of Mechanical Engineering
Politecnico di Milano
Milan, Italy

Federico Zanelli
Department of Mechanical Engineering
Politecnico di Milano
Milan, Italy

Stefano Giappino
Department of Mechanical Engineering
Politecnico di Milano
Milan, Italy

ISSN 2196-7806 ISSN 2196-7814 (electronic)
Research Topics in Wind Energy
ISBN 978-3-032-17422-2 ISBN 978-3-032-17423-9 (eBook)
https://doi.org/10.1007/978-3-032-17423-9

This Springer imprint is published by the registered company Springer Nature Switzerland AG
The registered company address is: Gewerbestrasse 11, 6330 Cham, Switzerland

Preface

Since 2010, Ph.D. and Master of Science students at Politecnico di Milano have had the opportunity to take a 6 ECTS course on Wind Engineering. This course was established due to the dedication and expertise of several professors and researchers who had been conducting experimental and numerical research on wind engineering topics in the preceding years.

A significant factor contributing to the development of world-class research in wind engineering at Politecnico di Milano is the synergic approach used to study various topics. This involves performing experimental analyses in the wind tunnel and developing numerical models. The wind tunnel facility, built in 2001, is one of the largest in the world, specifically designed for wind engineering applications. This facility has enabled the research group to collaborate with the international wind engineering community, enhancing their experimental and numerical modeling skills.

Originally presented as slides, the course content was later transformed into lecture notes and is now presented as a comprehensive book. The motivation behind creating the wind engineering course and writing this book is to impart the core knowledge of the research group, preparing a new generation of wind engineers. Such professionals are in high demand in the international marketplace due to the increasing frequency and complexity of wind engineering problems.

The book's structure reflects this aim, beginning with introductory chapters covering fundamental aspects of wind engineering, followed by specific research topics developed at Politecnico di Milano over the last decades:

- Chapter 1: Basics of wind climate physics, understanding wind genesis and main windy phenomena.
- Chapter 2: Engineering modeling of the atmospheric boundary layer and its statistical characteristics, including normative considerations.
- Chapter 3: Experimental wind tunnel techniques, scaling laws, aeroelastic and rigid replica models, and instrumentation for models and wind flow. The subsequent chapters delve into specific research topics unique to this book.

- Chapter 4: Wind-induced vibrations on conductors of High Voltage Transmission Lines, including Aeolian Vibrations, Sub-span Oscillations, and Ice Galloping.
- Chapter 5: Wind effects on long span bridges, covering aerodynamic static and dynamic (aeroelastic) problems with analytical methods and wind tunnel tests.

For each of these topics, the authors aimed to create a reference text to help readers gain a fundamental understanding of the involved phenomena and grasp the experimental and numerical engineering tools commonly used during the design phase of projects. In conclusion, the authors express their gratitude to all the individuals who contributed to the creation of this book through their comments, suggestions, ideas, and support to the research activities. This includes colleagues, researchers, professors, and both Ph.D. and Master students.

Suggestions for Instructors

The book is intended to be a reference textbook for the wind engineering Master and Ph.D. level for mechanical, civil and aerospace engineers. It is assumed that students already have a basic knowledge of mechanical vibrations and fluid mechanics. Each chapter of the book is stand-alone and can be read independently of the content of the other chapters.

Milan, Italy
October 2025

Giorgio Diana
Daniele Rocchi
Tommaso Argentini
Federico Zanelli
Stefano Giappino

Competing Interests The authors have no competing interests to declare that are relevant to the content of this manuscript.

Contents

Chapter 1
Wind Climate

1.1 The Atmosphere

1.1.1 Definition and Basic Properties

The Earth's atmosphere is an extensive mixture of gases and suspended particles that forms a thin shell around the planet. Although it extends for hundreds of kilometres above the surface, about 99.99997% of its total mass is found below an altitude of 100 km. Figure 1.1 shows the atmosphere as seen from the International Space Station, appearing as a faint luminous layer enveloping the Earth.

The variation of temperature with altitude under average conditions, shown in Fig. 1.2, serves as the basis for dividing the atmosphere into four principal layers: the *troposphere*, *stratosphere*, *mesosphere*, and *thermosphere*. The upper boundaries of these layers are referred to collectively as *pauses* (e.g. tropopause, stratopause).

The lowest region, the *troposphere* (from the Greek *tropos*, meaning "turning" or "mixing"), extends from the ground up to roughly 12–16 km, depending on latitude and season. It contains more than 80% of the total atmospheric mass and nearly all of its water vapour, clouds, and precipitation; in other words, almost all weather phenomena occur here.

Air properties within the troposphere vary markedly with both height and latitude. Table 1.1 lists the standard air properties at sea level, while Fig. 1.3 illustrates how pressure, temperature, and density decrease with altitude within the lowest 2 km of the standard atmosphere.

The rate at which temperature decreases with altitude is known as the *lapse rate* (Γ). Its mean value in the troposphere is about 6.5 °C per kilometre,[1] though it can vary in space and time. Occasionally, thin layers form in which temperature increases with height (i.e. negative lapse rate). These *temperature inversions* tend to suppress vertical mixing and strongly influence the dispersion of pollutants and

[1] The dry adiabatic lapse rate, representing the theoretical upper limit for unsaturated air, is approximately 10 °C/km.

D. Rocchi et al., *Wind Engineering*, Research Topics in Wind Energy 9,
https://doi.org/10.1007/978-3-032-17423-9_1

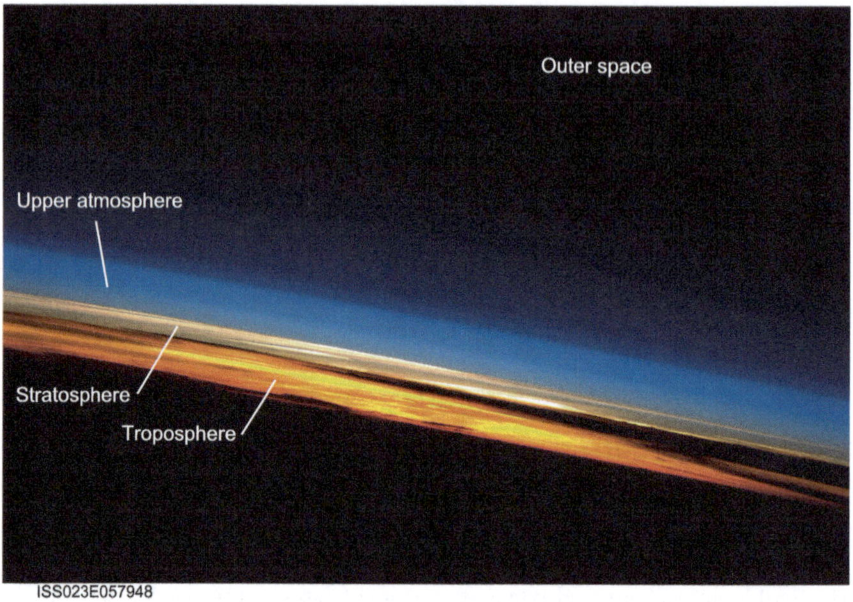

Fig. 1.1 Sunset seen from the International Space Station (NASA, Mission ISS023, https://eol.jsc.nasa.gov/SearchPhotos/photo.pl?mission=ISS023&roll=E&frame=57948)" The image presents an edge-on, or limb view, of the Earth's atmosphere as seen from orbit. The Earth's curvature is visible along the horizon line, that extends across the image from center left to lower right. Above the darkened surface of the Earth, a brilliant sequence of colors roughly denotes several layers of the atmosphere. Deep oranges and yellows appear in the troposphere. The pink to white region above the clouds appears to be the lower stratosphere. Above the stratosphere, blue layers likely mark the transition between the middle and upper atmosphere as it gradually fades into the blackness of outer space"

moisture near the ground. For reference, in the standard atmosphere over the first 2000 m, temperature decreases from 15 °C to about 2 °C, pressure from 1013 hPa to 795 hPa, and air density from 1.22 to roughly 1.00 kg/m^3.

Above the troposphere lies the *stratosphere* (from *strato*, meaning "layered"), where vertical mixing is again inhibited because temperature increases with altitude. This temperature rise is mainly caused by the absorption of ultraviolet radiation by ozone molecules. The resulting heat input is responsible for the temperature maximum at about 50 km, defining the *stratopause*. The ozone concentration in this layer is crucial for life on Earth, as it filters out biologically harmful ultraviolet radiation.

Above the ozone layer extends the *mesosphere* (from *meso*, meaning "middle"), where temperature again decreases with height, reaching a minimum at the *mesopause*. The overlying *thermosphere* shows a marked increase in temperature with altitude due to the absorption of high-energy solar radiation, which dissociates nitrogen and oxygen molecules and ionizes atoms. Temperatures in the outer ther-

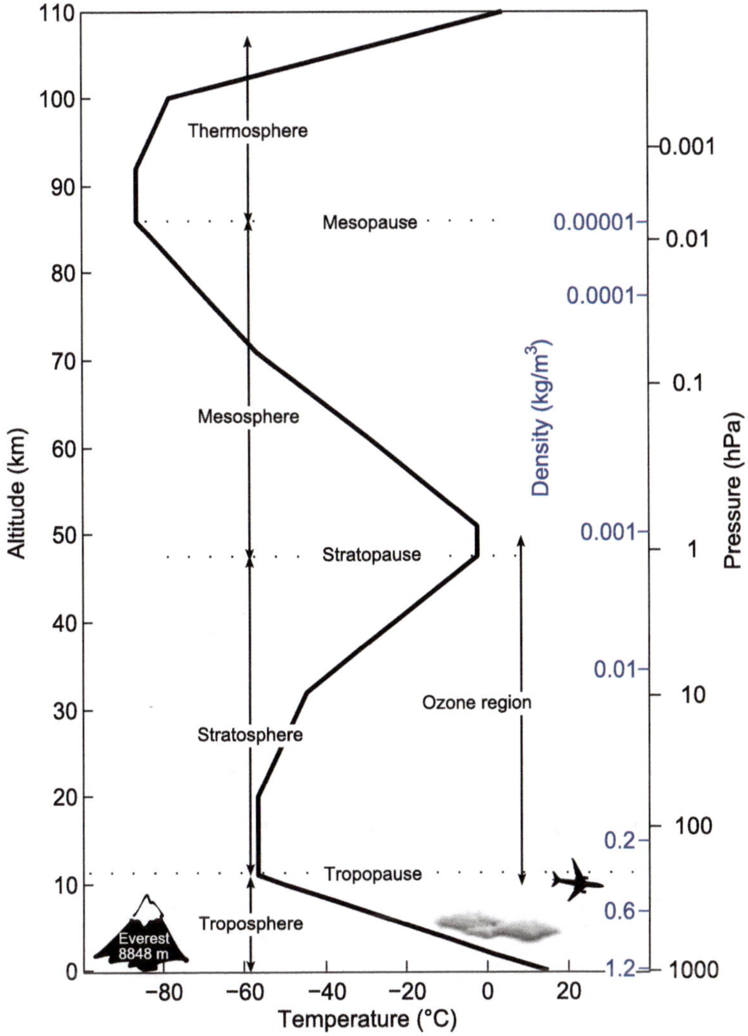

Fig. 1.2 Vertical structure of the atmosphere: temperature layers

mosphere vary greatly depending on solar activity and the emission of ultraviolet and X-ray radiation from the Sun's outer atmosphere.

1.1.2 Scales of Atmospheric Motion

Atmospheric motions responsible for wind generation and weather phenomena span an enormous range of spatial and temporal scales. Each scale can be characterized by

Table 1.1 Properties of Standard atmosphere at sea level (International Organization for Standardization, Standard Atmosphere, ISO 2533:1975, 1975)

Property	Value
Temperature, T_0	288.15 K (15 °C)
Pressure, p_0	1013.25 hPa
Density, ρ_0	1.225 kg/m^3
Viscosity, μ	$1.789 \cdot 10^{-5}$ Ns/m^2

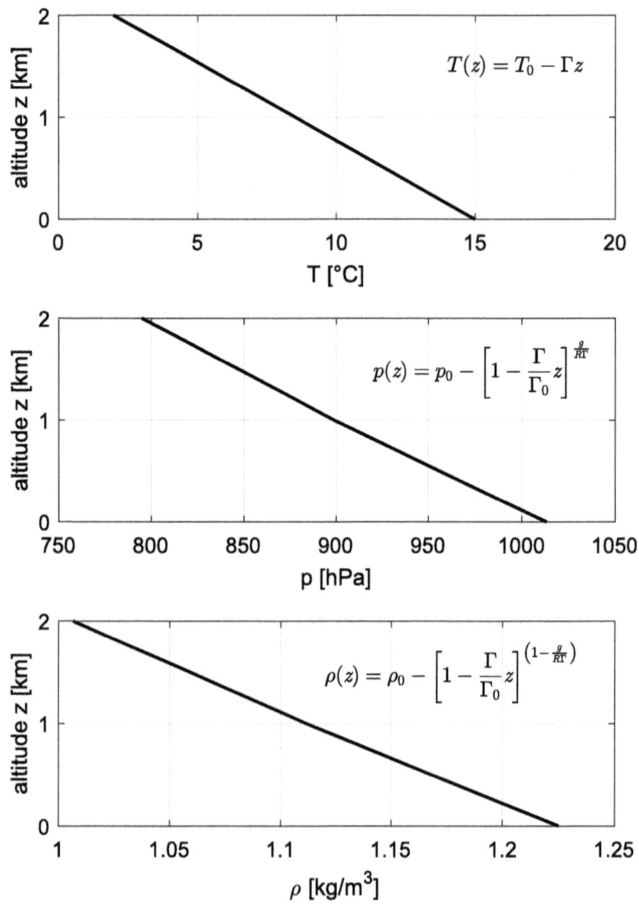

Fig. 1.3 Vertical profiles of temperature, pressure, and density for the standard atmosphere near ground ($\Gamma = 6.5$°C/km, $R = 287$ J/(K ·kg), g= 9.8065 m/s^2)

its typical horizontal extent and duration. This concept is intuitive if one considers a few common examples: a windy afternoon, a travelling storm, or a tropical cyclone.

A windy day is usually accompanied by brief gusts that last only a few seconds and affect regions a few metres across-small-scale turbulent motions that shake leaves and branches. In contrast, a mid-latitude cyclone, such as those often displayed on weather charts, extends over roughly 1000 km and persists for several days. Still larger and longer-lasting are the global circulation patterns that steer hurricanes originating over tropical oceans and transport them westward toward continents, where they can produce severe damage near the surface.

These examples illustrate that longer time scales generally correspond to larger spatial scales. For this reason, meteorologists classify atmospheric processes primarily by their horizontal dimension, recognizing three broad categories: the *microscale*, *mesoscale*, and *macroscale*.

The study of fine-scale motions close to the surface–those with horizontal dimensions from a few metres to a few kilometres and lasting seconds to minutes–belongs to *micrometeorology*. At the opposite extreme, phenomena extending beyond about 1000 km and lasting several days, such as synoptic-scale pressure systems, are part of *macrometeorology*. Between these two domains lies the *mesoscale*, which includes convective storms, sea breezes, and local wind systems of tens to hundreds of kilometres in size and lasting from hours to about a day.

Figure 1.4 schematically represents these characteristic scales of motion, while Table 1.2 reports typical spatial and temporal ranges for representative atmospheric processes.

When we approach the study of meteorology, it is important to recognize that atmospheric motion occurs simultaneously at all spatial and temporal scales. Furthermore, interactions between these scales are both significant and often complex.

As illustrated in Fig. 1.4, energy is introduced into the atmosphere mainly at the macroscale. Through the so-called *energy cascade*, this energy is transferred progressively to smaller scales–from the planetary scale down to the microscale–where it is eventually dissipated, primarily through viscous friction and re-radiation to space.

1.1.3 Equations Governing Air Motion

The motion of the atmosphere is described by the fundamental physical laws that express the conservation of mass, momentum, and energy. In addition, because air is a mixture of gases, an equation of state is required to relate pressure, density, and temperature.

These governing principles can be summarized as follows (Holton 2004):

- **Momentum conservation (Newton's second law)**: The rate of change of momentum of a fluid parcel is equal to the sum of the forces acting upon it:

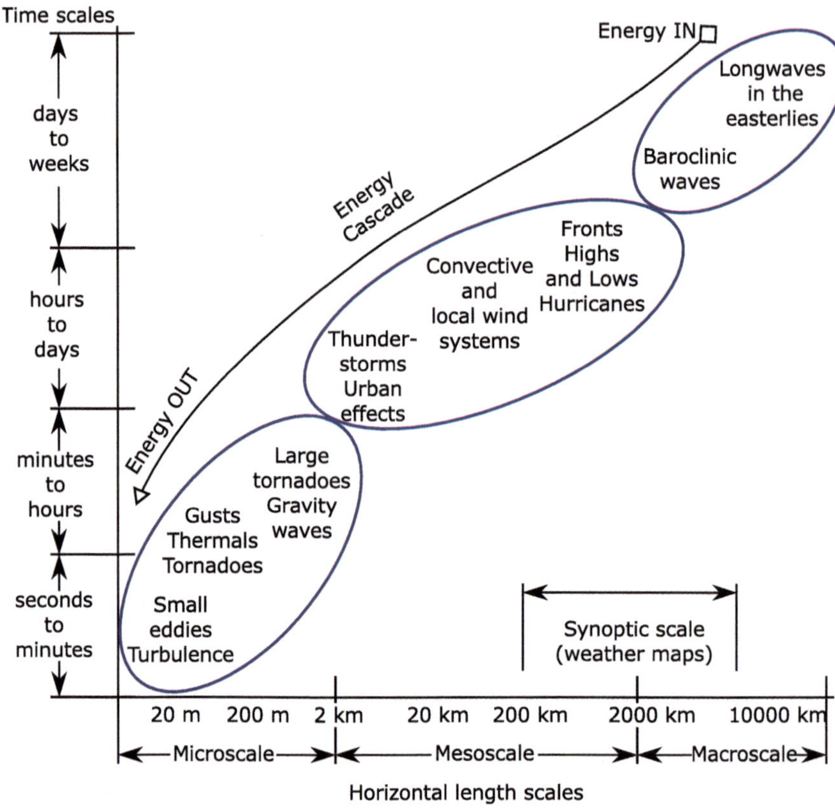

Fig. 1.4 Scales definitions and characteristic time and horizontal length scales of a variety of atmospheric processes (adapted from Orlanski 1975)

Table 1.2 Scales of atmospheric motion (Holton 2004)

Type of motion	Horizontal scale [m]
Molecular mean free path	10^{-7}
Minute turbulent eddies	10^{-2}–10^{-1}
Small eddies	10^{-1}–1
Dust devils	1 – 10
Gusts	10–10^{2}
Tornadoes	10^{2}
Cumulonimbus clouds	10^{3}
Fronts, squall lines	10^{4}–10^{5}
Hurricanes	10^{5}
Synoptic cyclones	10^{6}
Planetary waves	10^{7}

$$\rho \frac{D\mathbf{V}}{Dt} = -\rho \mathbf{g} - \nabla p + \operatorname{div} \boldsymbol{\tau} \tag{1.1}$$

where ρ is the air density, $\frac{D}{Dt}$ is the total (material) derivative with respect to time t, $\mathbf{V}$ is the velocity vector, $\mathbf{g}$ the gravitational acceleration vector, ∇p the pressure gradient, and $\boldsymbol{\tau}$ the stress tensor. The pressure gradient represents the primary driving mechanism for atmospheric motion: winds arise from horizontal differences in pressure, which in turn are established by thermodynamic processes governed by the equation of state and the first law of thermodynamics.

- **Equation of state**: For most atmospheric applications, air can be treated as an ideal (perfect) gas. The relationship between pressure, density, and temperature is therefore given by:

$$p = \rho R T \tag{1.2}$$

where $R = 287\ \mathrm{m^2\,s^{-2}\,K^{-1}}$ is the specific gas constant for dry air, and T is the absolute temperature (in kelvins).

- **Energy conservation (First law of thermodynamics)**: The rate of change of internal energy equals the sum of the heat added to, and the mechanical work done on, a fluid element:

$$\rho C_v \frac{DT}{Dt} = -p \operatorname{div} \mathbf{V} + \rho Q + \mu \chi \tag{1.3}$$

where $\rho C_v T$ represents the internal energy of the air parcel, $-p \operatorname{div} \mathbf{V}$ the rate of mechanical work, ρQ the rate of heat addition per unit volume (where Q may include conductive, latent, or radiative heat sources, e.g. $Q = \operatorname{div}(k\nabla T)$), and $\mu \chi$ the rate of viscous dissipation of kinetic energy into heat.

- **Additional relations**: To close the system of equations, further physical relationships must be included, such as the continuity equation (mass conservation), and, where relevant, chemical, radiative, and electrical equilibrium equations.

1.2 Wind as an Energy Vector

To understand why pressure gradients form, it is helpful to consider the atmosphere as a quasi-closed thermodynamic system that exchanges energy with both the Earth's surface and outer space.

Figure 1.5 illustrates a schematic representation of the Earth's mean annual energy budget, averaged spatially over the globe and temporally over several years. In the following, we briefly examine the main energy fluxes: the incoming solar radiation, the outgoing terrestrial radiation, and the resulting global energy balance

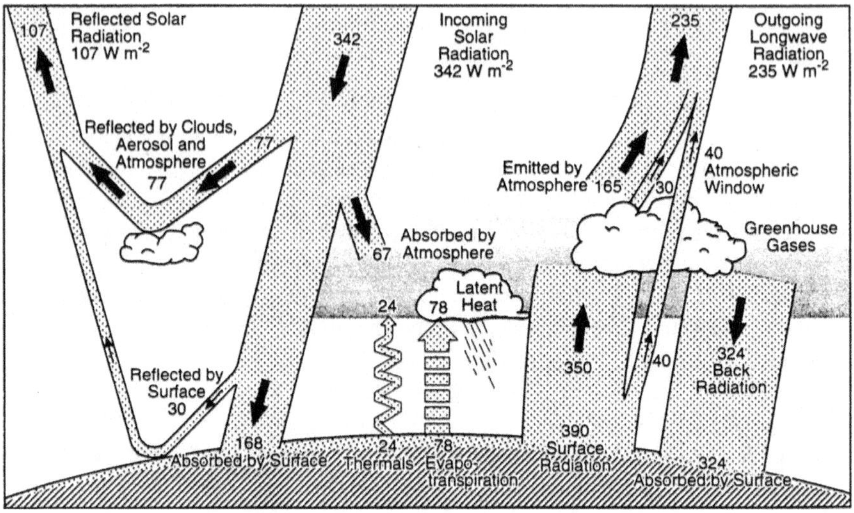

Fig. 1.5 Schematic representation of Earth's global mean energy balance (from J.T. Kiehl and K.E. Trenberth, 1997: "Earth's annual global energy budget", *Bull. Am. Meteorol. Soc.*, 78, 197–208).©American Meteorological Society. Used with permission

1.2.1 Incoming Solar Radiation

The Sun emits most of its radiation in the visible band (400–700 nm) and, to a lesser extent, in the ultraviolet (10–400 nm). When averaged globally and annually, the total solar irradiance reaching the top of the atmosphere is approximately 342 Wm^{-2}.

Of this, around 19% (67 Wm^{-2}) is absorbed directly by the atmosphere. Short-wavelength solar radiation, particularly in the ultraviolet range, is strongly absorbed by gases such as ozone in the stratosphere, which acts as an effective shield against biologically harmful radiation.

Approximately 49% (168 Wm^{-2}) of the incident solar energy is absorbed by the Earth's surface, while the remainder is reflected back to space: about 8% (30 Wm^{-2}) directly from the surface and about 23% (77 Wm^{-2}) by clouds, aerosols, and atmospheric gases. The ratio between the reflected and incident radiation is known as the *albedo*. On a planetary scale, the mean albedo is approximately 0.31, indicating that roughly one-third of incoming solar radiation is reflected, mainly by clouds.

1.2.2 Outgoing Terrestrial Radiation

The Earth's surface, atmosphere, and clouds emit energy predominantly in the infrared (IR) band. The mean upward longwave emission from the surface is about 390 Wm^{-2}. A large fraction of this energy, roughly 350 Wm^{-2}, is absorbed by atmospheric constituents such as water vapor, carbon dioxide, and ozone–collectively known as *greenhouse gases*. These gases absorb radiation selectively within specific wavelength intervals.

The absorbed infrared energy is then re-emitted in all directions, including downward toward the surface. The flux of longwave radiation returning to the surface is about 324 Wm^{-2}. This back-radiation plays a crucial role in maintaining the planet's mean surface temperature near 15 °C, whereas without an atmosphere the global mean temperature would be approximately -18 °C.

Only a small fraction, about 40 Wm^{-2}, of the surface-emitted radiation escapes directly to space through the so-called *atmospheric window*–a spectral region between 8 and 11 μm in which absorption is weak. Overall, the energy radiated to space as terrestrial longwave emission amounts to roughly 235 Wm^{-2}.

1.2.3 Global Energy Balance

When averaged globally and over long periods, the Earth-atmosphere system is approximately in radiative equilibrium. At the top of the atmosphere, the incoming solar flux (342 Wm^{-2}) is balanced by the sum of reflected shortwave (107 Wm^{-2}) and outgoing longwave radiation (235 Wm^{-2}), yielding a near-zero net flux:

$$342 - 107 - 235 \approx 0 \text{ Wm}^{-2}.$$

At the surface, the balance involves the absorbed shortwave flux (168 Wm^{-2}), the absorbed atmospheric back-radiation (324 Wm^{-2}), the emitted longwave flux (390 Wm^{-2}), and the upward turbulent heat fluxes associated with convection and evaporation (102 Wm^{-2}):

$$168 + 324 - 390 - 102 \approx 0 \ Wm^{-2}.$$

Hence, when globally averaged, the net radiative and turbulent energy exchanges are balanced. However, if we isolate the atmosphere from the surface, it exhibits a net radiative cooling. This cooling is offset by the sensible and latent heat supplied from the surface.

Spatial and temporal variations of radiative fluxes are also significant. Figure 1.6 shows the global distribution of net surface radiation for January and July (multi-year averages). Here, positive values correspond to a net gain of energy by the surface, while negative values indicate a net loss.

The pattern shows that the tropics and subtropics experience an excess of absorbed solar radiation, while middle and polar latitudes experience a deficit. This equator-to-pole gradient in radiative heating represents the fundamental driver of both atmospheric and oceanic circulations. On long timescales, when the global mean temperature is steady, this imbalance is compensated by the meridional transport of energy by winds and ocean currents.

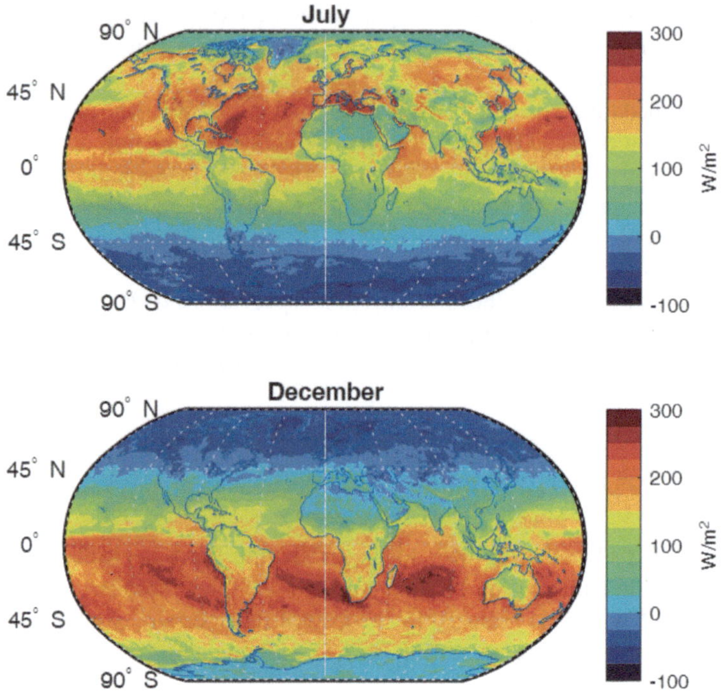

Fig. 1.6 Average net radiation at the Earth's surface in January and July. *Data source* NASA GEWEX surface radiation budget

1.3 Global Circulation

As outlined in the previous section, the large-scale circulation of the atmosphere arises from fundamental energetic imbalances between low and high latitudes. The earliest conceptual framework for this motion, known as the *single-cell model*, was formulated by George Hadley in 1735 based on simple thermodynamic arguments. With the advancement of meteorological science, particularly in the twentieth century, this model was refined into the *three-cell model*, which incorporates additional physical processes such as the Coriolis effect. Although still idealized, this model reproduces the main features of the observed global circulation with remarkable accuracy.

1.3.1 Single-Cell Model (Hadley, 1735)

A simplified representation of the atmospheric circulation can be developed under a set of idealized assumptions:

1. the Earth does not rotate;
2. the Earth's surface is uniform and homogeneous;
3. the differential heating between equator and poles creates a temperature gradient, with warm air at the equator and cold air at the poles.

In the schematic of Fig. 1.7, the upper diagram depicts an isobaric distribution decreasing with height from 1000 to 500 millibars, without any horizontal temperature or pressure gradient–hence, no motion occurs.

In the second diagram, the equatorial region receives greater solar input, heating the surface and warming the air above through conduction and convection. Consequently, the spacing between isobars increases aloft as the air expands and rises. Toward the poles, reduced solar input leads to cooling and denser air. This temperature contrast produces an upper-level pressure gradient, which drives air flow from the equator toward the poles.

In the third diagram, this process generates a complete circulation cell: air transported poleward aloft descends near the poles and returns equatorward at the surface, completing the loop.

The resulting motion, illustrated in Fig. 1.8, would form a single convective cell in each hemisphere. Warm air rises near the equator and moves poleward in the upper troposphere, while cold, dense air sinks at the poles and flows back toward the equator at the surface.

1.3.2 Three-Cell Model (1920s)

When the effects of Earth's rotation are introduced, the simple single-cell scheme must be replaced by a more complex structure. Due to the Coriolis effect, moving air is deflected to the right of its trajectory in the Northern Hemisphere and to the left in the Southern Hemisphere, as shown in Fig. 1.9. The strength of this deflection depends on both the wind velocity and latitude–it is zero at the equator and reaches a maximum at the poles.

Taking the Coriolis force into account, global circulation can be described by a three-cell system in each hemisphere (Fig. 1.10): the *Hadley cell* in the tropics, the *Ferrel cell* in midlatitudes, and the *polar cell* at high latitudes. Although simplified, this conceptual framework effectively captures the main features of planetary-scale motion.

- **Tropical (Hadley) cell**: Low-latitude air moves toward the equator, rises due to surface heating and convection, and diverges poleward aloft before subsiding near 30 ° latitude, forming the subtropical highs. This cell dominates tropical and subtropical climates.
- **Midlatitude (Ferrel) cell**: A secondary, thermally indirect circulation in which surface winds flow poleward and eastward, while upper-level air flows equatorward and westward.

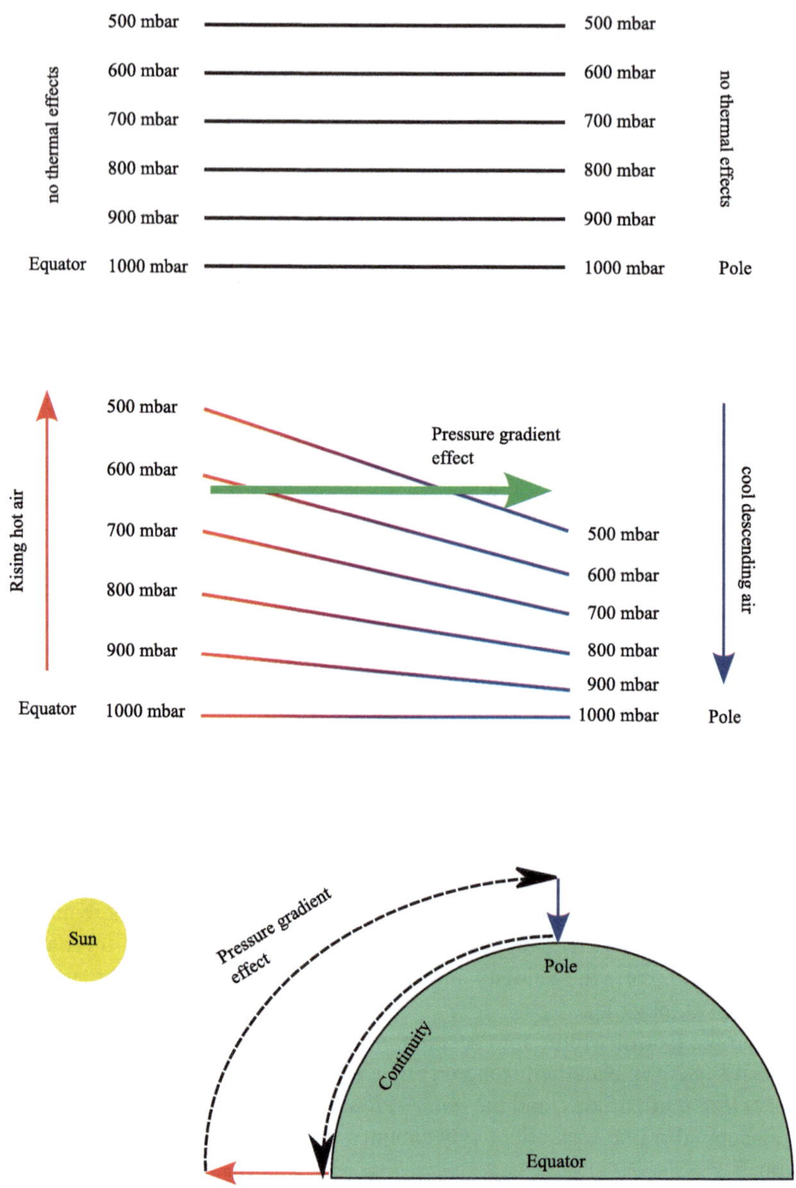

Fig. 1.7 Schematic mechanism leading to the single-cell model

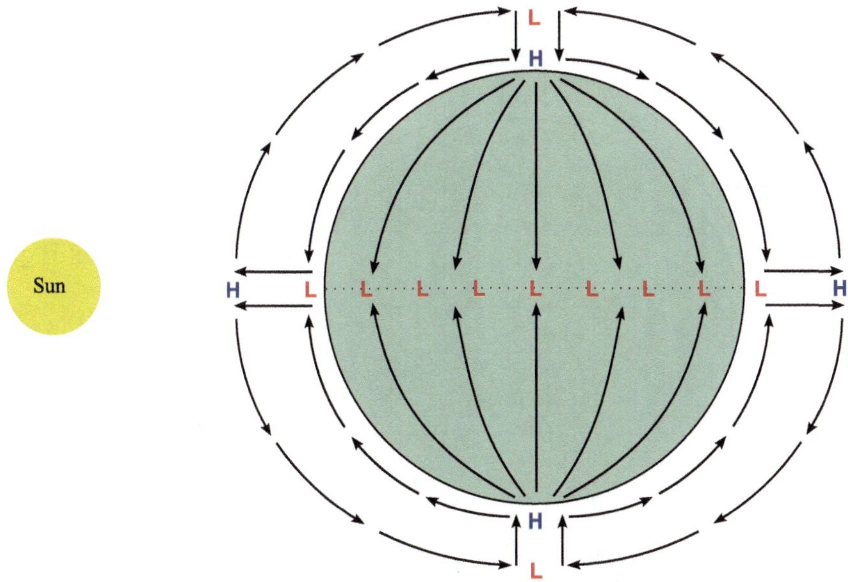

Fig. 1.8 Theoretical single-cell model of atmospheric circulation

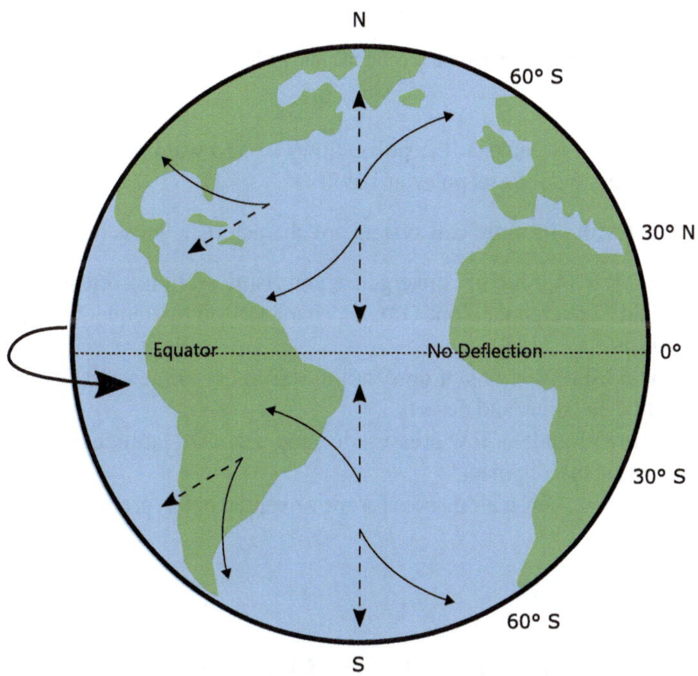

Fig. 1.9 Schematic representation of the Coriolis deflection on moving air masses

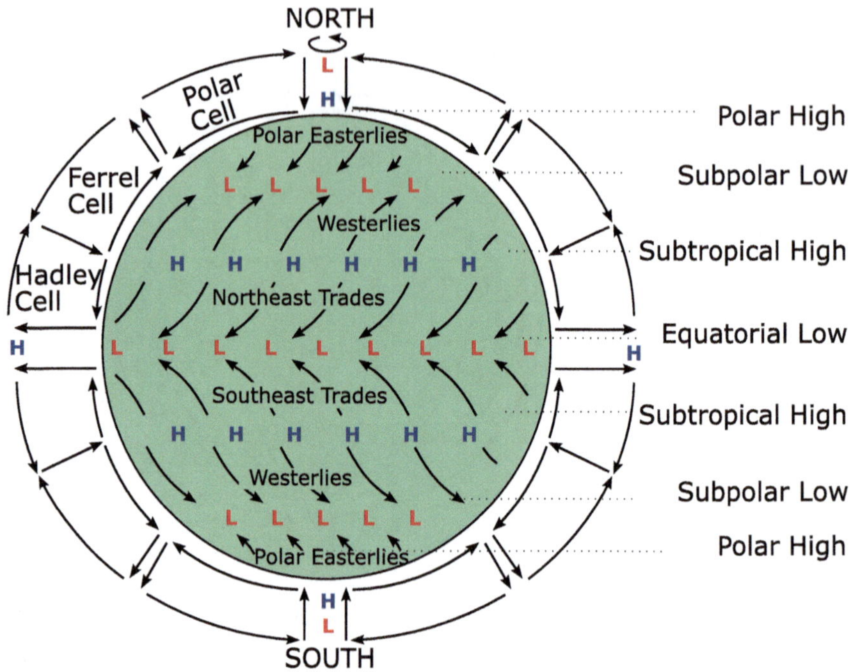

Fig. 1.10 Idealized three-cell atmospheric circulation model

- **Polar cell**: Cold air sinks at the poles, flows equatorward at the surface, rises at the polar front, and returns poleward aloft.

Associated with this three-cell system are characteristic pressure belts:

- **Equatorial low**: A zone of convergence and rising motion, corresponding to the Intertropical Convergence Zone (ITCZ), characterized by frequent cloudiness and precipitation.
- **Subtropical high**: Regions of descending warm, dry air near 30 ° latitude, associated with clear skies and deserts.
- **Subpolar low**: Bands of low pressure located near 60 ° latitude where polar and midlatitude air masses meet.
- **Polar high**: Cold, dense air masses forming over the poles, producing high surface pressure.

1.3.3 Global Circulation: Observed Wind and Pressure Belts

In reality, the global circulation is far more intricate due to Earth's rotation, axial tilt, and the unequal distribution of land and ocean between hemispheres. Figure 1.11

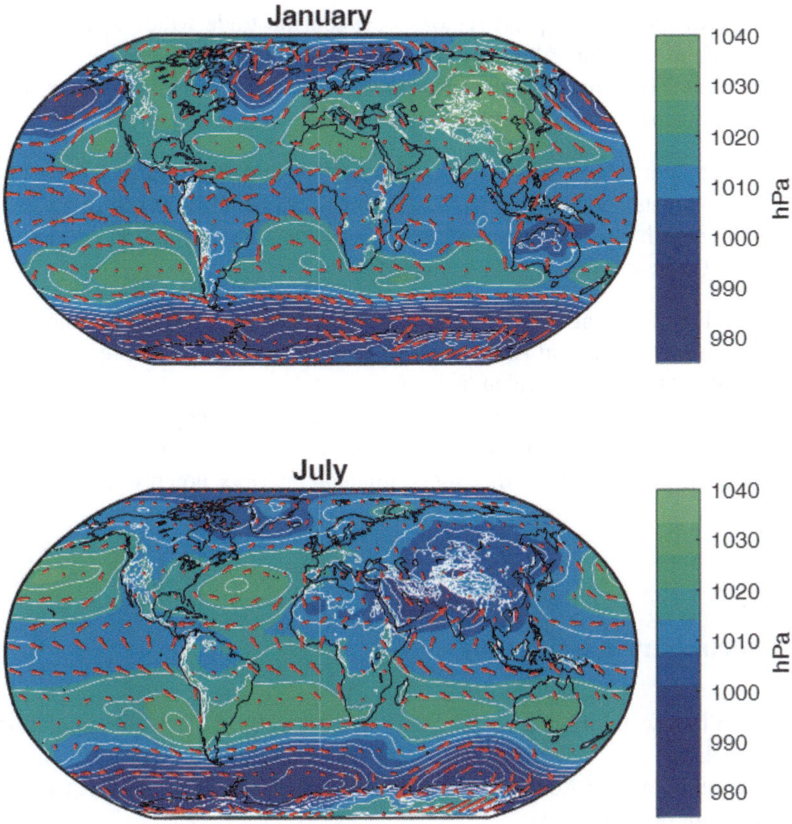

Fig. 1.11 Mean surface pressure and 10 m-wind velocity for January and July 2024. *Data source*: Copernicus climate change service, climate data store, (2023): ERA5 hourly data on single levels from 1940 to present

illustrates the observed average surface pressure and prevailing wind patterns for January and July, highlighting the seasonal migration of major pressure belts.

Despite seasonal and regional variations, the general structure of the three-cell model is evident. The principal wind belts include:

- The **trade winds** in the tropics,
- The **prevailing westerlies** in midlatitudes,
- The **polar easterlies** at high latitudes.

These wind systems define three distinctive regions:

- **Doldrums**: A low-latitude zone of weak winds and high humidity near the ITCZ, marked by frequent rainfall and convective activity. Its position oscillates seasonally north and south of the equator.

- **Horse latitudes**: Subtropical regions between the trade winds and westerlies, characterized by calm or light winds and arid conditions.
- **Polar front**: The boundary between the polar easterlies and midlatitude westerlies, a zone of strong temperature gradients and active weather systems.

The interaction between continents and oceans introduces further complexity. Land-sea contrasts and topographic effects break up idealized pressure zones into *semi-permanent* high- and low-pressure centers that vary in position and intensity throughout the year. Major examples include:

- **Winter**: Development of polar highs over Siberia and Canada; strong subtropical highs over the Pacific and Atlantic (Azores High); and deep low-pressure centers such as the Aleutian and Icelandic Lows.
- **Summer**: The Azores High shifts westward and strengthens into the Bermuda High, while the Pacific High also intensifies and migrates westward. Polar highs weaken, and thermal lows form over continental interiors, notably across southern Asia.

1.4 Secondary Circulation

In addition to low and high pressure zones generated by the general circulation of the troposphere, there exist low and high pressure zones that are not permanent, but local, temporary, and moving: they are called cyclonic (low pressure) and anti-cyclonic (high pressure) areas.

These temporary pressure zones have a daily or weekly time scale, and a length scale of 100–1000 km. Their passage is generally associated to a weather change. Usually anticyclones are associated to nice weather, while cyclones bring rain, clouds, and high wind speeds.

The pressure field is represented on weather charts in terms of a set of isobars (i.e., lines or contours along which the pressure is equal to a constant value) on a horizontal surface, such as sea level (see Fig. 1.12). Isobars are usually plotted at uniform increments: for example, every 4 hPa on a sea-level pressure chart (e.g., ...996, 1000, 1004 ...hPa).

Local maxima in the pressure field are referred to as high pressure centers or simply highs, denoted by the symbol H, and minima as lows (L). At any point on a pressure chart the local horizontal pressure gradient is oriented perpendicular to the isobars and is directed from lower toward higher pressure. The strength of the horizontal pressure gradient is inversely proportional to the horizontal spacing between the isobars in the vicinity of that point.

As it will be discussed in the following these pressure zones are responsible for winds that can reach moderate to relevant speeds. These atmospheric processes, that are in the mesoscale, form the secondary circulation of atmosphere.

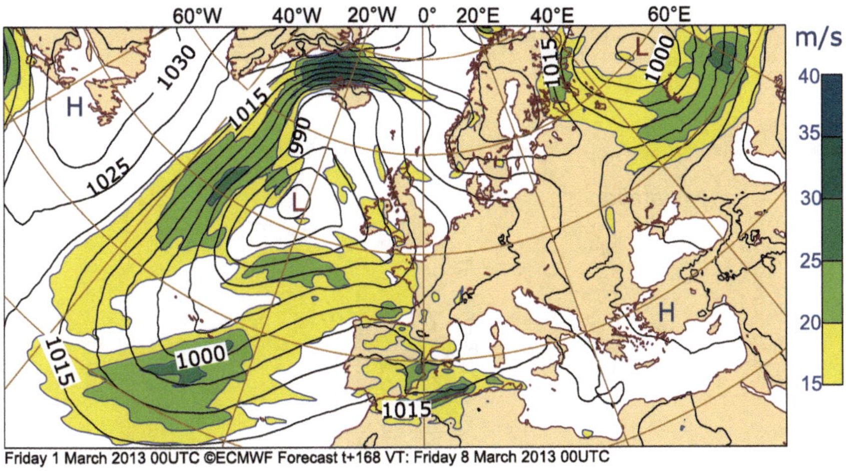

Fig. 1.12 Isobars plotted on sea-level pressure chart *Source* European Centre for Medium-Range Weather Forecasts (ECMWF) https://www.ecmwf.int

1.4.1 A Simple Framework for Wind Kinematics

In order to understand which are factors that drive the secondary circulation, we introduce a simple framework for the analysis of the kinematics and the dynamics of air masses.

Earth can be simplistically modeled as a rigid sphere that rotates about a fixed axis passing through the poles at a constant speed $\Omega = 7.27 \cdot 10^{-5}$ rad/s (one revolution in 24 h). With reference to Fig. 1.13, we define an inertial reference frame $OXYZ$, with versors $\hat{\mathbf{i}}$, $\hat{\mathbf{j}}$, and $\hat{\mathbf{k}}$. In this frame, the angular velocity of Earth is

$$\hat{\mathbf{\Omega}} = \begin{bmatrix} 0 \\ 0 \\ \Omega \end{bmatrix} = \Omega\hat{\mathbf{k}} \tag{1.4}$$

However, to describe the movement of the atmosphere it is convenient to use a local moving reference frame $oxyz$, fixed to the Earth's surface. This is a convenient choice since anemometric stations measure wind speeds directly in this reference frame. The sketch in Fig. 1.13 depicts the traditional choice for a local Cartesian framework of reference: the x-axis is oriented eastward, the y-axis, northward and the z-axis, upward. ϕ indicates the latitude and λ the longitude of the origin o. The angular velocity of Earth in this reference system is (Fig. 1.14)

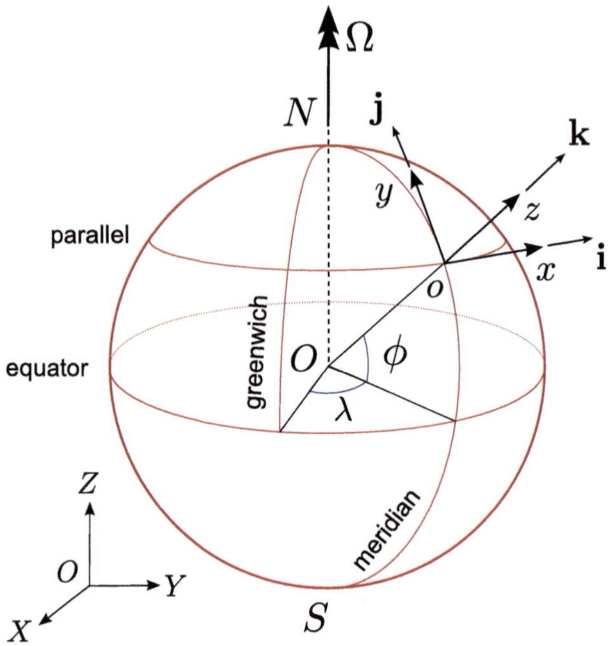

Fig. 1.13 Inertial and non inertial reference systems

Fig. 1.14 Angular velocity of Earth in the inertial and non-inertial reference systems

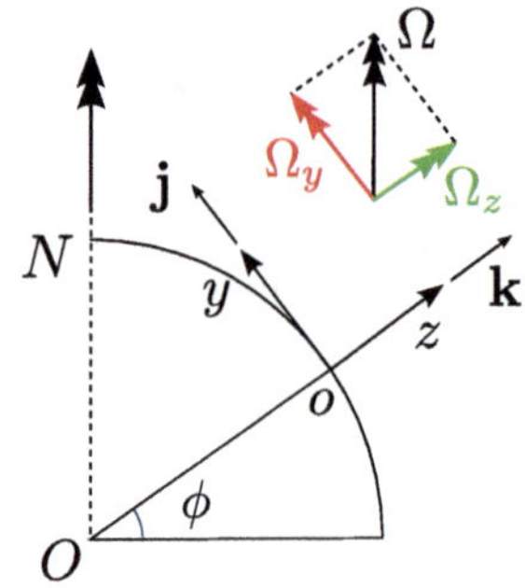

$$\boldsymbol{\Omega} = \begin{bmatrix} 0 \\ \Omega_y \\ \Omega_z \end{bmatrix} = \begin{bmatrix} 0 \\ \Omega \cos \phi \\ \Omega \sin \phi \end{bmatrix} = \Omega \cos \phi \mathbf{j} + \Omega \sin \phi \mathbf{k} \tag{1.5}$$

We describe the kinematics of a particle of air P (representative of a wind system) in the local reference system $oxyz$. The velocity of P relative to an observer standing on the Earth's surface is[2] (see Fig. 1.15)

[2] This is an approximation, since the curvature of the Earth is not negligible, and we should use a curvilinear coordinate for x and y to describe the three-dimensional motion of P (see Holton

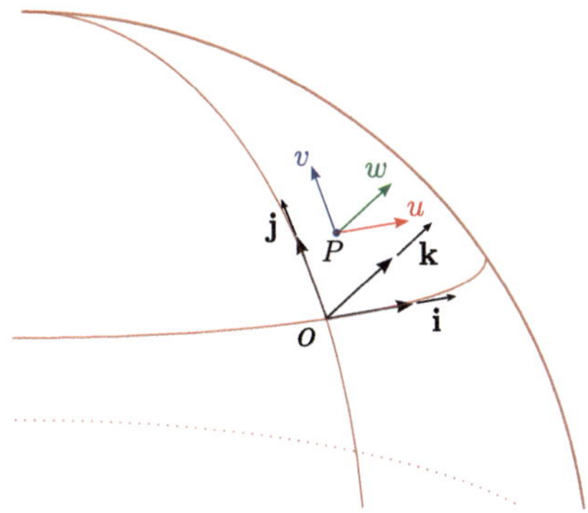

Fig. 1.15 Velocity of particle P in the non inertial system $oxyz$

$$\mathbf{v}_r = \begin{bmatrix} u \\ v \\ w \end{bmatrix} = u\mathbf{i} + v\mathbf{j} + w\mathbf{k} \tag{1.6}$$

The $oxyz$ reference system is moving due to Earth's rotation, therefore the absolute acceleration of P (and the inertial forces that governs its dynamics) can be computed as the sum of three terms: the relative acceleration $\mathbf{a}_r$, the entrainment acceleration $\mathbf{A}_s$, and the complementary (or Coriolis) acceleration $\mathbf{A}_c$. Therefore the absolute acceleration of P is:

$$\mathbf{A} = \mathbf{a}_r + \mathbf{A}_s + \mathbf{A}_c \tag{1.7}$$

The relative acceleration $\mathbf{a}_r$ express the acceleration relative to the moving frame $oxyz$:

$$\mathbf{a}_r = \frac{du}{dt}\mathbf{i} + \frac{dv}{dt}\mathbf{j} + \frac{dw}{dt}\mathbf{k} \tag{1.8}$$

The entrainment acceleration $\mathbf{A}_s$ is given by:

$$\mathbf{A}_s = \boldsymbol{\Omega} \wedge (\boldsymbol{\Omega} \wedge (P - O)) \tag{1.9}$$

where $(P - O)$ is the vector describing the position of P in the absolute reference frame

$$(P - O) = (P - o) + (o - O) \approx 0\mathbf{i} + 0\mathbf{j} + R\mathbf{k} \tag{1.10}$$

with R the mean radius of Earth (6371 km).

2004). However we will restrict the application of these equations to a limited bi-dimensional surface, which allows for this simplification.

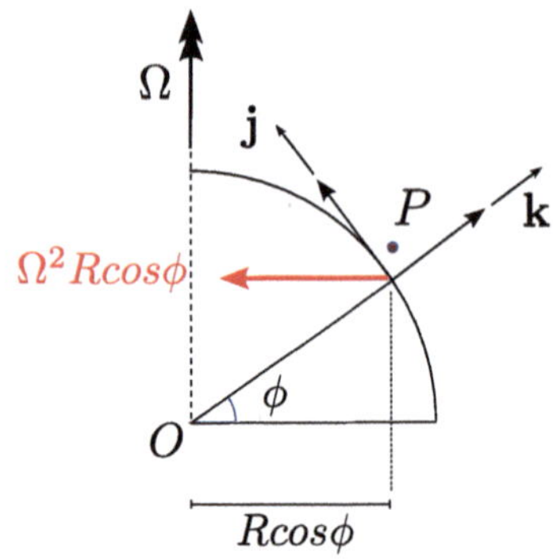

Fig. 1.16 Entrainment acceleration

The entrainment acceleration $\mathbf{A}_s$ results as

$$\mathbf{A}_s = \begin{bmatrix} 0 \\ \Omega^2 R \cos\phi \sin\phi \\ -\Omega^2 R\cos^2\phi \end{bmatrix} \tag{1.11}$$

As shown in Fig. 1.16 the entrainment acceleration is the centripetal acceleration due to the rotation of Earth projected onto **j** and **k**.

The complementary (or Coriolis) acceleration $\mathbf{A}_c$ is given by the vector product of the angular velocity and the relative velocity as (see also Fig. 1.17):

$$\mathbf{A}_c = 2\Omega \wedge \mathbf{v}_r \tag{1.12}$$

$$\begin{aligned} \mathbf{A}_c &= 2 \begin{vmatrix} \mathbf{i} & \mathbf{j} & \mathbf{k} \\ 0 & \Omega\cos\phi & \Omega\sin\phi \\ u & v & w \end{vmatrix} \\ &= \begin{bmatrix} w2\Omega\cos\phi - v2\Omega\sin\phi \\ u2\Omega\sin\phi \\ -u2\Omega\cos\phi \end{bmatrix} = \begin{bmatrix} wf^* - vf \\ uf \\ -uf^* \end{bmatrix} \end{aligned} \tag{1.13}$$

where $f = 2\Omega\sin\phi = 1.45 \cdot 10^{-4} \sin\phi$ and $f^* = 2\Omega\cos\phi$ are named respectively Coriolis parameter and Complementary Coriolis parameter. Notice that Coriolis parameter is maximum at Poles and null at Equator, and change sign according to the considered hemisphere.

Therefore, recalling Eq. (1.7), the absolute acceleration of particle P is

$$\mathbf{A} = \begin{bmatrix} \frac{du}{dt} \\ \frac{dv}{dt} \\ \frac{dw}{dt} \end{bmatrix} + \begin{bmatrix} 0 \\ \Omega^2 R \cos\phi \sin\phi \\ -\Omega^2 R\cos^2\phi \end{bmatrix} + \begin{bmatrix} wf^* - vf \\ uf \\ -uf^* \end{bmatrix} \tag{1.14}$$

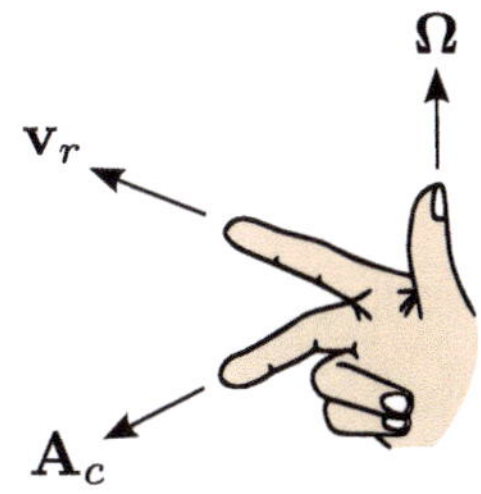

Fig. 1.17 Right hand rule for Coriolis acceleration

The centripetal acceleration, that appears in the entrainment part, together with the acceleration of gravity, that is not reported in the expression, mainly influence the hydrostatic equilibrium of the atmosphere and it will not be considered in the following analysis that is focused on a simplified description of the dynamics of air masses. For this reason, we will consider the acceleration **A** as:

$$\mathbf{A} = \mathbf{a}_r + \mathbf{A}_c \tag{1.15}$$

Relative and Coriolis acceleration have the same order of magnitude and must be accounted for in the dynamic of air masses.

1.4.2 Pressure Zones and Air Dynamics

If we analyze a limited region, we can neglect curvature of Earth and the thickness of the troposphere. This simplification allows to approach the study of air dynamics considering the problem to be 2-dimensional, as sketched in Fig. 1.18. Notice that even if we consider a 2-D problem there is a component of the Earth's angular velocity Ω_z, normal to the considered plane, that depends on the latitude.

Applying Newton's second law to the particle P, which is representative of a limited volume of air, the equations of motion per unit volume can be written as:

$$\rho\mathbf{A} = \rho(\mathbf{a}_r + \mathbf{A}_c) = -\nabla p \tag{1.16}$$

where ρ is the air density, **A** the absolute acceleration, and ∇p the pressure gradient. In this equilibrium we have neglected friction forces, and body forces like buoyancy, gravitational effects and centrifugal forces.

For an observer fixed to the Earth's surface, interested in the relative velocity $\mathbf{v}_r$, the Coriolis term can be interpreted as an external force, and the equilibrium can be written as:

$$\rho\mathbf{a}_r = -\rho\mathbf{A}_c \quad - \nabla p \tag{1.17}$$

In the following sections, we will use this equilibrium equation to develop the concepts of geostrophic and gradient winds.

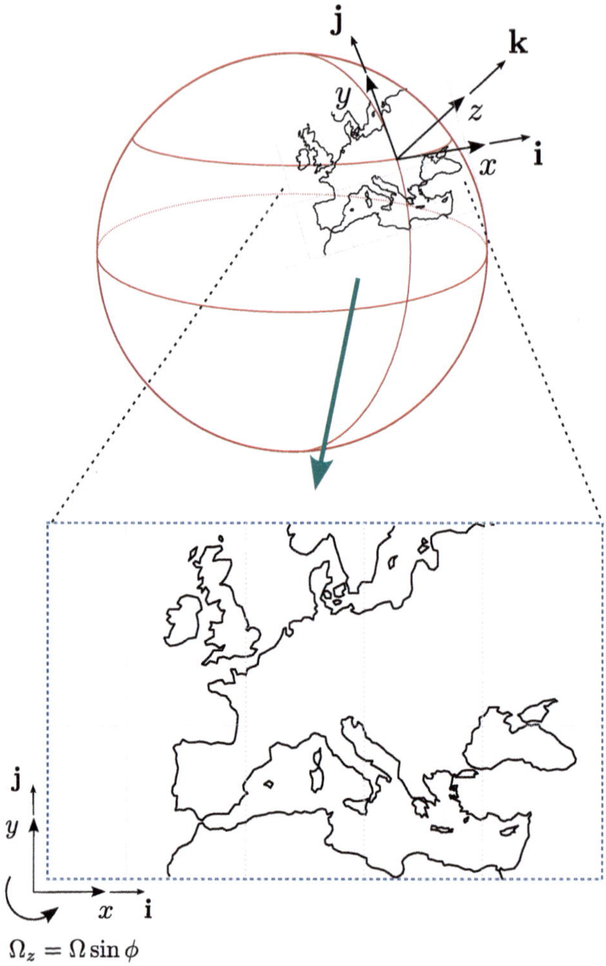

Fig. 1.18 From 3D to 2D problem (thickness of troposphere ≈ 12 km, mean radius of Earth 6371 km)

1.4.3 Geostrophic Wind Approximation

Let us consider a 2D case in which isobars are straight lines (i.e. negligible curvature).

Let us try to perform a qualitative analysis of the dynamics that governs air motion in the northern hemisphere. In case of negligible curvature of isobars, pressure gradient $\nabla p = \partial p/\partial r$ is normal to the isobars, directed from lower pressure p_1 to higher pressure p_2, as shown in Fig. 1.19a.

Therefore, the particle P will be subjected to a pressure gradient force directed from high to low pressure. Due to this force, the particle P starts moving toward low pressure and the relative velocity $\mathbf{v}_r$ rises. Contextually, since there is an angu-

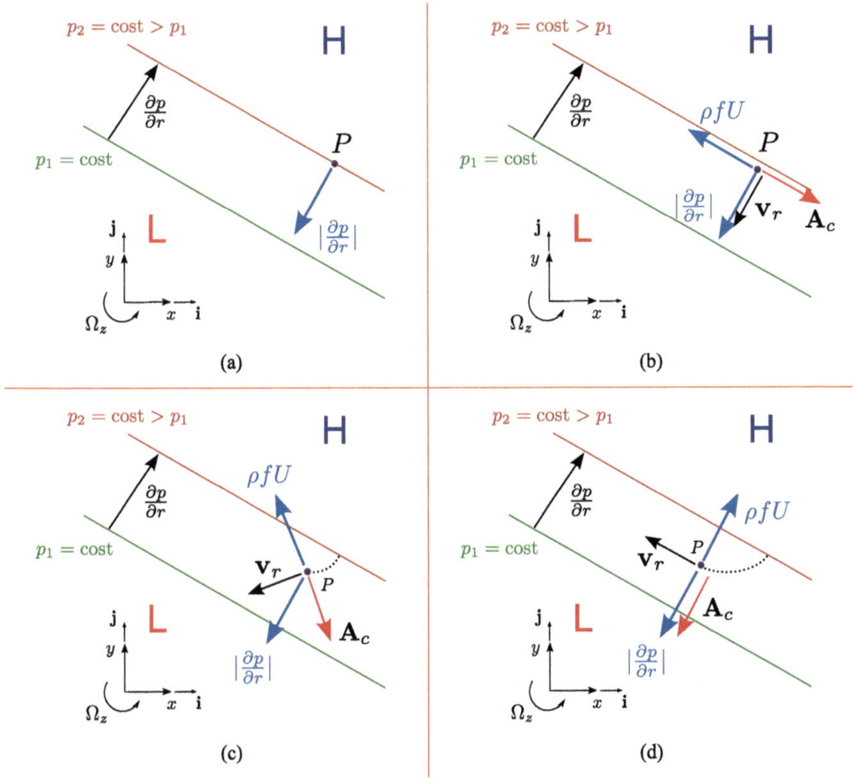

Fig. 1.19 Geostrophic wind

lar velocity Ω_z, the complementary acceleration $\mathbf{A}_c$ appears, and consequently the Coriolis force with magnitude $\rho f|\mathbf{v}_r| = \rho f U$, as shown in Fig. 1.19b.

The air particle P, under the action of pressure gradient and Coriolis forces, turns right, as shown in Fig. 1.19c, and keeps turning right until the velocity vector is parallel to the isobars (Fig. 1.19d). In this situation, the pressure gradient force is equilibrated by the Coriolis force, and the system has a stationary solution.

Therefore, in the geostrophic approximation, wind blows parallel to the isobars. If the Earth did not rotate there would be a flow normal to the isobars and non stationary.

As a rule of thumb, if wind blows on your back low pressure is on the left (on Northern hemisphere). The difference between Northern and Southern hemisphere is the sign of the Coriolis parameter f.

This result can be also verified analytically. For a 2D flow (w=0) the equation of motions (Eq.(1.17)) are:

$$\begin{cases} \rho \frac{du}{dt} = +\rho v f - \frac{\partial p}{\partial x} \\ \rho \frac{dv}{dt} = -\rho u f - \frac{\partial p}{\partial y} \end{cases} \tag{1.18}$$

where the pressure gradient is

$$\frac{\partial p}{\partial r} = \left(\frac{\partial p}{\partial x}\right)\mathbf{i} + \left(\frac{\partial p}{\partial y}\right)\mathbf{j} \tag{1.19}$$

We look for a stationary solution ($d/dt = 0$) that equilibrates the system

$$\begin{cases} 0 = +\rho v f - \frac{\partial p}{\partial x} \\ 0 = -\rho u f - \frac{\partial p}{\partial y} \end{cases} \tag{1.20}$$

$$\begin{cases} u = -\frac{1}{\rho f}\frac{\partial p}{\partial y} \\ v = \frac{1}{\rho f}\frac{\partial p}{\partial x} \end{cases} \tag{1.21}$$

If we perform the scalar product between $\mathbf{v}_r$ and the pressure gradient, the result is zero. This means that the two vectors are orthogonal, therefore wind blows parallel to the isobars.

$$\mathbf{v}_r \cdot \nabla p = \begin{bmatrix} u \\ v \end{bmatrix} \cdot \begin{bmatrix} \frac{\partial p}{\partial x} \\ \frac{\partial p}{\partial y} \end{bmatrix} = -\frac{1}{\rho f}\frac{\partial p}{\partial y}\frac{\partial p}{\partial x} + \frac{1}{\rho f}\frac{\partial p}{\partial x}\frac{\partial p}{\partial y} = 0 \tag{1.22}$$

This is a little counterintuitive, since one would expect a wind normal to the isobars, and this would be the case if there were not the Coriolis effect. Geostrophic balance means that horizontal pressure gradients are largely balanced by Coriolis forces associated with horizontal winds.

The wind speed magnitude in stationary conditions is

$$U = \sqrt{u^2 + v^2} = \left|\frac{1}{\rho f}\right| \sqrt{\left(\frac{\partial p}{\partial x}\right)^2 + \left(\frac{\partial p}{\partial y}\right)^2} = \left|\frac{1}{\rho f}\frac{\partial p}{\partial r}\right| \tag{1.23}$$

1.4.4 Gradient Wind Approximation

Now let us consider a 2D case in which isobars have a curvature. For simplicity we consider circular concentric isobars.

In analogy with the results obtained for the geostrophic approximation, we assume a uniform circular motion of P, and we verify whether the system can be in equilibrium. With reference to Fig. 1.20, let us consider first the case of a cyclone, that is the case in which the low pressure is in the center of the pressure system. In analogy with the geostrophic wind approximation, if wind blows on your back low pressure is on the left (on Northern hemisphere).

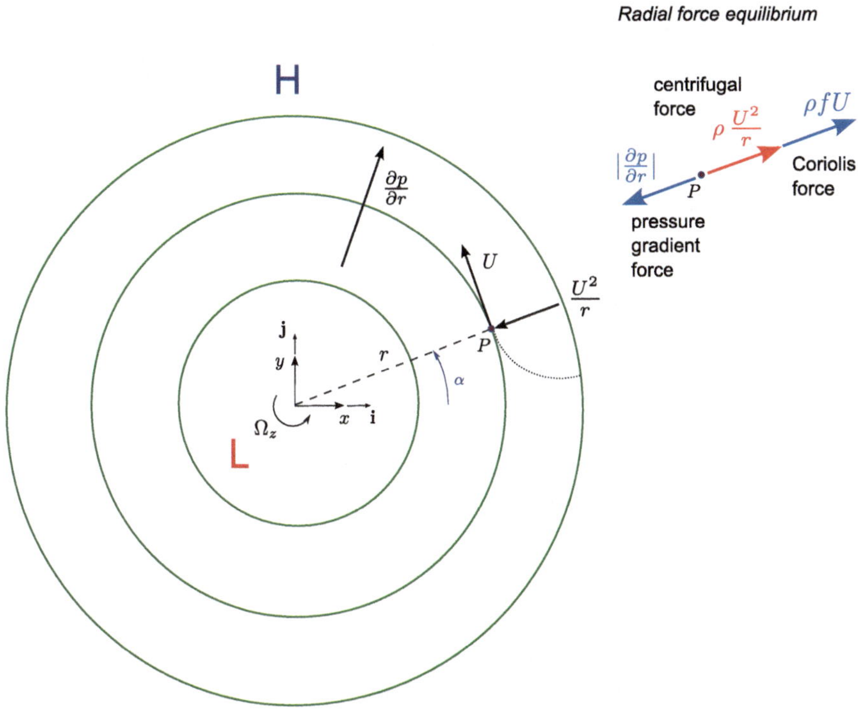

Fig. 1.20 Gradient wind approximation: cyclone

Considering the particle P in uniform circular motion, we can write its kinematics using a reference system centered in the center of the cyclone and using a complex notation as:

$$\begin{aligned} (P - o) &= r e^{i\alpha(t)} \\ \mathbf{v}_r &= r\dot{\alpha} e^{i(\alpha(t)+\pi/2)} = U e^{i(\alpha(t)+\pi/2)} \\ \mathbf{a}_r &= -r\dot{\alpha}^2 e^{i\alpha(t)} = -\frac{U^2}{r} e^{i\alpha(t)} \end{aligned} \tag{1.24}$$

where r is the radius of the isobar, α the angular position, and U the magnitude of wind speed.

In this flow conditions, the particle P is subjected to a centripetal pressure gradient force $\left|\frac{\partial p}{\partial r}\right|$, a centrifugal force $\rho \frac{U^2}{r}$, and a centrifugal Coriolis force $\rho f U$ (see Fig. 1.20).

Therefore we can write a dynamic equilibrium along r as

$$\left(\frac{U^2}{r} + Uf\right) - \frac{1}{\rho}\left|\frac{\partial p}{\partial r}\right| = 0 \tag{1.25}$$

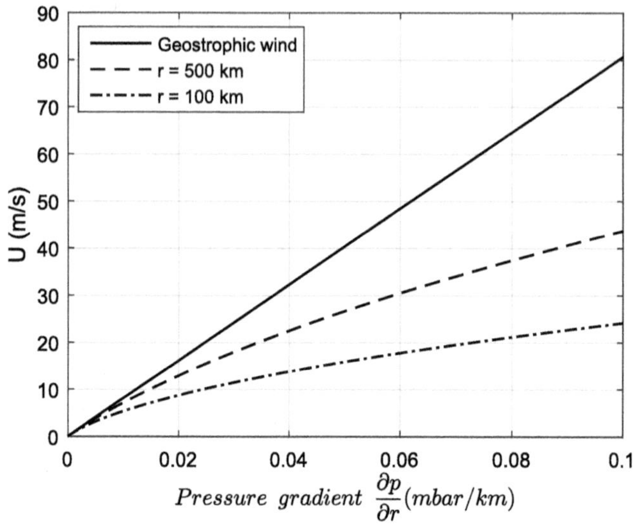

Fig. 1.21 Cyclone wind speed as a function of pressure gradient and isobar curvature $\phi = 45°$

from which we can compute the wind speed U necessary to balance the dynamic system. Solving the quadratic equation in U we get:

$$U = -\frac{rf}{2} + \sqrt{\left(\frac{rf}{2}\right)^2 + \frac{r}{\rho}\left|\frac{\partial p}{\partial r}\right|} \tag{1.26}$$

This expresses gradient wind balance and is a better approximation than geostrophic balance (which neglects the centrifugal force). The "minus" sign root is not considered since this solution would not have a physical meaning (there would be a wind speed in absence of a pressure gradient).

In this case the velocity U can be very high if large pressure gradients are present, as shown in Fig. 1.21.

The previous solution holds if pressure gradient points outward (cyclone). If we consider a pressure system with negative pressure gradient, we must consider a negative pressure gradient and a point P which rotates clockwise. This is called anticyclone.

Following the same considerations as done for the cyclone, with reference to Fig. 1.22, we can write the dynamic equilibrium along r as:

$$\left(\frac{U^2}{r} - Uf\right) + \frac{1}{\rho}\left|\frac{\partial p}{\partial r}\right| = 0 \tag{1.27}$$

from which we can compute the wind speed U necessary to balance the dynamic system. Solving the quadratic equation in U we get:

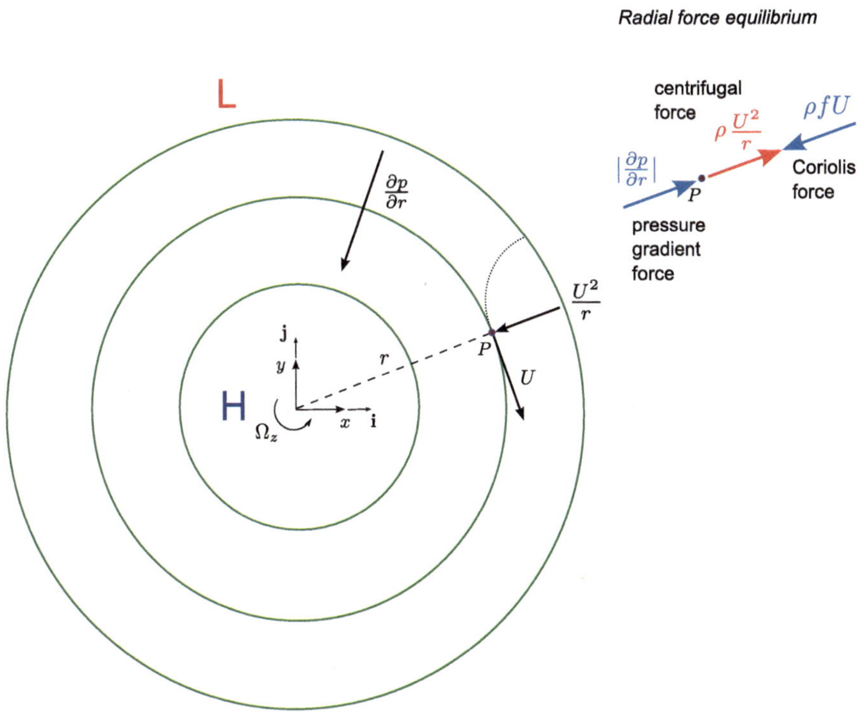

Fig. 1.22 Gradient wind approximation: anticyclone

$$U = +\frac{rf}{2} - \sqrt{\left(\frac{rf}{2}\right)^2 - \frac{r}{\rho}\left|\frac{\partial p}{\partial r}\right|} \tag{1.28}$$

This expresses gradient wind balance for anticyclones. In this case the "plus" sign root is not considered since this solution would not have a physical meaning (there would be a wind speed in absence of a pressure gradient). In this case the velocity U is limited to $fr/2$, as shown in Fig. 1.23.

Using this simplistic model, we can understand why high wind speeds are associated to cyclones and not to anticyclones. Note that in Europe, cyclones are by far the most important source of high wind velocities.

The gradient wind is a very simplistic model, in reality due to the 3-dimensionality and to friction effect, cyclones are three-dimensional in-spirals, while anticyclones are three-dimensional out-spirals. Moreover the way of rotation of the spirals depends on the sign of the Coriolis parameter (which depends on the hemisphere), as sketched in Fig. 1.24.

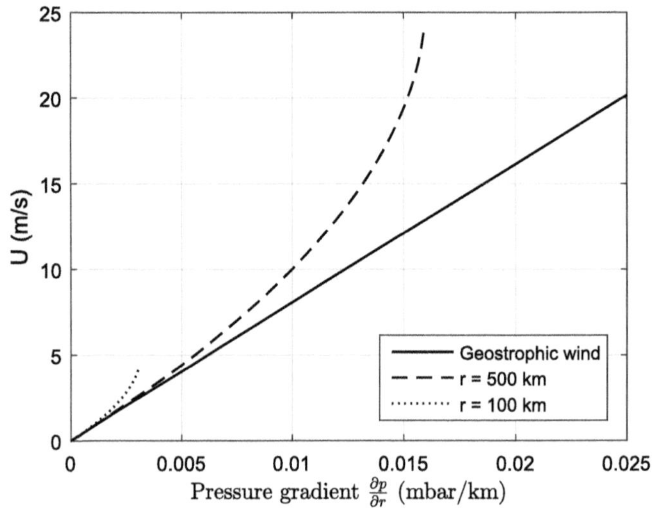

Fig. 1.23 Anticyclone wind speed as a function of pressure gradient and isobar curvature $\phi = 45°$

1.4.5 Tropical and Extra-Tropical Cyclones

Two main categories of cyclones can be identified according to their origin and dynamics: *tropical* and *extra-tropical*. Extra-tropical, or mid-latitude, cyclones develop from the dynamic interaction between warm tropical and cold polar air masses along the polar front. In contrast, tropical cyclones–commonly known as hurricanes–derive their energy primarily from the latent heat released during condensation processes.

Although mid-latitude cyclones are widespread and meteorologically significant, their intensity and destructive potential are generally much lower than those of tropical cyclones, which involve far greater exchanges of atmospheric energy. Moving away from the Equator, the available energy to sustain such systems decreases with the diminishing input of solar radiation and surface heating. Figure 1.25 illustrates the typical tracks followed by both mid-latitude and subtropical cyclones, which are steered by the prevailing winds of the large-scale atmospheric circulation.

1.4.5.1 Mid-Latitude Cyclones

Mid-latitude, or frontal, cyclones are large-scale traveling low-pressure systems that typically extend up to 2000 km in diameter. In intense cases, surface pressures may drop to about 970 mb, compared to the mean sea-level pressure of 1013 mb. These storms generally last from three to ten days, moving predominantly from west to east, and they represent the dominant weather systems in the mid-latitudes.

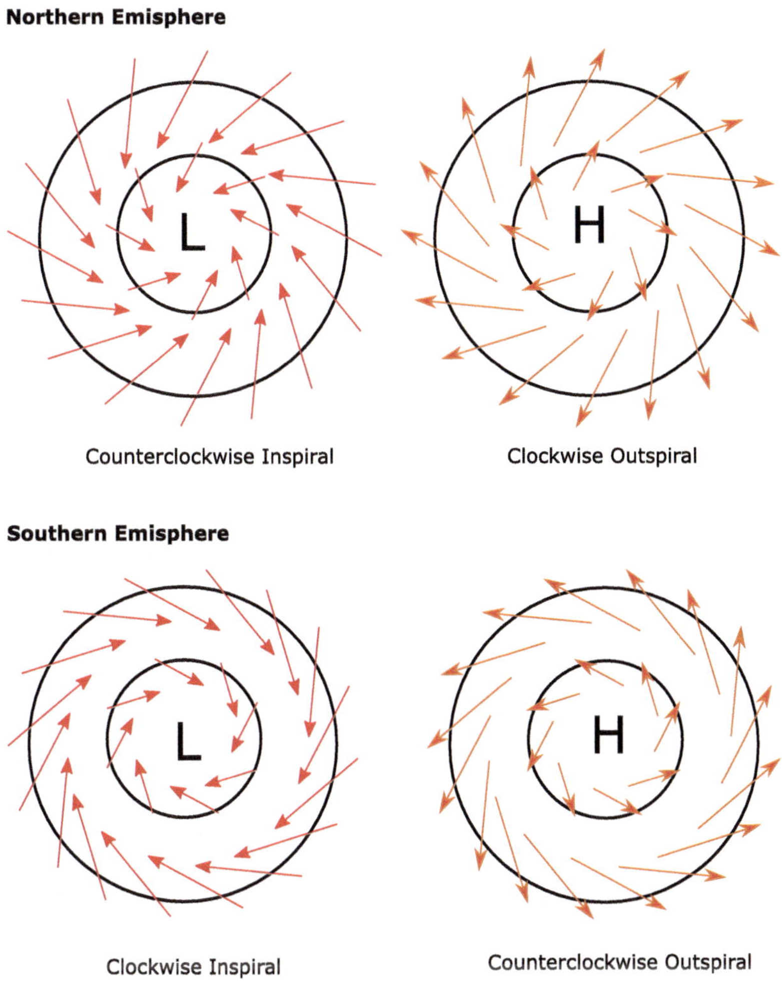

Fig. 1.24 Effect of friction and N/S hemisphere on cyclones and anticyclones

These cyclones form along the polar front where cold polar air meets warmer tropical air. The interaction between these air masses promotes uplift of warm air, which cools adiabatically and interacts with the overlying cold air, leading to extensive cloud formation and precipitation. In this process, large amounts of energy are transferred from lower to higher latitudes, contributing to the global redistribution of heat.

Precipitation associated with mid-latitude cyclones can take multiple forms–rain, snow, sleet, freezing rain, or hail–depending on local temperature conditions. Some cyclones undergo rapid intensification, known as *explosive cyclogenesis*. A notable example is the so-called "Great Storm of 1987" that affected the United Kingdom,

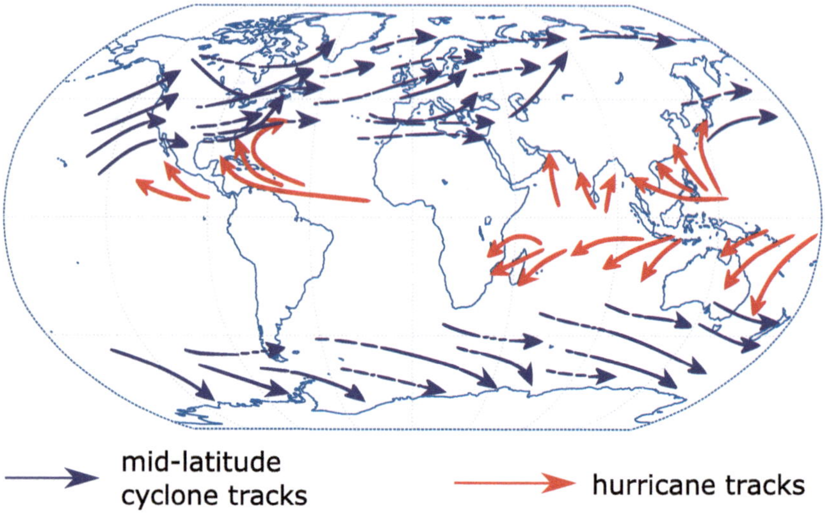

Fig. 1.25 Typical paths of mid-latitude cyclones are represented by black arrows. This image also shows the typical paths traveled by subtropical hurricanes (green arrows)

where surface pressure dropped to 953 mb and wind speeds reached 220 km/h (about 60 m/s).

The evolution of a mid-latitude cyclone, or *cyclogenesis*, is depicted in Fig. 1.26. The process begins with a weak disturbance along a stationary front where contrasting air masses meet (Stage 1). Warm air begins to rise over the denser cold air, forming a cyclonic circulation around a developing low-pressure center (Stages 2–3). As the system matures, the pressure at the center drops and distinct cold and warm fronts form (Stages 4–5). The region of warm air between these fronts is known as the *warm sector*. Since cold fronts advance more rapidly than warm fronts, the system eventually occludes when the cold front overtakes the warm front, lifting the warm air aloft (Stages 6–7). After occlusion, the system gradually weakens and dissipates (Stages 8–9).

1.4.5.2 Hurricanes

Hurricanes are intense cyclonic storms that develop over the warm oceans of the tropics (they can exist between 5 ° and 40 ° of latitude, and they reach full strength between 10 and 30 °). These tropical storms go by other names in the various parts of the world: India/Australia - cyclones; western North Pacific - typhoons; Philippines - baguio. By international agreement, the term tropical cyclone is used by most nations to describe hurricane-like storms that originated over tropical oceans.

Surface atmospheric pressure in the center of a hurricane tends to be extremely low. The lowest pressure reading ever recorded for a hurricane (typhoon Tip, 1979) is 870

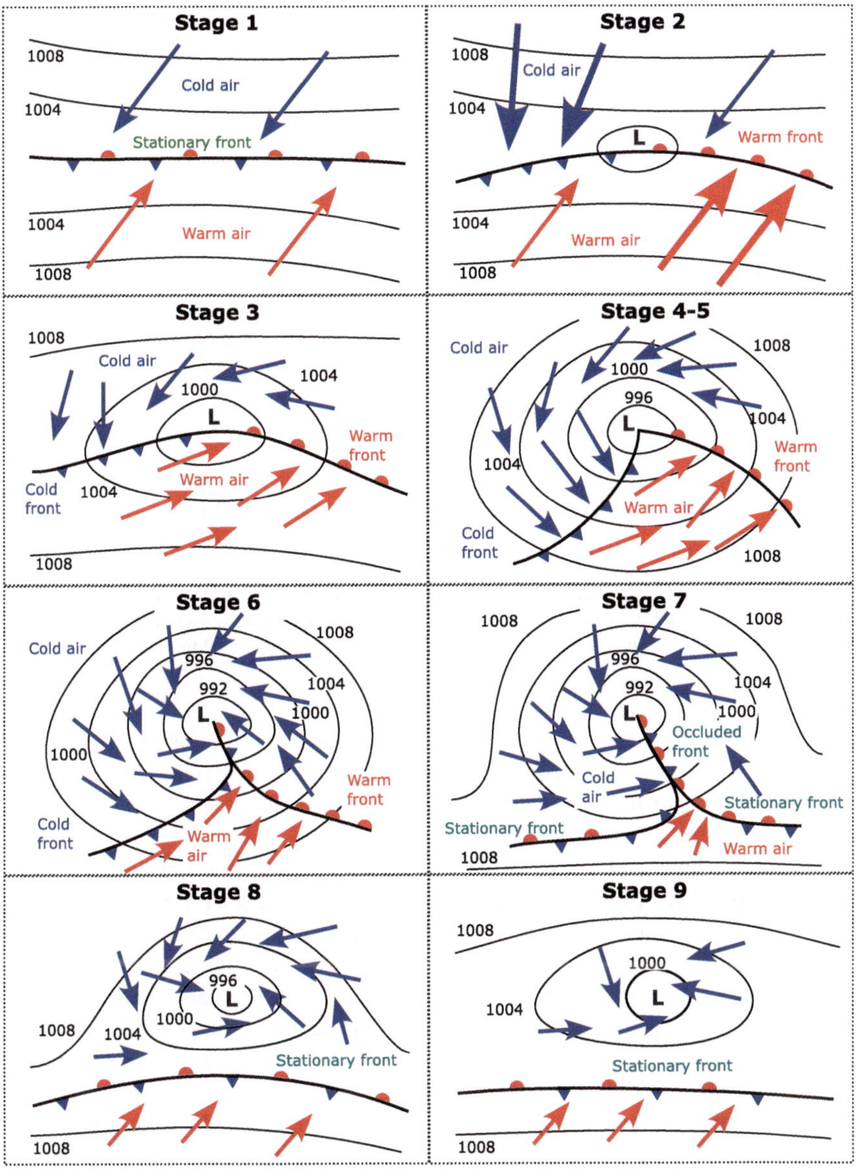

Fig. 1.26 Life cycle (cyclogenesis) of a mid-latitude cyclone

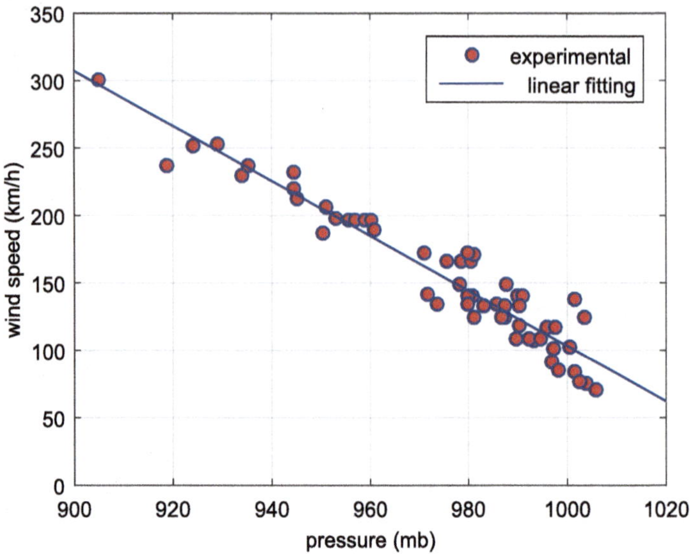

Fig. 1.27 Relationship between surface pressure and wind speed for a number of tropical low pressure systems. Tropical low pressure systems are classified as hurricanes when their pressure is 980 millibars or lower, and sustained wind speeds are greater than 118 kilometres per hour

millibars (mb). However, most storms have an average pressure of 950 millibars. To be classified as a hurricane, sustained wind speeds must be greater than 118 kilometres per hour at the storm's center. Wind speed in a hurricane is directly related to the surface pressure of the storm (see Fig. 1.27).

Hurricanes have no fronts associated with them like the mid-latitude cyclones of the polar front. They are also smaller than the mid-latitude cyclone, measuring on average 550 kilometres in diameter. One of the largest hurricanes ever measured was Typhoon Tip (October 12, 1979) which had a diameter of about 2100 kilometres. Mature hurricanes usually develop a cloud-free eye at their center. In the eye, air is descending creating clear skies. The eye of the hurricane may be 20 to 50 kilometres in diameter. Surrounding the eye are bands of organized thunderstorm clouds formed as warm air moving in and up into the storm.

As described in the previous sections, Coriolis force strongly influence cyclones whose way of rotation depends on the Coriolis parameter, as shown in Fig. 1.28 where the picture of one cyclone in the northern hemisphere is compared with that of a cyclone in the southern hemisphere.

Fig. 1.28 Effect of Coriolis force for N and S hemisphere: Gilbert (North, counterclockwise) and Catarina (South, clockwise)

1.4.6 Monsoon Winds

Monsoons are regional scale wind systems that predictably change direction with the passing of the seasons. Like land/sea breezes (discussed in the next section), these wind systems are created by the temperature contrasts that exist between the surfaces of land and ocean. However, monsoons are different from land/sea breezes both spatially and temporally. Monsoons occur over distances of thousands of kilometres, and their two dominant patterns of wind flow act over an annual time scale.

During the summer, monsoon winds blow from the cooler ocean surfaces onto the warmer continents. In the summer, the continents become much warmer than the oceans because of a number of factors. These factors include:

- Specific heat differences between land and water.
- Greater evaporation over water surfaces.
- Subsurface mixing in ocean basins which redistributes heat energy through a deeper layer.

Precipitation is normally associated with the summer monsoons. Onshore winds blowing inland from the warm ocean are very high in humidity, and slight cooling of these air masses causes condensation and rain. In some cases, this precipitation can be greatly intensified by orographic uplift. Some highland areas in Asia receive more than 10 m of rain during the summer months. In the winter, the wind patterns reverse as the ocean surfaces are now warmer. With little solar energy available, the continents begin cooling rapidly as longwave radiation is emitted to space. The ocean surface retains its heat energy longer because of water's high specific heat and subsurface mixing. The winter monsoons bring clear dry weather and winds that flow from land to sea.

1.5 Local Wind Systems

Local wind systems are atmospheric phenomena that occur over spatial scales ranging from approximately 1 to 10 km and evolve over time scales of a few hours. Their development is strongly influenced by local geographical features, atmospheric stratification, or by the interaction between the two ESDU 1987. These systems are typically associated with moderate to occasionally strong wind speeds, which can have significant local effects.

1.5.1 Thunderstorms

Thunderstorms may develop when warm, moist air is forced to rise within the atmosphere. This upward motion can occur in association with frontal systems, in which case the resulting flows are referred to as *frontal winds*. As the ascending air cools and condensation takes place, latent heat is released, providing additional energy that sustains and intensifies the convective motion. Precipitation within the cloud system subsequently triggers downdrafts near the storm center, establishing a circulation of rising and descending air.

Typical horizontal dimensions of individual thunderstorms are on the order of 10 km, with lifetimes of about one hour. Maximum wind speeds commonly reach values between 15 and 30 m/s.

1.5.2 Tornadoes

A tornado is a violently rotating column of air that extends from the base of a deep convective cloud to the Earth's surface (see Fig. 1.29). Tornadoes represent one of the most extreme and destructive forms of atmospheric motion.

Although their local frequency of occurrence is relatively low, tornadoes can cause catastrophic damage where they strike. Historical examples include those that hit Topeka (Kansas, 1966) and Lubbock (Texas, 1970), each producing damage exceeding $100 million. A major tornado outbreak in April 1974 affected eleven U.S. states, resulting in over 4000 injuries, 329 fatalities, and approximately $700 million in losses.

Tornadoes typically have diameters of 100–300 m at ground level, though their size can range from just a few meters to nearly 500 m. Track lengths vary from about 1 km up to several hundred kilometers. The most intense events, mostly recorded in the central United States, exhibit maximum tangential wind speeds approaching 150 m/s, although such estimates remain uncertain due to the difficulty of obtaining direct measurements. More typical wind speeds are in the range of 40–80 m/s. Outside

Fig. 1.29 Category F5 tornado (upgraded from initial estimate of F4) viewed from the southeast as it approached Elie, Manitoba, on June 22, 2007. *Author* Justin Hobson https://en.wikipedia.org/wiki/File:F5_tornado_Elie_Manitoba_2007.jpg

North America, tornadoes generally show comparable frequency but lower intensity. In the most severe cases, wind speeds can exceed 100 m/s.

A key factor in the destructive potential of tornadoes is the substantial pressure deficit within the vortex, often amounting to about 0.1 atm relative to the surrounding environment. This pressure drop can induce major structural damage even in the absence of direct wind impact.

Tornado formation is associated with environments of strong atmospheric instability and vigorous convection, such as those found in severe thunderstorms or occasionally within hurricanes. Tornado-genesis begins when intense updrafts within a convective cell generate localized cyclonic rotation near the surface. If the vortex extends upward into a region of strong vertical wind shear–such as near the jet stream–air is evacuated from the core, leading to further pressure reduction and intensification of rotation. The resulting low-pressure funnel can extend down to the surface, forming a tornado over land or a waterspout over the sea.

Tornadoes occur in many parts of the world where severe convective storms develop. In the United States, the high frequency of events in the central plains ("Tornado Alley") arises from the convergence of warm, moist air from the Gulf of Mexico and cool, dry air descending from the Rocky Mountains. Globally, tornadoes are most frequent in the latitude bands between 20 and 50 ° in both hemispheres. Their seasonal occurrence depends on regional climate, typically peaking from late spring to early autumn, following the annual migration of the jet stream and the polar front FUJITA 1973.

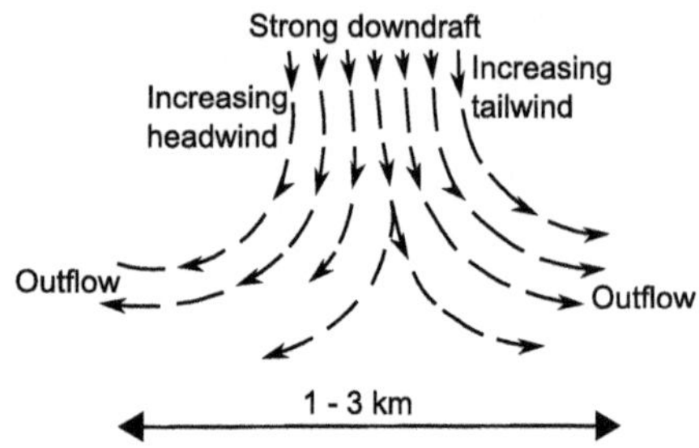

Fig. 1.30 Schematic of a downburst showing descending air and divergent outflow near the surface

Urban characteristics, such as increased surface roughness and the urban heat island effect, may reduce tornado formation or weaken storms as they move into built-up areas. Nevertheless, damaging urban tornadoes have been documented, including the event that struck northwest London in December 1954. Tornadoes are rare in mountainous terrain, where surface irregularities inhibit the organized development of strong vortices.

1.5.3 Downbursts

Downbursts are intense, localized downdrafts of air that descend rapidly from convective clouds and spread outward upon reaching the ground (see Fig. 1.30). These phenomena can generate strong and damaging surface winds over areas several kilometers wide and pose major hazards to aircraft during takeoff and landing.

In contrast to tornadoes, where the flow pattern is rotational, the wind field within a downburst is primarily divergent–air spreads out radially from the point of impact at the surface. When the descending air mass reaches the ground, it produces a sudden and violent surge of horizontal wind, which can reach or exceed the intensity of a weak tornado but over a shorter duration and more localized area.

The strong negative buoyancy of the descending air is mainly due to evaporative cooling and, in some cases, to the melting of hail within the downdraft. As the air cools and becomes denser than the surrounding environment, it accelerates downward under gravity. Upon hitting the ground, this air mass is forced to diverge horizontally, producing a ring of strong outflow winds that can cause significant structural damage in their path.

Depending on their horizontal extent, downbursts are classified as *microbursts* (less than 4 km in diameter) or *macrobursts* (greater than 4 km). Microbursts, in particular, are of critical concern for aviation safety, as they can generate abrupt wind shear conditions—a rapid change in wind speed and direction over a short distance—that can severely affect aircraft performance.

1.5.4 Sea and Land Breezes

Sea and land breezes are diurnal wind systems that arise due to the differential heating between land and sea surfaces. During the day, the land warms more rapidly than the adjacent water, leading to lower surface pressure over land and relatively higher pressure over the sea. This contrast drives a gentle onshore flow known as the *sea breeze*, which often brings cooler, moist air inland.

After sunset, the situation reverses: land surfaces cool faster than the sea, resulting in higher pressure over land and lower pressure over the ocean. Air thus flows seaward, generating the *land breeze*. These alternating circulations can be viewed as a small-scale analogue of the monsoon system. Figure 1.31 schematically illustrates this daily cycle.

The vertical circulation associated with sea breezes promotes the development of cumulus clouds and, occasionally, convective storms along coastal regions, particularly where topography enhances upward motion. The inland penetration of a sea breeze depends on local geography and the intensity of surface heating, but typically extends tens of kilometers inland.

1.5.5 Mountain and Valley Winds

Mountain and valley winds are local circulations generated by diurnal temperature variations along mountain slopes. During the day, solar radiation heats the slopes more rapidly than the adjacent air in the valley. This causes air to rise upslope, generating an *anabatic* (valley) wind that transports moisture and heat upward.

At night, radiative cooling of the slopes makes the air denser and colder than the surrounding atmosphere, producing a downslope flow known as a mountain wind. These nocturnal winds can be strong, particularly in regions with steep terrain, and may contribute to the formation of temperature inversions and localized fog within valleys.

1.5.6 Katabatic and Foehn Winds

A *katabatic wind* is a broader term for downslope flows driven by the cooling and consequent densification of air near elevated terrain. When the air mass accelerates under gravity, it can produce strong gusts, particularly in polar regions or high mountains where surface cooling is intense. Famous examples include the *Mistral* in southern France and the *Bora* along the Adriatic coast.

In contrast, the *foehn* (or *Chinook*) wind is a warm, dry downslope flow that occurs when moist air ascends a mountain barrier, loses moisture through condensation and precipitation on the windward side, and descends adiabatically on the leeward side.

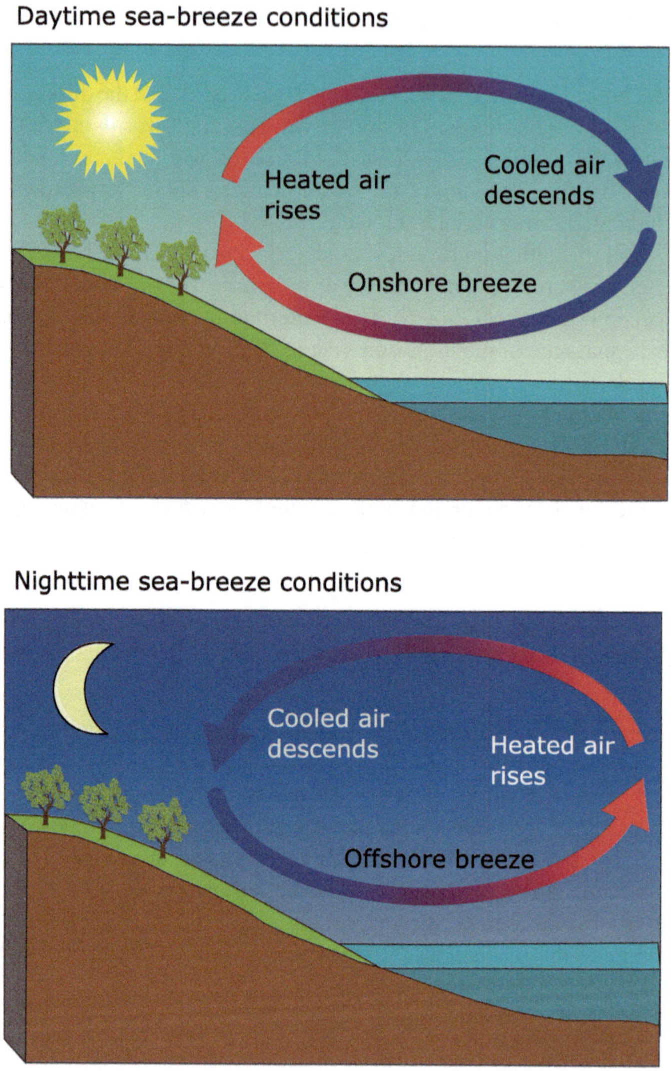

Fig. 1.31 Schematic representation of sea and land breeze circulation

As descending air warms at the dry adiabatic lapse rate, temperatures can rise sharply, producing abrupt thaws or snowmelt (see Fig. 1.32). The foehn effect is common in the Alps, the Rockies, and the Andes.

Local wind systems, though small in spatial scale compared to planetary circulation, play an essential role in redistributing heat and moisture near the Earth's surface. They also influence boundary-layer stability, pollutant dispersion, and the microclimates of valleys and coastal zones.

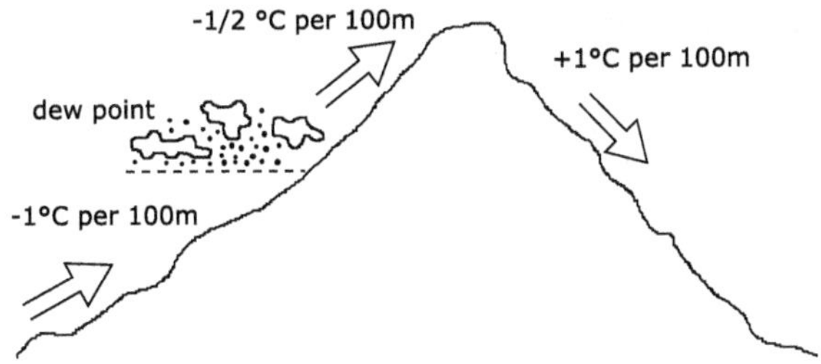

Fig. 1.32 Mechanism of the foehn (Chinook) wind

1.6 Conclusions

The study of atmospheric motion reveals a complex and interconnected system, governed by both global-scale dynamics and local interactions. From the global circulation of the troposphere to localized wind phenomena, each process contributes to the continuous redistribution of energy and momentum within the atmosphere. The main points may be summarized as follows:

- Most weather processes occur within the lowest layer of the atmosphere, the *troposphere*, where temperature decreases with altitude and vertical motions are significant.
- Atmospheric motion occurs across a broad range of spatial and temporal scales. Much like a meandering river filled with eddies of various sizes, large-scale atmospheric flows contain smaller and smaller embedded vortices, each interacting with and influencing the others.
- Wind systems result from a combination of factors, including global and local energy imbalances, seasonal variations, synoptic-scale weather systems, and surface characteristics such as topography and land-sea contrasts.
- Atmospheric circulation is typically classified into three interrelated categories: *primary circulation* (planetary-scale motions associated with global energy balance), *secondary circulation* (synoptic and mesoscale weather systems), and *local winds* (microscale thermally and topographically driven flows).
- Because of the inherently chaotic and multiscale nature of the atmosphere, wind conditions at any specific location cannot be predicted deterministically. For this reason, wind engineering treats wind as a *statistical phenomenon*, analyzed through probabilistic and stochastic approaches rather than fixed deterministic models.

References

ESDU (1987) World-wide extreme wind speeds. Part 1: origins and methods of analysis. Item 87034. ESDU International, London

FUJITA T (1973) Tornadoes around the world. Weatherwise 26:56–62 and 78–83

Holton JR (2004) An introduction to dynamic meteorology. Int Geophys Series, vol 88, 4th edn. Elsevier Academic Press, ISBN:0-12-354015-1

Orlanski I (1975) A rational subdivision of scales for atmospheric processes. Bull Amer Meteor Soc 56:527–530

Chapter 2
The Atmospheric Boundary Layer

2.1 Introduction

As the wind approaches the Earth's surface, frictional forces caused by the terrain become more important. The wind becomes a turbulent flow that varies in a random way in both space and time. The height z_g where the surface friction effects are zero is called *gradient height*. The lowest part of the atmosphere, below the gradient height, is named *Atmospheric Boundary Layer* (ABL) and typically it extends up to 500–1000 m, while the remainder of the air in the troposphere is called *free atmosphere*. The near-earth characteristics are different from what we observe in the rest of the atmosphere, because of the dominating influence of the ground on the lowest layers of the air Stull 1997; Cook 1990. This region is extremely important for wind engineering studies, since most of the structures are immersed in the ABL.

The surface shear stress modifies the geostrophic balance, and the mean wind vector turns from being parallel to the isobars to having a component towards the low pressure. This results in a gradual change in both the wind speed and the direction with the height above the ground, named *Ekman Spiral* (Fig. 2.1). The typical turn is about 20° through the boundary layer, but it is small (a few degrees) and can be neglected in the height range of typical civil structures Dyrbye and Hansen 1999. This chapter deals with the wind characteristics typical of strong winds produced by gales in the higher latitudes. Extreme winds caused by tropical cyclones, thunderstorms and tornadoes can also be important in specific regions worldwide. A description of these phenomena can be found in Holmes 2007.

The thermal state of the atmosphere can also influence the wind properties in the boundary layer. Thermal effects cannot be ignored if wind velocity is less than 10 m/s, but, in strong winds, surface shear stress effects are the most important. In some cases dynamic phenomena on the structures are caused by moderate wind speeds as in vortex induced vibrations, and thermal conditions may be important Dyrbye and Hansen 1999.

D. Rocchi et al., *Wind Engineering*, Research Topics in Wind Energy 9,
https://doi.org/10.1007/978-3-032-17423-9_2

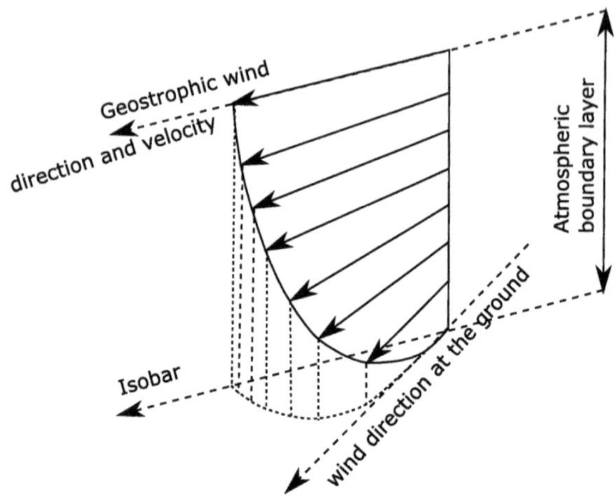

Fig. 2.1 The ekman spiral. The figure shows a simplified and intentionally exaggerated picture of these directional changes Dyrbye and Hansen 1999

2.2 Wind Characteristics

The knowledge of wind characteristics in the atmospheric boundary layer that we have nowadays is due mainly to the analysis of large databases of anemometric records performed in different locations of the world for many years. Long term weather stations are present in the airports, in the air force locations and many interesting points distributed on the ground. A weather station is a facility, either on land or on sea, with instruments and equipment for observing atmospheric conditions to provide information for weather forecasts and to study weather and climate (Fig. 2.2). The measurements include temperature, barometric pressure, humidity, wind speed, wind direction, and precipitation amounts. Wind measurements are usually taken at a standard height of 10 m, in locations far from obstacles that could affect the measurement, while temperature and humidity measurements are performed avoiding direct solar radiation or insulation (Fig. 2.3).

Repetition of measurements at different times in the same place lead to different results. Nevertheless, there are characteristics that are common among a long-term set of measurements in the boundary layer, suggesting that we can develop empirical relationships for the variables of interest.

Figure 2.4 shows a recorded time history of a near-surface wind speed, measured during a 6 h period on the Messina strait at the height of 128 m from the ground. A number of features stand out Stull 1997:

- The wind speed varies in an irregular pattern, revealing the turbulent nature of the wind;
- We can identify a change in the typical (mean) value of the velocity over the day duration;
- The wind speed varies in a limited range over the mean value: this is measurable and it is characteristic of turbulence;
- Fluctuations cover a wide range of frequencies (time scales).

Fig. 2.2 Locations of weather stations all over the world *Source* NOAA national climatic data center www.ncdc.noaa.gov)

Fig. 2.3 Anaemometric station. Author: Famartin *Source* https://commons.wikimedia.org/wiki/File:2008-07-01_Elko_ASOS_viewed_from_the_south_cropped.jpg

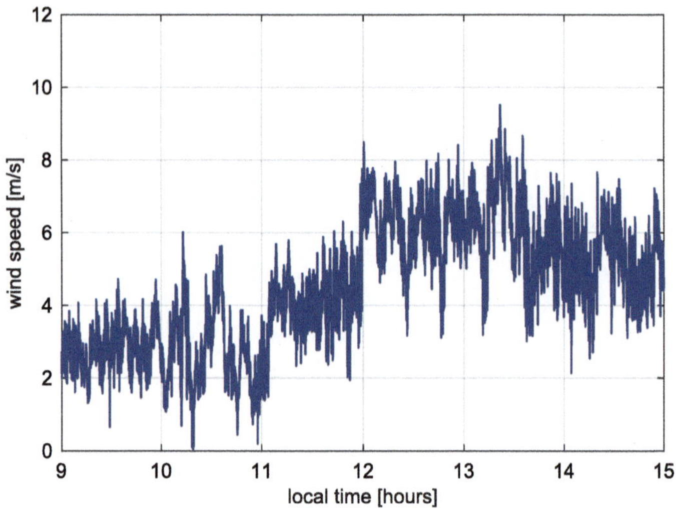

Fig. 2.4 Trace of wind speed on Messina Strait, measured at 128 m height

Turbulence is characterized by a randomness that makes difficult to describe it in a deterministic way: statistical methods must be used. Moreover, if we compare different wind traces measured simultaneously at different heights in the same place (Fig. 2.5), we can also observe:

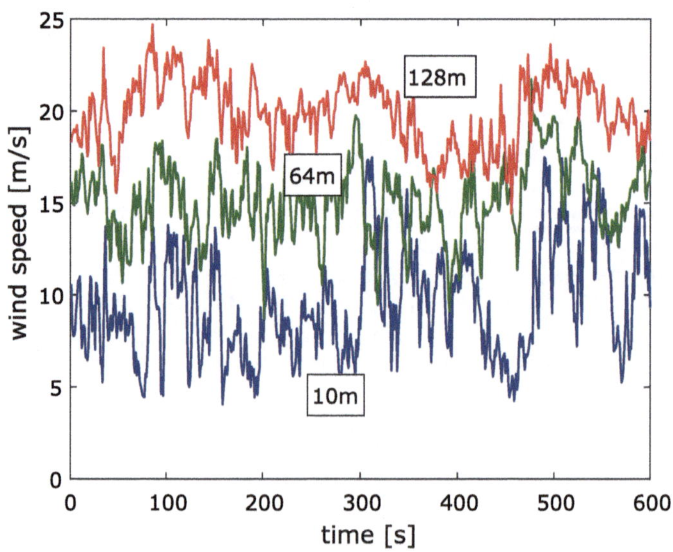

Fig. 2.5 Wind speed at different heights on Messina Strait, 10 min period

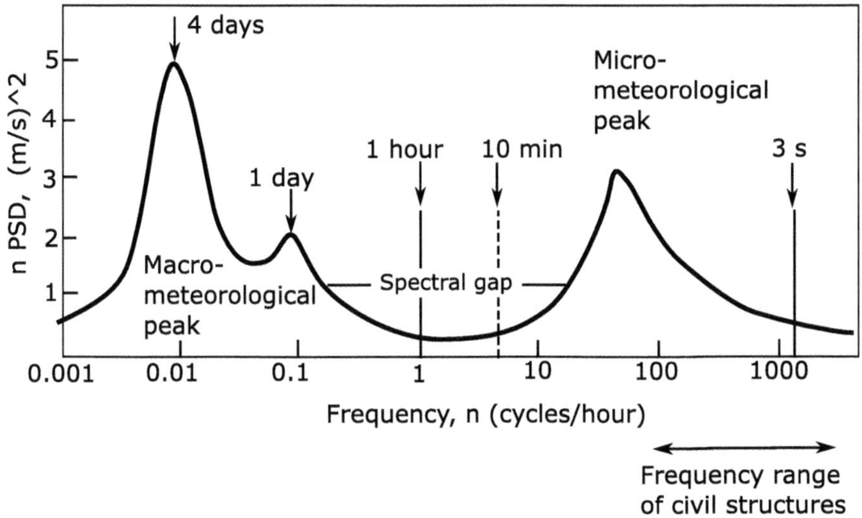

Fig. 2.6 Spectrum of horizontal wind speed after Van der Hoven Van der Hoven 1957

- The average wind velocity increases with the height from ground;
- There is some similarity between the patterns of gusts at all heights, that is indicative of a spatial correlation of the wind turbulence.

Data measured by anemometers contains information both on wind climate and on the atmospheric boundary layer: these two contributions belong to different characteristic scales of the phenomena. Van der Hoven in 1957 compiled the first wind spectrum, that was based on data collected at Brookhaven, NY, USA and it is reproduced in Fig. 2.6 Van der Hoven 1957. In abscissa we have the frequency of the wind-speed variation that can be associated to a characteristic size of eddies: small eddies have shorter time periods (i.e. higher frequencies) than large eddies. The ordinate is a measure of the energy associated with each frequency. We can recognize different important features:

1. The leftmost peak with a period of about 100 h (4 d) corresponds to wind speed variations associated with the transit of fully developed weather systems that is named *macrometeorological peak*;
2. The next peak, at 24 h shows the diurnal increase of wind speed during day and decrease at night;
3. The rightmost peak comprises a range of higher frequencies that are associated to the turbulence of the ABL and that span from a period of 10 min to less than 3 s. This is named *micrometeorological peak*;
4. There is a distinct lack of wind speed variation having periods between 10 min and 1 h. This has been called *spectral gap*.

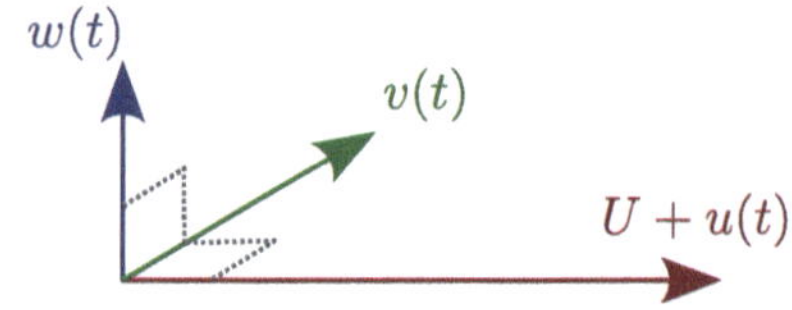

Fig. 2.7 Wind components: longitudinal, lateral and vertical direction *u*, *v* and *w*

Mean and turbulent wind components

The spectral gap (Fig. 2.6) represents a separation between the scales associated with the wind climate and the ones associated with the atmospheric turbulence. Air motions on the left of the gap are associated with the **mean flow**, while motions on the right constitute the **turbulence**. We can define the wind mean velocity U as:

$$\begin{aligned} U(z,t) &= U(z) + u(z,t) \\ U(z) &= \frac{1}{T}\int_0^T U(z,t)dt \quad \text{with} \quad \text{T} = \text{10 min - 1 hour} \end{aligned} \tag{2.1}$$

Generally wind speed is averaged over a period of 10 min or 1 h.
Turbulence is a three dimensional phenomenon so that we have variations in longitudinal, lateral and vertical direction named respectively *u*, *v* and *w*, as visible in Fig. 2.7. The longitudinal direction is defined as the mean wind direction so that the turbulence components have a zero mean value.
Considering typical wind engineering problems we can treat slow wind speed fluctuations as variations in the mean wind velocity that produce only static effects on the structures. On the contrary the high frequency fluctuations (micrometeorological), the wind turbulence, produce dynamic effects on the structures since the typical structural frequencies lay in this range. Moreover the lower is the structure frequency the higher is the wind energy at the same frequency (see also Sect. 2.4).

Taylor's hypothesis

Turbulence, the gustiness superimposed on the mean wind, can be visualized as consisting of irregular swirls of motion called *eddies*. Usually turbulence consists of many different size eddies superimposed on each other Stull 1997. We often need information on the size of eddies and on the scales of motions in the boundary layer. It is difficult to have a picture of the wind motion in many points simultaneously but, on the contrary, it is easier to make measurements at one point in space over a long time period.
In 1938 Taylor suggested that the turbulence can be considered to be *frozen*; than the mean wind is used to translate turbulence measurements as a function of time to their corresponding measurements in space (see Stull 1997). Using this assumption spatial variations of the wind velocity field could be based on temporal variations and vice versa. If an eddy of diameter L is advected at mean wind speed U, the time

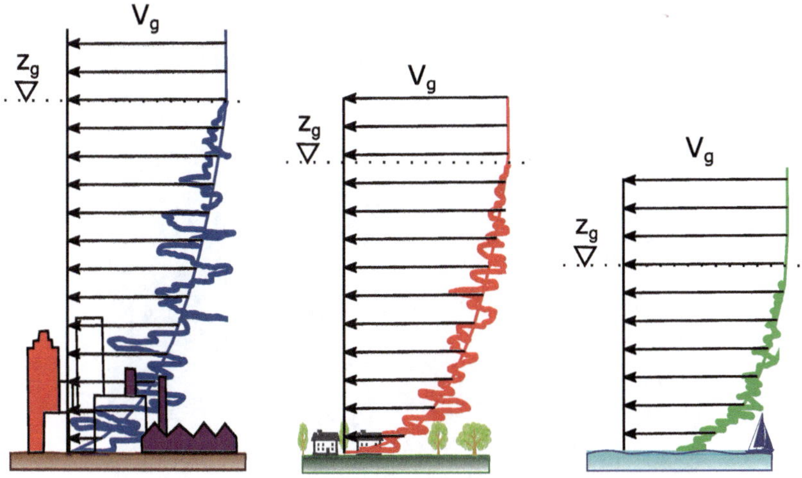

Fig. 2.8 Wind profiles on different terrain roughness CNR-DT 207/2008 2009

period T for it to pass by a stationary sensor is $T = L/U$.
Even if eddies are not really frozen Taylor's hypothesis is a good approximation of natural wind.

2.3 Mean Wind Velocity Profile

Since the boundary layer is dominated by the effects of the surface roughness we expect that the mean wind velocity vertical profile depends on the "types" of roughness elements on the ground, as visible in Fig. 2.8. Frictional drag causes the wind speed to become zero close to the ground, while the pressure gradient forces cause the wind to increase with height. The surface roughness can be characterized by the *aerodynamic roughness length* parameter z_0 [m]. It is related to the dimensions and spacing of roughness elements. Although the roughness length is not equal to the height of the individual roughness elements there is a correspondence between the roughness elements and the aerodynamic roughness length Stull 1997: higher roughness elements (trees, houses, tall buildings) are associated to larger aerodynamic roughness lengths As an example, ESDU 82026 2002 suggests a value of $z_0 \approx 10$ m for centres of large cities, $z_0 \approx 0.5$ m for farmlands, and $z_0 \approx 0.001$ m for calm open sea. Pictures of different terrains are shown in Figs. 2.9, 2.10, 2.11, 2.12.

The logarithmic profile

Figuer 2.13 shows a typical vertical profile of 10 min mean wind speed for two different terrain types: open country ($A:\ z_0 = 0.05$) and town ($B:\ z_0 = 0.3$). It is

Fig. 2.9 Aerial view of a coastal area $z_0 = 0.01$ m *Source* google maps

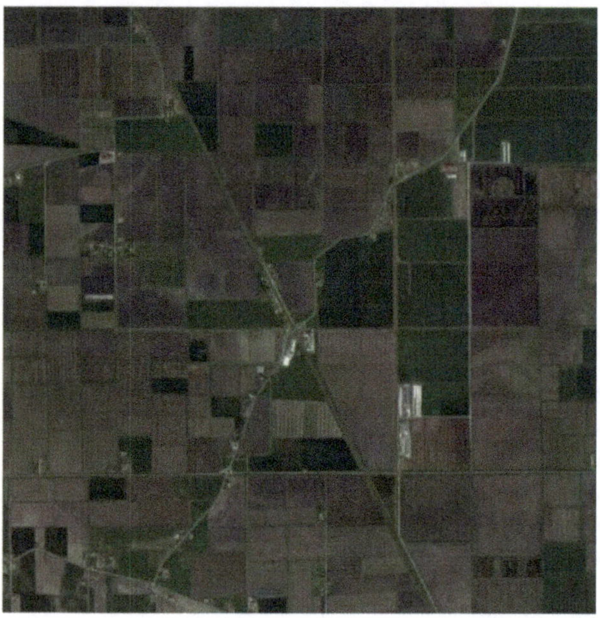

Fig. 2.10 Aerial view of a open country area (farmland) $z_0 = 0.05$ m *Source* google maps

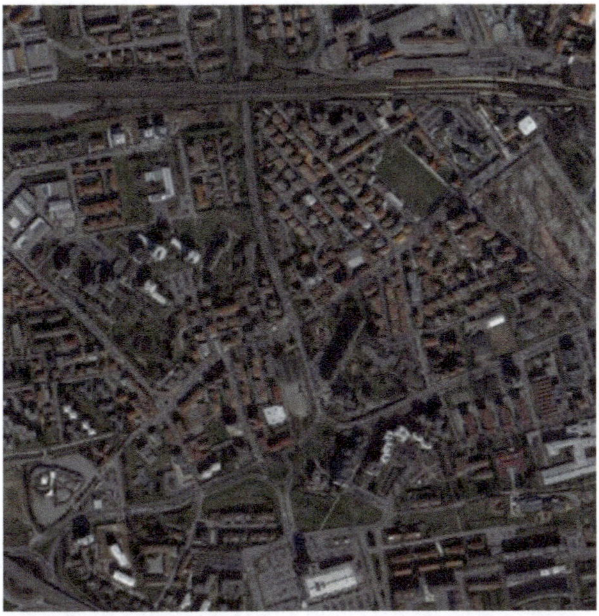

Fig. 2.11 Aerial view of a suburban area (small town) $z_0 = 0.3$ m *Source* google maps

Fig. 2.12 Aerial view of city centre area, with many high buildings $z_0 = 0.7$ m *Source* google maps

possible to observe that in the terrain with higher roughness length (B: town) the velocity is more retarded close to the ground.

It is useful to plot the same graph of Fig. 2.13 but with logarithmic height axis and linear speed axis. We can observe that the lower part of the boundary layer (up to 100–200 meters) seems to follow quite well a linear relationship. To draw the mean wind speed as a function of the height z above the ground we apply a simplified method based on the Buckingham Pi Theory. We suppose that only the following physical variables are relevant: the wind velocity U [m/s], the height above the ground z [m], the roughness length z_0 [m] and the surface stress τ_0. It is useful to replace the surface stress by the friction velocity u^* [m/s]: this is not a physical velocity but is an abstraction introduced for expressing the shear stress in the dimension of velocity ($\tau_0 = \rho_a\, u^{*2}$ being ρ_a the air density).
Upon applying the Pi Theory we found two dimensionless groups:

$$\pi_1 = \frac{U}{u^*} \qquad \pi_2 = \frac{z}{z_0} \tag{2.2}$$

Based on data plotted in Fig. 2.14 we might expect a logarithmic relationship between these two groups:

$$\frac{U}{u^*} = \frac{1}{\kappa} \ln\left(\frac{z}{z_0}\right) \tag{2.3}$$

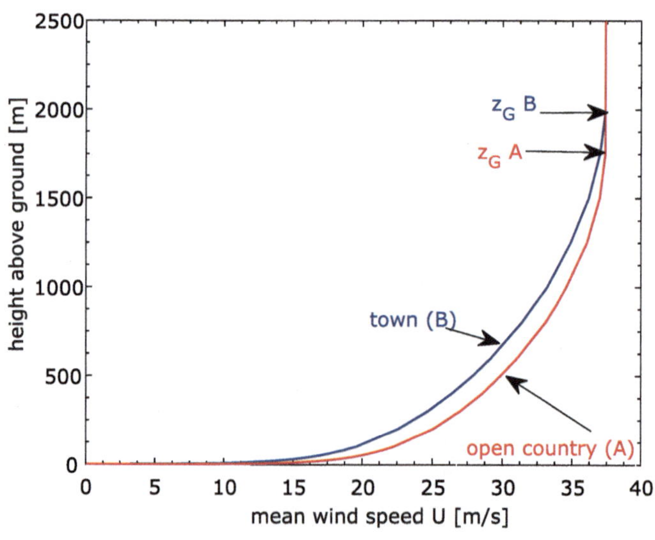

Fig. 2.13 Mean wind velocity vertical profile up to the gradient height for two different terrain types

where $1/\kappa$ is a constant of proportionality. κ is named the von Karman constant and is supposed a universal constant that is not a function of the flow nor of the surface. The value of the constant is near $\kappa = 0.4$ Stull 1997.
This gives the form of the logarithmic wind profile often seen in literature:

$$U(z) = \frac{u^*}{\kappa} \ln\left(\frac{z}{z_0}\right) \tag{2.4}$$

Comparing the log-profile with the real one we see that the log-profile is a good approximation up to 200 m. The logarithmic relationship has also a physical justification and it is based on the Prandtl boundary layer theory Stull 1997; Cook 1990. It is to note that the roughness length z_0 can also be interpreted as the height where the mean velocity is zero ($U(z_0) = 0$), as visible in Fig. 2.14.
More sophisticated analytical models are available to extend the equations through all the ABL up to the gradient height z_g. The one proposed by Deaves and Harris can be found in Cook 1990. It is also useful to use Eq. (2.5) to relate the mean wind speed at two different heights:

$$\frac{U(z_1)}{U(z_2)} = \frac{\ln\left(\frac{z_1}{z_0}\right)}{\ln\left(\frac{z_2}{z_0}\right)} \tag{2.5}$$

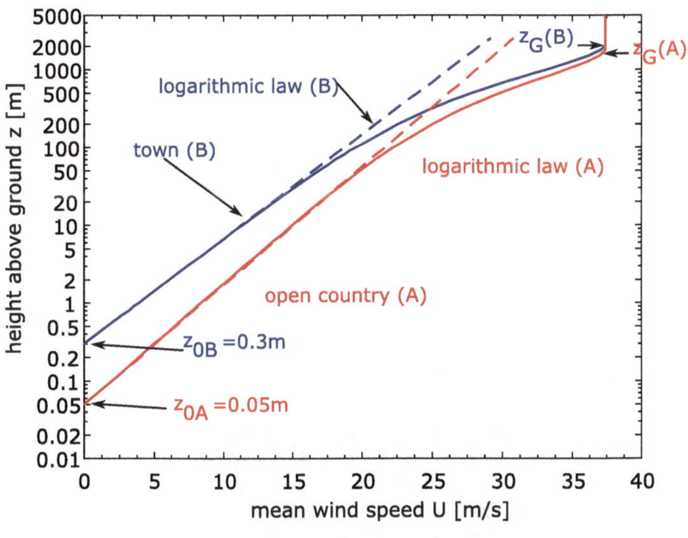

Fig. 2.14 Mean wind velocity vertical profile with log-linear axes

The zero-plane displacement

Over land, if obstacles are very close each other, the flow will be raised such that the top of the obstacles can act like a new displaced surface. Mathematically this can be taken into account in the logarithmic profile introducing a *zero-plane displacement* d [m]:

$$U(z) = \frac{u^*}{\kappa} \ln\left(\frac{z-d}{z_0}\right) \tag{2.6}$$

The zero-plane displacement can be taken as about 75% of the general rooftop height Holmes 2007 (Fig. 2.15).

The power law profile

Another empirical formulation that is used to describe the wind profile is the *power law*:

$$U(z) = U_{ref}\left(\frac{z}{z_{ref}}\right)^{\alpha} \tag{2.7}$$

where U_{ref} is the wind velocity at the reference height z_{ref} and α is the power law exponent and is related to the terrain type. Typical values of α range from 0.12 to 0.30. The exponent α can be related to the roughness length z_0 by the relationship Holmes 2007:

$$\alpha \cong \left(\frac{1}{\ln\left(z_{ref}/z_0\right)}\right) \tag{2.8}$$

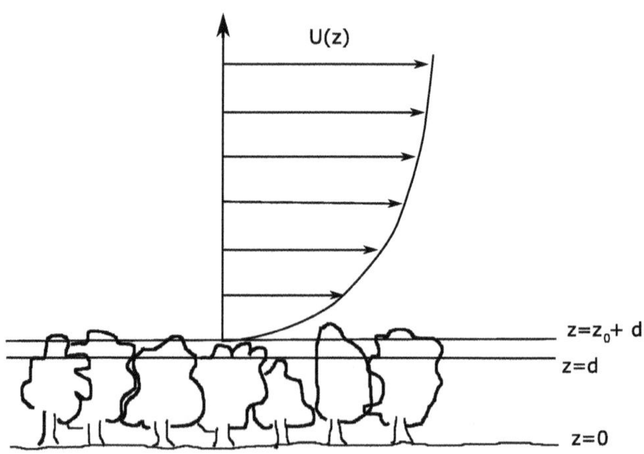

Fig. 2.15 Zero-plane displacement "d" Dyrbye and Hansen 1999

Inhomogeneous terrain

The logarithmic law model has been derived assuming uniform terrain conditions: when the wind has travelled a long distance over the same roughness the flow in the ABL settles down to equilibrium. Up to 50 km of uniform roughness may be necessary to reach a complete equilibrium ESDU 84011 1993. However there are situations where the 'ideal' conditions are not satisfied:

- Changes in surface roughness
- Directionality effects
- Topography effects

Changes in surface roughness

When the air moves from one terrain with roughness z_{01} over a second one with roughness z_{02} the adjustment to a new roughness is not instantaneous. A process of adjustment develops, starting at ground level and gradually moves upward like visible in Fig. 2.16. Even if the new shape of the boundary layer below the *interface* is correct, the magnitude of the wind velocity continues to change until the new boundary layer does not reach equilibrium, that is until the interface reaches the gradient height z_g. Numerical model are available to compute the transition and they can be found in ESDU 82026 2002 and Cook 1990; Dyrbye and Hansen 1999 (see for example Fig. 2.17).

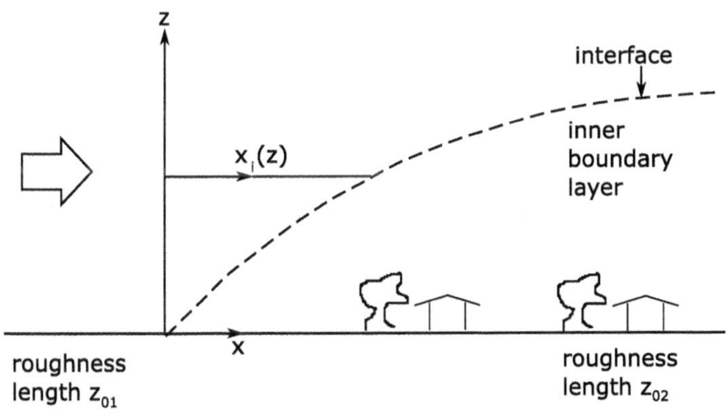

Fig. 2.16 Development of the new boundary layer after a change of terrain roughness (Holmes 2007)

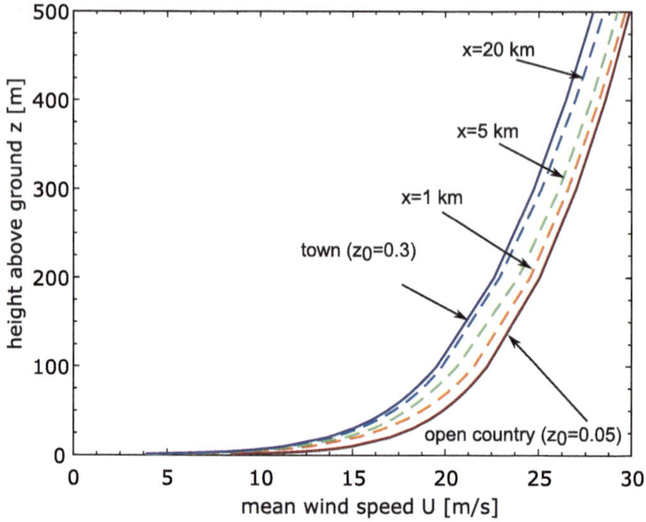

Fig. 2.17 Modification of mean wind profiles after a roughness change from open country ($z_0 = 0.05$) to town ($z_0 = 0.3$) at different upwind distances x from roughness change

Directionality effects

If a structure is not located in a uniform roughness terrain often happens that terrain roughness and roughness changes depend on the incoming wind direction. In this case directionality effects may be considered and the wind profile varies according to wind direction: terrain roughness and roughness changes should be investigated upwind of the site considering sectors of about 30°.

Topography

Topography is a large displacement of the ground surface that is different from the surface roughness and affects the atmospheric boundary layer in different ways:

- Gradual change in altitude;
- Escarpments and ridges;
- Hills and valleys.

All these kind of terrain changes can cause acceleration, deceleration or separation bubbles in the flow and so modify the velocity profile. Engineering models are available to estimate flow modification and can be found in codes, as Eurocode 1 2010 or in bibliography Cook 1990; ESDU 82026 2002.

Table 2.1 Terrain categories and parameters according to Eurocode 1 2010

#	Terrain category	z_0 [m]	$z_{\min}$ [m]
0	Sea or coastal area exposed to the open sea	0.003	1
I	Lakes or flat and horizontal area with negligible vegetation and without obstacles	0.01	1
II	Area with low vegetation such as grass and isolated obstacles (trees, buildings) with separations of at least 20 obstacle heights	0.05	2
III	Area with regular cover of vegetation or buildings or with isolated obstacles with separations of maximum 20 obstacle heights (such as villages, suburban terrain, permanent forest)	0.3	5
IV	Area in which at least 15 % of the surface is covered with buildings and their average height exceeds 15 m	1.0	10

2.3.1 *Mean Wind Profiles According to EUROCODE*

Eurocode 1 2010 contains a formulation based on the logarithmic law to describe the velocity profile:

$$\begin{cases} U(z) = V_b k_r \ln\left(\frac{z}{z_0}\right) & z_{\min} \le z \le 200\,\mathrm{m} \\ U(z) = U(z_{\min}) & z \le z_{\min} \end{cases} \tag{2.9}$$

where a *basic wind velocity* V_b and a terrain factor k_r are introduced. The basic wind velocity depends on the wind climate and it is evaluated at 10 m above the ground of terrain category II (see Fig. 2.1). The methods for the evaluation of V_b are described in Section 2.5. Below a $z_{\min}$ height the velocity profile is considered constant. The terrain factor is related to the roughness length by the expression:

$$k_r = 0.19\left(\frac{z_0}{0.05}\right)^{0.07} \tag{2.10}$$

In Eurocode 1 the roughness length z_0 is simplified to 5 different representative terrain categories that are indicated in Fig. 2.1.

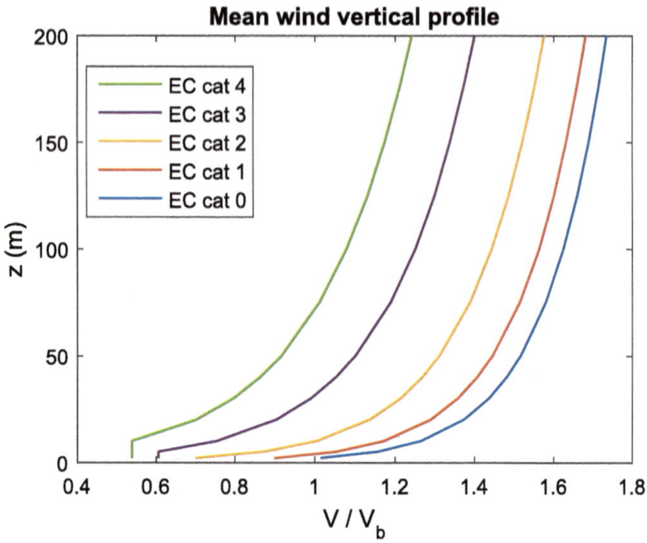

Fig. 2.18 Wind velocity profiles according to (Eurocode 1 2010)

The velocity profile for the five different terrain categories is illustrated in Fig. 2.18. In the graph the velocity is normalized with respect to the *basic wind velocity* V_b value.

2.4 Wind Turbulence

Wind in the atmospheric boundary layer is always turbulent, that is a chaotic and irregular motion. Turbulence has to be described using a stochastic approach.
The properties of atmospheric turbulence can be characterized by a variety of statistical quantities with a single point analysis (time) as follows:

- Turbulence intensity (energy);
- Time scales and integral length scales (vortex size);
- Power spectral distribution of wind velocity fluctuation (energy);
- Peak estimation (gust factor);

and with a multi points analysis (space), as follows:

- Cross spectral density (correlation of wind fluctuations in different points at different frequencies);
- Integral length scales (vortex size);

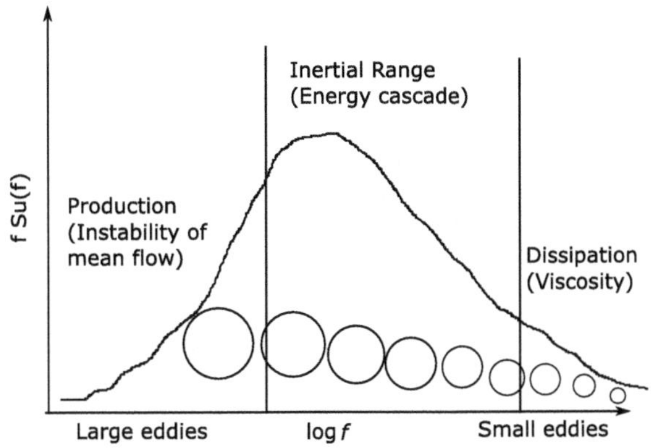

Fig. 2.19 Energy cascade

Kolmogorov's similarity theory

On the basis of the eddy model of turbulence we can introduce the power spectrum that describes the distribution of the turbulence with frequency: a low frequency fluctuation implies a large eddy and a high frequency fluctuation a small eddy. The typical shape of atmospheric spectra is shown in Fig. 2.19 expressed in the format of $f S_u$ versus $\log f$ (Cook 1990). The area under the curve represents the variance of the turbulence and can be divided in three ranges by frequency:

- A production range at low frequency where the turbulence is generated as large eddies;
- An *inertial range* over the middle range of frequency where large eddies break up transferring their energy to smaller scales, than smaller and smaller ones forming the so-called **energy cascade**;
- A dissipation range where eddies are very small so that viscosity is important and energy is dissipated by heat.

A well known verse by the meteorologist L.F. Richardson (1922) describes these effects:

Big whirls have little whirls
That feed on their velocity,
And little whirls have lesser whirls
And so on to viscosity

Turbulence intensity

In homogeneous terrain conditions, the standard deviation of wind velocity slowly decreases with height and is close to zero at the gradient height. Nevertheless up to the height of 200 m it is possible to assume that the standard deviation is almost constant. It is then useful to define a dimensionless parameter that is named *turbulence intensity* defined as (for the along wind turbulence component u):

$$I_u(z) = \frac{\sigma_u(z)}{U(z)} \tag{2.11}$$

where $\sigma_u(z)$ is the standard deviation of the turbulence component u and $U(z)$ is the mean wind velocity, both at height z. In the same way it is possible to define the lateral and vertical turbulence intensities:

$$I_v(z) = \frac{\sigma_v(z)}{U(z)} \qquad I_w(z) = \frac{\sigma_w(z)}{U(z)} \tag{2.12}$$

For instance in Fig. 2.20 it is possible to observe the trace of the wind turbulence measured at Messina strait and the corresponding turbulence intensity indexes. For flat terrain and homogeneous roughness the turbulence intensity is approximately given by (Dyrbye and Hansen 1999):

$$I_u(z) = \frac{1}{\ln(z/z_0)} \tag{2.13}$$

where z_0 is the roughness length.
The relationship between the three indices is approximately:

$$I_v = 0.75\, I_u \qquad I_w = 0.50\, I_u \tag{2.14}$$

Eurocode 1 2010 proposes the same formulation for the along wind turbulence intensity. The turbulence profiles for the different terrain categories are shown in Fig. 2.21. It is possible to observe that the turbulence increases as the roughness increases and decreases with the height above the ground.

Integral length scales

Integral length scales are a measure of the sizes of the vortices in the wind, or in other words the average size of gust in a given direction Dyrbye and Hansen 1999. As an example, L_u^x is the integral length scale for turbulence component u measured in the longitudinal direction x. It is formally defined trough the cross correlation function:

$$L_u^x = \int_0^\infty R_u(r_x)dr_x \qquad [m] \tag{2.15}$$

where $R_u(r_x)$ is the cross correlation function between two different points separated longitudinally (along the mean wind speed direction) by a distance r_x and measured simultaneously.

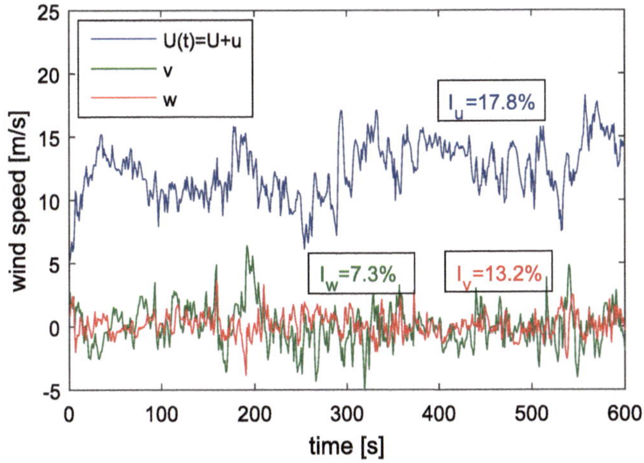

Fig. 2.20 Signature of the wind turbulence measured at the Messina strait over a period of 10 min, measured at 64 m height

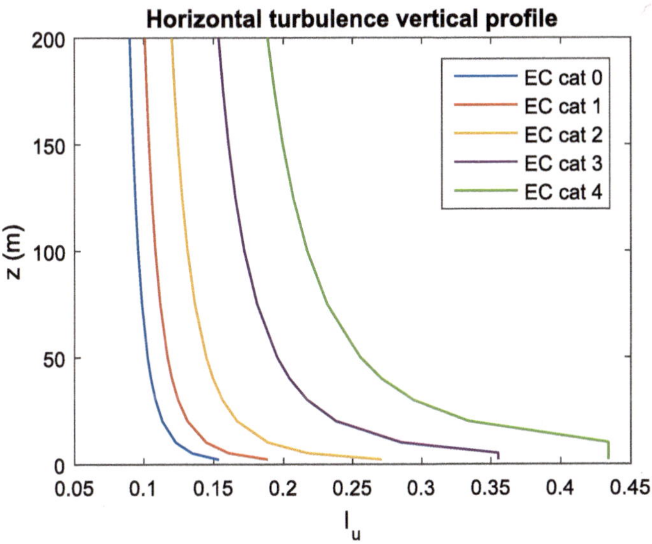

Fig. 2.21 Wind turbulence profiles according to (Eurocode 1 2010)

Similarly, nine integral length scales can be defined:

$$\begin{array}{llll} \text{for longitudinal turbulence component } u & L_u^x & L_u^y & L_u^z \\ \text{for lateral turbulence component } v & L_v^x & L_v^y & L_v^z \\ \text{for vertical turbulence component w} & L_w^x & L_w^y & L_w^z \end{array} \tag{2.16}$$

Assuming the Taylor's hypothesis we can calculate the longitudinal scale of turbulence based on the measurement performed in one point. We use the autocorrelation function $R_u(\tau)$ to compute the time scale T:

$$T_u(z) = \int_0^{\infty} R_u(z, \tau)\, d\tau \tag{2.17}$$

since the "frozen" turbulence is transported by the mean wind speed U the integral length scale is equal to:

$$L_u^x(z) = T_u(z)\, U(z) \tag{2.18}$$

for instance Fig. 2.22 shows the autocorrelation function of the longitudinal turbulence component u. It is possible to calculate a time scale $T = 16.2$ s, than multiplying by the mean wind velocity $U = 12.3$ m/s we obtain a length scale $L_u^x = 200$ m. Full scale measurements are used to estimate integral length scales, but with a large scatter of results. The integral length scales depend on the height from the ground and the terrain roughness. Empirical formulations are suggested like the one proposed in Eurocode 1 2010:

$$\begin{aligned} L_u^x(z) &= 300\left(\frac{z}{200}\right)^{\alpha} \\ \alpha &= 0.67 + 0.05 \ln(z_0) \end{aligned} \tag{2.19}$$

The integral length scales depend on height z above ground and on roughness length z_0.

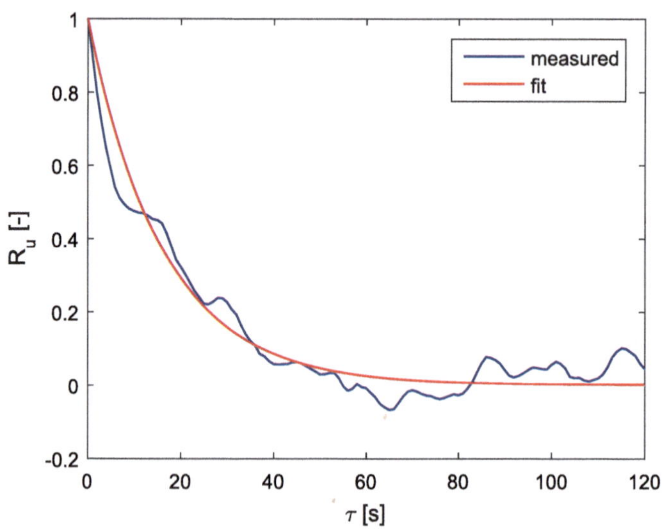

Fig. 2.22 Autocorrelation function of the longitudinal turbulence component of Fig. 2.20

Measures of the spatial correlation in lateral direction indicates an exponential decay behaviour that is represented by the function Dyrbye and Hansen 1999

$$R_u\left(r_y\right) = \exp\left(-\frac{r_y}{L_u^y}\right) \tag{2.20}$$

The remaining integral length scales are often expressed as a function of the longitudinal one:

$$\begin{aligned} L_v^x &\approx 0.25 L_u^x \quad L_w^x \approx 0.1\, L_u^x \\ L_u^y &\approx 0.3\, L_u^x \quad L_u^z \approx 0.2\, L_u^x \end{aligned} \tag{2.21}$$

Power spectral density

The wind spectrum has been yet introduced describing the mechanism generating turbulence. Many equations have been proposed in the literature to describe quantitatively the wind spectrum. Von Karman proposed the following equation:

$$\frac{f \cdot S_u(f)}{{\sigma_u}^2} = \frac{4\left(\frac{f L_u^x}{U}\right)}{\left[1 + 70.8\left(\frac{f L_u^x}{U}\right)^2\right]^{5/6}} \tag{2.22}$$

where the frequency distribution is described by the non-dimensional power spectral density ($S_u(f)$ is the power spectral density of the velocity component u)ESDU 85020 2001. The frequency is normalized introducing a non-dimensional frequency:

$$f^* = \frac{f L_u^x}{U} \tag{2.23}$$

that is calculated using the integral length scale L_u^x. A similar expression is proposed in Eurocode 1 2010:

$$\frac{f \cdot S_u(f)}{{\sigma_u}^2} = \frac{6.8\left(\frac{f L_u^x}{U}\right)}{\left[1 + 10\frac{f L_u^x}{U}\right]^{5/3}} \tag{2.24}$$

The two different spectra are compared in Fig. 2.23. It is to note that in the high frequency range the spectrum of Eq. 2.22 has an almost linear trend with a fixed slope:

$$\frac{f \cdot S_u(f)}{{\sigma_u}^2} = 0.115 f^{*-2/3} \tag{2.25}$$

Basing on the relationship between the frequency at which the curve $f \cdot S_u(f)$ reaches a maximum and the integral length scale L_u^x we can estimate the integral length scale by Simiu and Scanlan 1996:

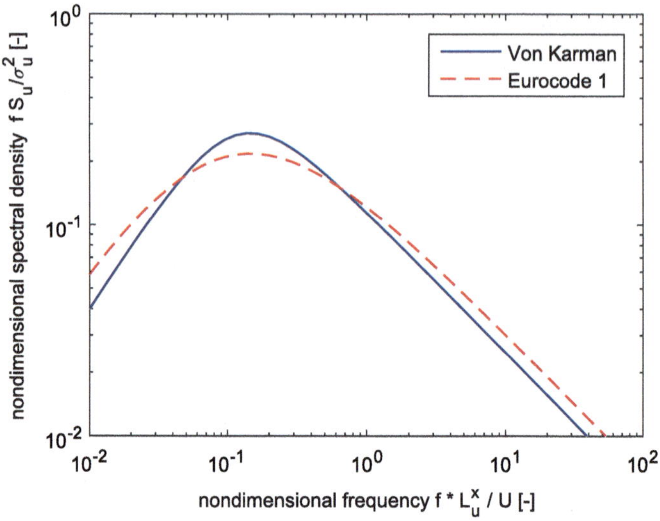

Fig. 2.23 Power spectral density function for the longitudinal turbulence component

$$L_u^x = \frac{1}{2\pi} \frac{U}{f_{peak}} \tag{2.26}$$

Von Karman model proposes also equations for the lateral and vertical turbulence components power spectra ESDU 85020 2001

$$\frac{f \cdot S_i(f)}{\sigma_i{}^2} = \frac{4\left(\frac{f^x L_i}{U}\right)\left(1 + 755.2\left(\frac{f^x L_i}{U}\right)^2\right)}{\left[1 + 283.2\left(\frac{f^x L_i}{U}\right)^2\right]^{11/6}} \quad i = v \quad \text{or} \quad w \tag{2.27}$$

Axes units of Fig. 2.23 are normalized in a dimensionless form. Figure 2.24 shows a real wind spectrum with axes in engineering units: it is possible to observe that the wind energy decreases as the frequency increases. As a consequence, considering the structures frequencies, dynamic amplifications in the wind induced response are more likely to happen if the structure frequency is low.

Cross spectral density and coherence

The normalized cross-spectrum describes the statistical dependence between two points (P an Q) at a frequency f. This dependence is due to the spatial dimension of the vortices in the wind field.

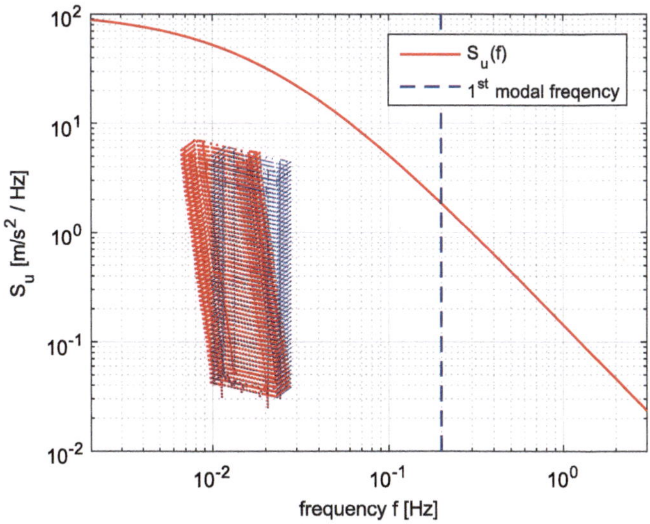

Fig. 2.24 Spectrum of the horizontal component of the wind velocity compared with the structural frequency of an high-rise building (vertical line)

The coherence function is defined as ESDU 86010 2001:

$$\gamma_{uu}^2 = \frac{|S_{uu}(P, Q, f)|^2}{S_{uu}(P, f) \cdot S_{uu}(Q, f)} \tag{2.28}$$

where $S_{uu}(P, f)$ and $S_{uu}(Q, f)$ are the single point spectral density functions at the two points P and Q, spaced in y direction, and $S_{uu}(P, Q, f)$ is the cross-spectral density function (see Bendat and Piersol 2000). Similar functions can be derived for the other turbulence components and spacings.

The coherence function is a real quantity, while the cross-spectral density function is a complex quantity and can be expressed in terms of a modulus and phase angle, or as the sum of real (in-phase) and imaginary (out-of-phase) components.

The simplest formulation for the root-coherence function is the one originally proposed by Davenport, on empirical basis Dyrbye and Hansen 1999

$$\gamma_{uu}(\Delta y, f) = \exp\left[-\left(C_y \frac{\Delta y \cdot f}{U}\right)\right] \tag{2.29}$$

where C_y is a non-dimensional decay constant (typical value is $C_y = 10$). Figure 2.25 shows experimental data of the root-coherence measured in a wind tunnel simulation of the atmospheric boundary layer. Coherence has been measured at different lateral separations.

The main limitation of this formulation is that it implies a close to unity value for small non-dimensional frequencies that is not true for large separations, larger than

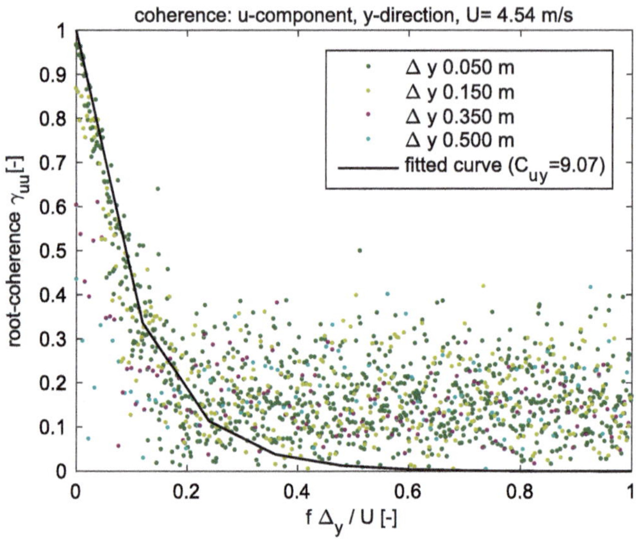

Fig. 2.25 Normalized cross-spectrum. Measured data in wind tunnel simulation of the atmospheric boundary layer

the typical size of vortexes. Different formulations, more sophisticated, are described in literature Dyrbye and Hansen 1999; ESDU 86010 2001.

Gust wind speed

The instantaneous wind speed $U(t)$ varies with time due to the randomness of the phenomenon. The gust wind speed is the expected maximum of the wind velocity, averaged over a period τ, occurring in the period T (generally 10 min or 1 h)

$$\hat{U}(\tau, z) = G(\tau, z) \cdot U(z) \tag{2.30}$$

where

$$G(\tau, z) = (1 + c(\tau) I_u(z)) \tag{2.31}$$

Specific formulations for the calculation of $c(\tau)$ and G can be found in ESDU 83045 2002. For $\tau = 3\,s, c(\tau) = 3$; in open country terrain G is about 1.5. Figure 2.26 shows values of G as a function of the average time τ in open country terrain conditions ($z_0 \approx 0.05$) and elevation $z = 10$ m Simiu and Scanlan 1996.

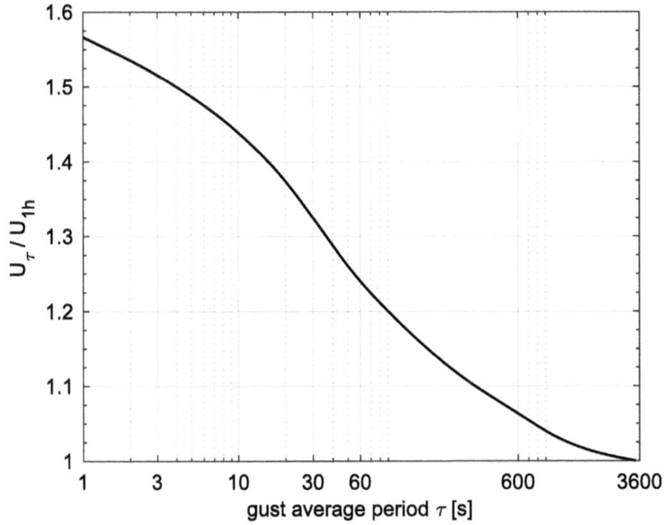

Fig. 2.26 Expected maximum of the wind velocity, averaged over a period τ to that averaged over 1 h Simiu and Scanlan 1996

2.5 Wind Statistics

In design stages, for wind loading calculations and risk analyses, we often need information about the probability of a certain wind speed to be exceeded. Such datum is required to the designer for the estimation of the wind loads that can affect the structure during its lifetime. The establishment of an appropriate design wind speed is a critical step in the calculation of the design wind loads. This procedure is based on the analysis of records of wind speed data over a number of years.

The statistical analysis of databases can define the properties of the parent probability distribution (complete population of wind speeds) or only a subset to define the probability distribution of extreme values (Fig. 2.27). Each element of the complete population is a 10-min average wind velocity, generally acquired by anemometric stations at regular intervals while each element of the population of annual maxima is the maximum value over each year (10-min average).

Return period and risk

In the definition of the design wind speed an important parameter is the **return period**. The return period R $[years]$ is defined as the inverse of the complementary cumulative distribution of the extremes:

$$R = \frac{1}{\text{Probability of exceedence}} = \frac{1}{1 - F_M(v)} \tag{2.32}$$

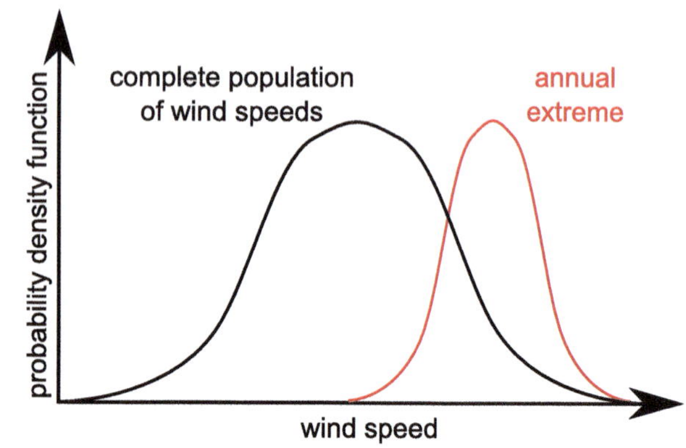

Fig. 2.27 Probability distribution of the complete population of wind speeds and of the extreme values

where $F_M(v)$ is the cumulative distribution of the annual maximum wind speed.

For instance, a 50-year return-period wind speed has a probability of exceedence of 0.02 $(1/50)$ in any one year or an average rate of exceedance of 1 in 50 years (it should not be interpreted as occurring regularly every 50 years). We refer to the return period also as *Mean Recurrence Interval.* We can now introduce the *Design life of the structure* L $[years]$ that is the nominal life of the structure defined in the design stage Holmes 2007. Figure 2.28, for instance, shows the indicative design lives for different types of structure, according to Eurocode Eurocode 2006. The probability that a wind speed having a return period R is exceeded over the design life of the structure L is named *risk* r Holmes 2007. Assuming that all years are statistically independent the risk is:

$$r = 1 - \left[1 - \left(\frac{1}{R}\right)\right]^L \tag{2.33}$$

If we set $R = 50$ and $L = 50$ we find $r = 0.636$. Therefore, there is a 64% chance that the 50-year return period wind speed is exceeded during 50-year lifetime. Loads

Design working life category	Indicative design working life (years)	Examples
1	10	Temporary structures [(1)]
2	10 to 25	Replaceable structural parts, e.g. gantry girders, bearings
3	15 to 30	Agricultural and similar structures
4	50	Building structures and other common structures
5	100	Monumental building structures, bridges, and other civil engineering structures
(1) Structures or parts of structures that can be dismantled with a view to being re-used should not be considered as temporary.		

Fig. 2.28 Example of design life of different types of structures Eurocode 2006

Table 2.2 Risk of exceedance for different return periods R, assuming a design life $L = 50$ years

Return period R [$years$]	50	100	200	500	1000
Risk of exceedance r (%)	63.6	39.5	22.2	9.5	4.9

calculated using this risk level are not generally suited for design purposes. In fact, in case of Ultimate Limit States loads they must be factored. Codes and standards include multiplier load factors that are applied to the wind action. Eurocode 2006 sets for ordinary structures a lifetime equal to 50 years to calculate the wind actions using a return reriod of 50 years and a safety factor of 1.5 is suggested to decrease the risk of failure.

In other design approaches, in order to reduce the risk, an higher return period is used, generally as large as about ten times the design life of the structure. ASCE standard 7–10 ASCE 7/10 2010 for the United States provides wind maps that are directly applicable rather than using importance factors and load factors: in this way the maps, providing the wind speed directly, more clearly inform about the storm intensities for which designs are performed. ASCE 7/10 2010 uses a 700-year return period wind speed for structures having a design life of 50 years. Loads calculated using this high return period wind speed are not factored when applied to the structure.

Table 2.2 shows the risk of exceedance as a function of the return period, assuming a design life $L = 50$ $years$. If we want a probability of exceedance of 5% during a design life of 50 years we need to use a wind velocity having a return period of 1000 years.

2.5.1 Extreme Winds

The statistical analysis of high return period winds is performed determining the probability distribution of extreme values. One of the most used model for the description of the distribution of the annual maximum is the **Type I Extreme value distribution** that is also known as *Gumbel distribution*. Theoretical methods of estimation of extremes were first presented by Fisher and Tippet (1928) and than by Gumbel (1958)Palutikof et al. 1999; Cook 1990.

The analysis attributes to the annual maximum the cumulative distribution

$$F_M(v) = \exp(-\exp(-(a(v-u)))) \tag{2.34}$$

where v is the annual maximum wind velocity and the two model parameter of the distribution are the *mode* u and the *dispersion* $1/a$. So $F_M(v)$ is the probability that the annual maximum is less than v. It is useful to introduce a reduced variable y as

$$y = a\,(v - u) \tag{2.35}$$

so that the cumulative distribution is

$$F_M\,(v) = \exp\left(-\exp\left(-y\right)\right) \tag{2.36}$$

Since the return period is directly related to the cumulative distribution of annual maxima (Eq. 2.32), once we know the model parameter a and u we can evaluate the maximum wind speed as a function of the return period:

$$\begin{aligned} v &= u - \frac{1}{a}\ln\left(-\ln\left(F_M\,(v)\right)\right) \\ &= u - \frac{1}{a}\ln\left(-\ln\left(1 - \frac{1}{R}\right)\right) \end{aligned} \tag{2.37}$$

It is also useful to remember the relationship (method of moments) between the mean μ and the standard deviation σ of the distribution and u and a:

$$\begin{aligned} \frac{1}{a} &= \left(\frac{\sqrt{6}}{\pi}\right)\sigma \\ u &= \mu - 0.450\sigma \end{aligned} \tag{2.38}$$

This method allows the evaluation of the parameters of the extreme distribution only by the calculation of μ and σ Holmes 2007.
Extreme value analysis can be performed both on 10 min and 1 h average wind speed or gust wind speed. The choice depends on the approach to be used in defining the wind action: 10 min average is used in Eurocode.

Gumbel's method

One simple method to calculate the model parameters a and u has been proposed by Gumbel and can be summarized in these steps Holmes 2007:

1. Extract the largest wind speed v in each year for M years;
2. Rank series from smallest to largest ($m = 1, 2, ...$to M)
3. Assign a probability of non-exceedance p to each value

$$p = \frac{m}{(M+1)} \tag{2.39}$$

4. Form the reduced variable $y = -\ln(-\ln(p))$
5. Plot v versus y in a graph and check that there is a linear relationship
6. Evaluate the model parameters a and u by a line fitting (for instance using a least-square fit). u is the intercept and $1/a$ the slope.

It is to point out that this method is biased Holmes 2007; ESDU 87034 1990 and gives distorted values of probability of non-exceedance. Alternative methods can be used such the Gringorten's method Holmes 2007; Palutikof et al. 1999 or the Lieblein BLUE estimators (best linear unbiased estimators) described in ESDU 87034 1990 where weighted coefficients are used in the evaluation of model parameters. Nevertheless the Gumbel's method, applied to extreme wind data, gives generally conservative results Cook 1982. Other authors Cook 1982; ESDU 87034 1990 suggest to perform the statistics on the wind pressure instead of the wind velocity, depending on the characteristics of the parent wind distribution. The topic is investigated in detail in (Lagomarsino et al. 1992; Cook and Harris 2004; Pagnini and Solari 2009). However, the differences in the results are generally small in practical design situations.

Case study: Torino Caselle Airport

The method previously described is here applied to a set of wind data. We use the annual maxima, averaged over a period of 10 min, measured by the anemometer placed at Torino Caselle airport for 20 years, in the period 1988–2007 (Table 2.3).

According to the steps previously described we sort the data in order of increasing magnitude and we assign a probability p according to equation (2.39) and than the value y as shown in Table 2.4. We plot the measures in a graph (Fig. 2.29) and we perform a straight line fitting using the least-squares method. The model parameter u is the intercept and $1/a$ the slope. Table 2.5 shows the model parameters calculated for the example using the Gumbel's method and the method of moments (Eq. 2.38).

In the end, we calculate the annual maximum wind velocity as a function of the return period R according to equation (2.37), as shown in Fig. 2.30.

Transformation of the data base

The extreme value analysis performed on the data measured at the airport is specific of the terrain conditions because it is influenced by possible obstacles close to the anemometer and the terrain roughness. The anemometer site can be far from the construction area of the structure we are studying so that it is needed to transport the analysis results from the anemometer site to the target area. It is than useful to transform the velocity value to reference conditions as defined in Sect. 2.5.3.1 (Fig. 2.31). We can define a conversion factor as:

$$K_{ref} = \frac{\text{velocity at reference site } (z = 10\,\text{m}, z_0 = 0.05)}{\text{velocity at anemometer site}} = \frac{V_{ref}}{V_a} \tag{2.40}$$

Table 2.3 Annual maxima (10 min mean) measured at Torino Caselle airport in the period 1988–2007

Year	Data	V [m/s]
1988	17-Mar-1988 07:20:00	16.5
1989	31-Jul-1989 17:20:00	9.3
1990	27-Feb-1990 21:00:00	19.5
1991	27-Jun-1991 16:20:00	14.4
1992	16-Feb-1992 17:50:00	15.4
1993	22-Feb-1993 12:20:00	13.3
1994	05-Jun-1994 04:50:00	11.8
1995	29-Mar-1995 15:50:00	15.9
1996	17-Feb-1996 13:50:00	11.3
1997	19-Mar-1997 12:00:00	13.4
1998	05-Mar-1998 12:50:00	13.4
1999	05-Feb-1999 19:50:00	19.0
2000	16-Feb-2000 20:50:00	10.3
2001	23-Nov-2001 03:50:00	11.8
2002	21-Mar-2002 21:50:00	13.4
2003	29-May-2003 18:20:00	11.3
2004	19-Nov-2004 12:50:00	13.4
2005	21-Jan-2005 10:20:00	14.9
2006	27-Jul-2006 17:50:00	12.3
2007	04-Jul-2007 16:20:00	13.4

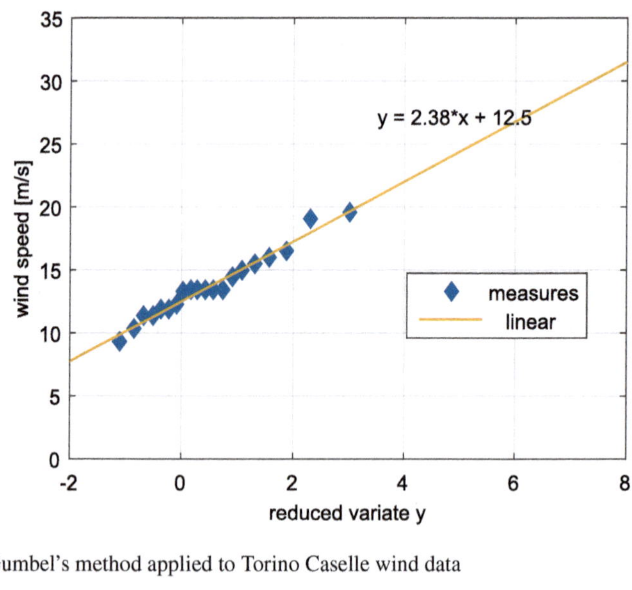

Fig. 2.29 Gumbel's method applied to Torino Caselle wind data

Table 2.4 Torino-Caselle airport: extreme value analysis

Rank	Year	V [m/s]	p	y
1	1989	9.3	0.0476	– 1.1133
2	2000	10.3	0.0952	– 0.8550
3	1996	11.3	0.1429	– 0.6657
4	2003	11.3	0.1905	– 0.5057
5	1994	11.8	0.2381	– 0.3612
6	2001	11.8	0.2857	– 0.2254
7	2006	12.3	0.3333	– 0.0940
8	1993	13.3	0.3810	0.0355
9	1997	13.4	0.4286	0.1657
10	1998	13.4	0.4762	0.2985
11	2002	13.4	0.5238	0.4360
12	2004	13.4	0.5714	0.5805
13	2007	13.4	0.6190	0.7349
14	1991	14.4	0.6667	0.9027
15	2005	14.9	0.7143	1.0892
16	1992	15.4	0.7619	1.3022
17	1995	15.9	0.8095	1.5544
18	1988	16.5	0.8571	1.8698
19	1999	19.0	0.9048	2.3018
20	1990	19.5	0.9524	3.0202

Table 2.5 Parameters of the Type I distribution

	Method of moments	Least squares
media	13.7	
std	2.63	
u	*12.52*	*12.45*
a	*0.49*	*0.42*
$V(50y)$ [m/s]	**20.5**	**21.7**

The value referred to standard condition can be used as basic wind velocity in the wind profiles formulations Ballio et al. 1991.

Further remarks

Wind climate studies can be more sophisticated if other factors are included in the analysis. Refined methods can foreseen that:

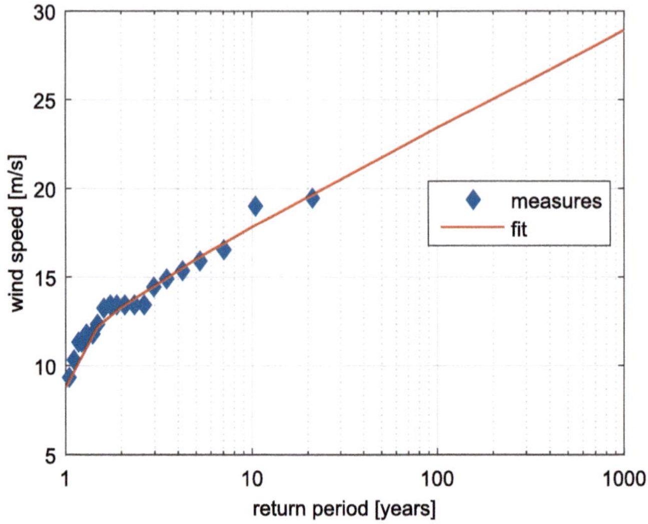

Fig. 2.30 Expected maximum wind speed as a function of the return period

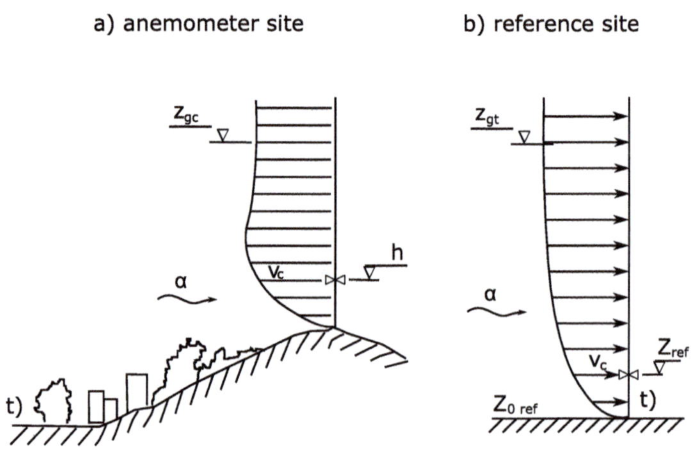

Fig. 2.31 Transformation of the wind data (adapted from Ballio et al. 1991)

- Records from several stations can be combined;
- Wind direction effects can be considered to take care of the typical incoming direction of high winds (Cook 1982, 1983);
- To increase the number of data available more than one maximum per year can be considered (method of independent storms) Cook 1998; ESDU 88037 1988;
- In locations where wind climate is dominated by more than one storm mechanism (thunderstorm or tropical storms), extreme value analysis must take care of the mixed wind climate condition Holmes 2007.

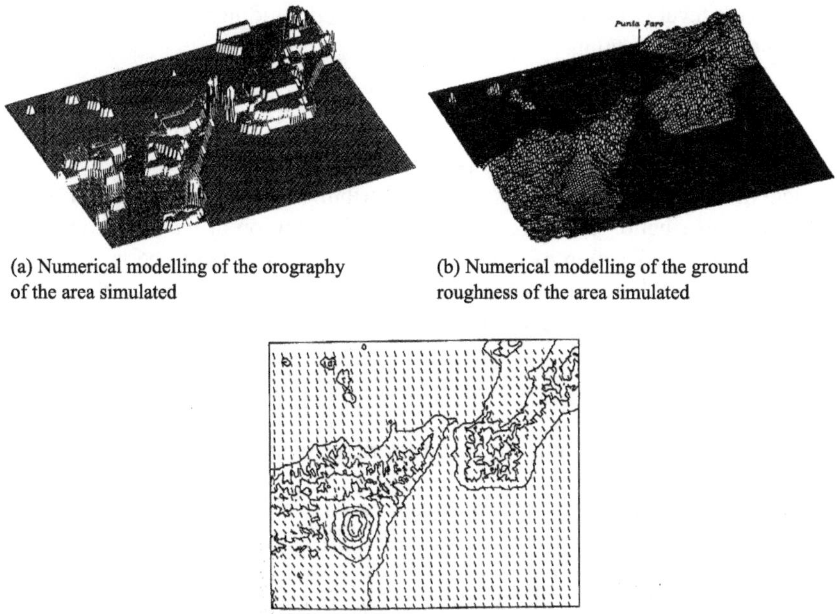

(a) Numerical modelling of the orography of the area simulated

(b) Numerical modelling of the ground roughness of the area simulated

(c) Wind field in the strait, at 10 m above the ground, gradient wind blowing from north

Fig. 2.32 Detailed wind study for the Messina strait Bocciolone et al. 1993. Reprinted with permission from Elsevier

In the design and construction of structures of outstanding importance and susceptibility to wind, an accurate analysis is generally achieved with an *ad hoc* installation of an anemometer in the construction area. After a short period of measures (1–2 years), it is possible to perform a correlation between the measures at construction site and measures performed by existing anemometric stations in a larger area (airports). This correlation analysis is used to evaluate the transformation of the data base with high accuracy (see paragraph 2.5.1). Moreover, in case of complicate terrain conditions, such as in presence of significant deviations of the flow due to orography, numerical simulations of the wind field can be performed Bocciolone et al. 1993, as shown in Fig. 2.32.

2.5.2 *Parent Wind Distribution*

In some design applications we need information on the distribution of the complete population of wind speeds at site:

- Fatigue damage. Fatigue is caused by repetitive cycles of loads.
- Wind comfort for occupants (comfort for people *inside* buildings).

- Pedestrian level wind (comfort for people *around* buildings).
- Wind energy production.
- High speed trains runnability.

In all these cases we are not interested in an extreme wind condition but the fraction of time that a certain level of wind speed is exceeded.
It is generally accepted that the parent wind distribution can be described by a Weibull probability distribution.The parent probability distribution of wind speed provides information on the fraction of time that wind can be expected to be greater than a specific value, but does not provide any information on how often a particular wind speed can be expected to be exceeded.

Weibull distribution

A statistical distribution commonly used for describing measured wind speed data is the Weibull distribution.
The probability density function is described by:

$$f_v(v) = \frac{k}{c}\left(\frac{v}{c}\right)^{k-1} \exp\left(-\left(\frac{v}{c}\right)^k\right) \qquad (v \geqslant 0) \tag{2.41}$$

while the complementary cumulative distribution (probability of exceedance) is:

$$P(V > v) = 1 - F_v(v) = \exp\left(-\left(\frac{v}{c}\right)^k\right) \tag{2.42}$$

where v is the wind velocity and the model parameters are the scale c and the shape k. Typical value of c are 2–10 m/s and k 1–2 Holmes 2007; Pagnini and Solari 2009. Different methods are available for the estimation of the model parameters on the basis of measured data and a review can be found in Conradsen et al. 1984. An open discussion is on the fact that the Weibull distribution imposes $f_v(0) = F_v(0) = 0$ while anemometric measurements shows long periods of calm. It is also to note that measuring errors are introduced by the use of wind speed measuring equipments with too high starting wind speed that introduce a high probability of calm. The Weibull distribution can be modified to take care of wind calms using the hybrid method that is reviewed in Conradsen et al. 1984.

$$P(V > v) = 1 - F_v(v) = A_0 \exp\left(-\left(\frac{v}{c}\right)^k\right) \tag{2.43}$$

where A_0 is the probability of $v > 0$.

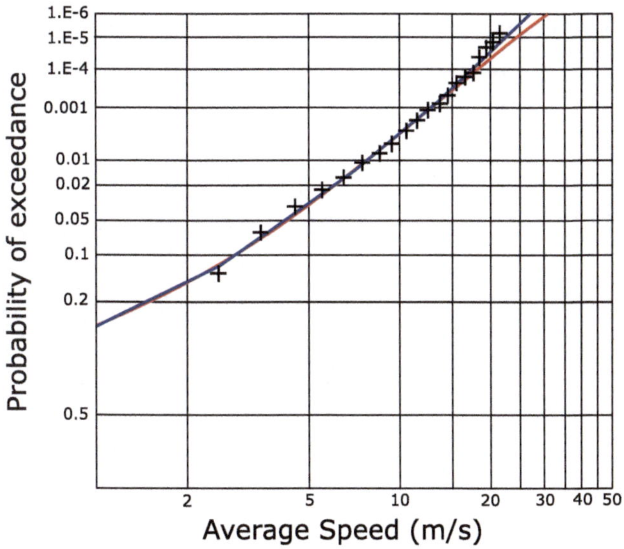

Fig. 2.33 Milano Malpensa airport. Probability of exceedance of a given wind speed irrespective on the wind direction Solari 2006

Case study: Milano Malpensa Airport

Weibull distribution has been fitted to wind measures performed at Milano Malpensa airport. Figure 2.33 shows the probability of exceedence as a function of the wind velocity.

Directionality effects

The Weibull distribution can be modified to take care of the wind direction. Measures are divided according to incoming wind sectors and the cumulative distribution becomes:

$$P\left(V > v, \theta\right) = A_{\vartheta} \exp\left(-\left(\frac{v}{c_{\vartheta}}\right)^{k_{\vartheta}}\right) \qquad \sum A_{\vartheta} = A_0 \tag{2.44}$$

where A_0 is the probability of $v > 0$ and θ is a specific sector of exposure. The results are generally expressed in terms of polar plot with lines having the same probability of exceedance as in Fig. 2.34. In the graph each curve corresponds to a given probability of exceedance in a 30° sector. We observe that prevailing winds blow from north (inner line) while extreme winds blow from a larger sector spanning from north-west to north-east (outer line).

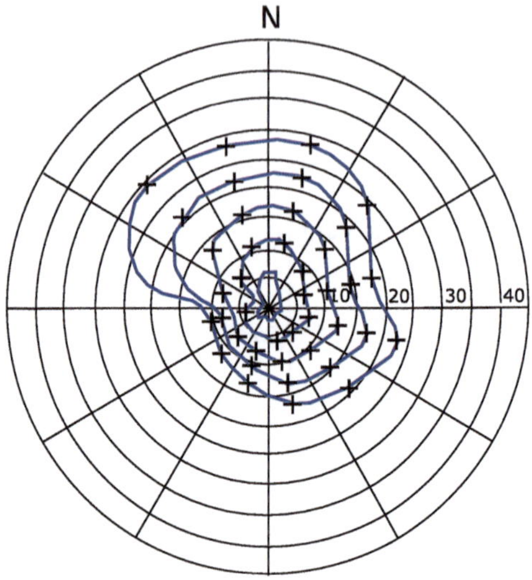

Fig. 2.34 Milano Malpensa airport. Probability of exceedance of a given wind speed depending on the wind direction Solari 2006. Lines corresponds to the probability of exceedance of (from inner to outer): 1%, 0.1%, 0.01%, 0.001%, 0.0001%

2.5.3 *Aeolian Wind Maps*

A territorial aeolian map is intended as an instrument to give to the designer, if specific studies are not available, a value of the reference wind velocity for a specific territory. A generic territory aeolian map may be represented by means of two alternative techniques. The first (used for instance in Great Britain) utilizes lines joining points characterized by equal reference wind speed (Fig. 2.35). The second, proposed for instance by Eurocode divides the territory into zones of equal reference wind speed (Fig. 2.36). The two methods possess both qualities and defects which are mutually exclusive. The first is congruent with the variation evidently continuous of V_{ref}; its application gives rise, on the other hand, to potentially controversial situations. The second eliminates every possible subjective interpretation when demarcations coincide with administrative limits that is very useful in rule-making sector. However, in concurrence with the demarcations, this method gives rise to clearly unrealistic conditions of discontinuity Ballio et al. 1991.

2.5.3.1 Basic Wind Velocity

Mean wind velocity profiles are often normalized with respect to a reference wind velocity in a nominal terrain roughness condition, as in Eurocode 1 2010 (see Sect. 2.3.1). Wind statistics is performed not on the measured wind velocities but on the transformed ones that take care of the terrain roughness at the anemometer location. Wind profile formulations are then used to define the wind speed value at the structure location, taking care of the local terrain roughness. Codes and standards

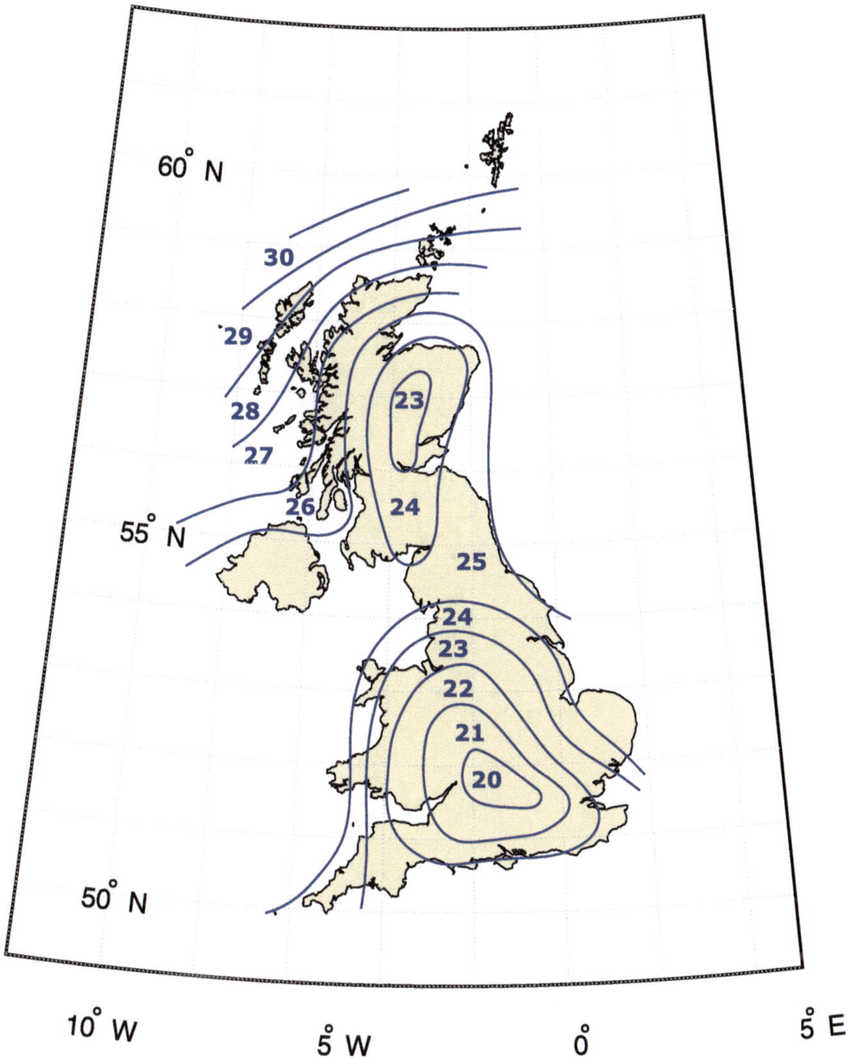

Fig. 2.35 UK wind map. Extreme hourly-mean wind speeds [m/s], with a 50-year return period assuming uniform open country terrain. Lines are characterized by equal reference speed

on wind action usually assign a reference value of the basic wind speed which represents the characteristic 10 min *mean wind velocity*, irrespective of wind direction and time of year, at 10 m *above ground level*, in open country terrain with low vegetation and isolated obstacles ($z0 = 0.05$), having a yearly exceedance probability of 0.02 (*Return period* 50 *years*).

Fig. 2.36 Italian wind map. Basic wind velocity [m/s], averaged over 10 min, at 10 m above ground level with a 50-year return period in terrain category II ($z_0 = 0.05$ m). Areas are characterized by equal reference speed NTC 2005

Change in return period

Wind maps generally refer to a return period of 50 years. In case of different return period we can use the formulation proposed for instance in Eurocode:

$$U_{b,p} = c_p \cdot U_{b,50y} \tag{2.45}$$

$$c_p = \left(\frac{1 - K \cdot \ln\left(-\ln\left(1 - p\right)\right)}{1 - K \cdot \ln\left(-\ln\left(0.98\right)\right)} \right)^n \tag{2.46}$$

where K is the shape parameter depending on the coefficient of variation of the extreme-value distribution and n is the exponent. If specific values are not given, we can assume $K = 0.2$ and $n = 0.5$.

For instance, if we have a 50-years return period wind velocity of 25 m/s and we want to calculate the value for a 100 years return period:

$$
\begin{aligned}
U_{b,50y} &= 25\,\text{m/s} \\
R &= 100y \\
p &= 1/100 = 0.01 \\
U_{b,100y} &= \left(\frac{1 - 0.2\ln(-\ln(1-0.01))}{1 - 0.2\ln(-\ln(0.98))} \right)^{0.5} \cdot 25 = 26\,\text{m/s}
\end{aligned} \tag{2.47}
$$

Different formulations can be used, depending also on the characteristics of the parent wind distribution. This topic is discussed in Lagomarsino et al. 1992; Pagnini and Solari 2009.

2.6 Conclusions

This section has described the main characteristics of the atmospheric boundary layer, in terms of mean wind velocity and turbulence. Formulations have been proposed for the estimation of the main important quantities. Then the problem of defining a design wind speed was presented. Since the wind is a random phenomenon, the information about the probability of a certain wind speed to be exceeded and risk analyses were introduced. The distribution of extreme wind, the return period and the risk of exceedance of a given wind speed in L years was presented. Then the parent distribution or the distribution of the mean wind was described since in some design applications the information of the distribution of the complete population of wind speeds at the site is also important.
We have generally proposed the most used and simplest formulations or the ones that are present in Eurocode 1 2010. These are also adequate for many wind engineering applications; nevertheless, in particular for structures of particular relevance or that are located in particular wind climate areas more sophisticated and refined methods must be used. A good reference is the ESDU Wind Engineering series whose items are quoted in the bibliography.

References

ASCE 7/10 (2010) Minimum design loads for buildings and other structures (ASCE/SEI 7/10)

Ballio G, Lagomarsino S, Piccardo G, Solari G (1991) A first step towards a map of Italian extreme winds. Part 2: results, ripercussion on standards, design implications. Costruzioni Metalliche 4

Bendat JS, Piersol AG (2000) Random data: analysis and measurement procedures. Wiley, New York

Bocciolone M, Gasparetto M, Lagomarsino S, Piccardo G, Ratto CF, Solari G (1993) Statistical analysis of extreme wind speeds in the straits of Messina. J Wind Eng Ind Aerodyn 48(2–3):359–377

CNR-DT 207/2008 (2009) Istruzioni per la valutazione delle azioni e degli effetti del vento sulle costruzioni. CNR

Conradsen K, Nielsen LB, Prahm LP (1984) Review of Weibull statistics for estimation of wind speed distributions. J Climate Appl Meteorol 23(8):1173–1183
Cook NJ (1982) Towards better estimation of extreme winds. J Wind Eng Ind Aerodyn 9(3):295–323
Cook NJ (1983) Note on directional and seasonal assessment of extreme winds for design. J Wind Eng Ind Aerodyn 12(3):365–372
Cook NJ (1990) The designer's guide to wind loading of building structures. Part 2. Static structures. Butterworths, London
Cook NJ (1998) Improving the Gumbel analysis by using m-th highest extremes. Wind Struct Int J 1(1):25–42
Cook NJ, Harris RI (2004) Exact and general ft1 penultimate distributions of extreme wind speeds drawn from tail-equivalent Weibull parents. Struct Saf 26(4):391–420
Dyrbye C, Hansen SO (1999) Wind loads on structures. J. Wiley, Chichester; New York
ESDU 82026 (2002) Strong winds in the atmospheric boundary layer. Part 1: hourly-mean wind speeds. ESDU International, London
ESDU 83045 (2002) Strong winds in the atmospheric boundary layer. Part 2: discrete gust speeds. ESDU International, London
ESDU 84011, (1993) Wind speed profiles over terrain with roughness changes. ESDU International, London
ESDU 85020 (2001) Characteristics of atmospheric turbulence near the ground. Part II: single point data for strong winds (neutral atmosphere). ESDU International, London
ESDU 86010 (2001) Characteristics of atmospheric turbulence near the ground. Part III: variations in space and time for strong winds (neutral atmosphere). ESDU International, London
ESDU 87034 (1990) World-Wide extreme wind speeds. Part 1: origin and methods of analysis. ESDU International, London
ESDU 88037 (1988) World-Wide extreme wind speeds. Part 2: examples using varius methods of analysis. ESDU International, London
Eurocode (2006) Eurocode: basis of structural design. CEN
Eurocode 1 (2010) Eurocode 1: actions on structures - Part 1–4: General actions-Wind actions. CEN
Holmes JD (2007) Wind loading of structures. Taylor and Francis, New York
Lagomarsino S, Piccardo G, Solari G (1992) Statistical analysis of high return period wind speeds. J Wind Eng Ind Aerodyn 41(1–3):485–496
NTC (2005) Nuove Norme Tecniche per le Costruzioni 14/09/2005
Pagnini LC, Solari G (2009) Preliminary elements for an innovative wind map of Italy. In: 5th European and African conference on wind engineering, EACWE 5
Palutikof JP, Brabson BB, Lister DH, Adcock ST (1999) A review of methods to calculate extreme wind speeds. Meteorol Appl 6(2):119–132
Simiu E, Scanlan RH (1996) Wind effects on structures fundamentals and applications to design. John Wiley, New York
Solari G (2006) Analisi del vento di progetto presso il quartiere storico dell'area della fiera di milano. Università di Genova, Tech. rep
Stull RB (1997) An introduction to boundary layer meteorology. Kluwer Academic Publ, Dordrecht
Van der Hoven I (1957) Power spectrum of horizontal wind speed in the frequency range from 0.0007 to 900 cycles per hour. J Meteorol 14(2), 160–164

Chapter 3
Wind Tunnel Testing

3.1 Similitude Requirements

The wind tunnel is intended primarily for scale model testing, and so we have to consider how to conduct scale tests and how the results can be applied to a real structure. The technique for doing this is a dimensional analysis based on a set of physical parameters assumed a priori to affect the phenomenon (Simiu and Scanlan 1996; White 1998; Munson et al. 2002). An alternative method is based on the analysis of the governing equations of the phenomenon and, even though they cannot be solved, they will reveal basic dimensionless parameters (White 1998; Munson et al. 2002) (Fig. 3.1).

3.1.1 Dimensional Analysis

When performing an experimental study it is useful to identify a set of governing dimensionless parameters. Maintaining the values of such groups between prototype and model will automatically ensure similarity. The method used to reduce a number of dimensional variables to a smaller number of dimensionless groups was proposed in 1914 by Buckingham and is now called the *Buckingham pi theorem* Buckingham (1914). The determination of the variables must be accomplished by the experimenter's knowledge of the problem and the physical laws that govern the phenomenon. We can now make a distinction between experiments on rigid models, where only flow similarity has to be achieved and aeroelastic models where also the dynamic properties of the structure have to be scaled (models will be described in detail in Sect. 3.3).

D. Rocchi et al., *Wind Engineering*, Research Topics in Wind Energy 9,
https://doi.org/10.1007/978-3-032-17423-9_3

Fig. 3.1 1 : 300 scaled aeroelastic wind tunnel model of the Messina Strait Bridge

3.1.1.1 Dimensional Analysis for Static Rigid Models

Suppose that we have to perform experiments to identify the forces acting on a cylindrical structure (a cable, a tall chimney, a street lamp ...) and to identify a possible vortex shedding frequency. We assume that the force F (mean and oscillating value) depends on: the flow speed U, the diameter D, the air density and viscosity ρ and μ and a frequency f for a total of 6 parameters.

$$F = h\,(U, D, \rho, \mu, f) \tag{3.1}$$

Applying the Pi Theorem a set of dimensionless numbers are obtained:

$$F \doteq \rho U^2 D^2 \left(\frac{Df}{U}\right)^{\delta} \left(\frac{\mu}{\rho U D}\right)^{\varepsilon} \tag{3.2}$$

The dimensionless force coefficient $F/(\rho U^2 D^2)$ that is analogous to the *nondimensional force coefficients* defined in the previous chapters, as (for drag force):

$$C_D = \frac{F_D}{\frac{1}{2}\rho U^2 D L} \tag{3.3}$$

where we recognize the dynamic pressure $\frac{1}{2}\rho U^2$ from the Bernoulli equation and a reference area DL replaces the square of the diameter D^2. A second number is the *Strouhal number*

$$St = \frac{Df}{U} \tag{3.4}$$

if f is a frequency of the flow (vortex shedding frequency). In case of f is a frequency of the structure the number is called *reduced frequency* or, the reciprocal $U/(Df)$, the *reduced velocity*. It is important in unsteady, oscillating flow problems. The third group is the inverse of the *Reynolds number* (after Osbourne Reynolds, 1883):

$$Re = \frac{\rho U D}{\mu} \tag{3.5}$$

It can be interpreted as the ratio between the inertia forces and the viscous forces acting on the flow After this analysis we can perform tests on a scaled model providing that these three parameters are satisfied.

3.1.1.2 Dimensional Analysis for Aeroelastic Models

In case of an aerolastic simulation also the mechanical properties of the structure has to be matched and the problem parameters number increases:

$$F = h\,(U, D, \rho, \mu, f, m, \zeta, g) \tag{3.6}$$

where we included the structure mass per unit length m, the damping ζ and the gravity acceleration g (the structure stiffness has not be included since it is related to the mass trough the frequency). It is to emphasize that any parameter can be related both to the fluid or to the structure so that the ratios must be maintained from prototype to model. For instance fluid density has to be scaled as the structure density or fluid dimensions (integral length scales) has to be scaled accordingly to geometry dimensions. After applying the Pi Theorem three new nondimensional products comes out: The first one is the *Froude Number* (after William Froude, 1870):

$$Fr = \frac{U^2}{gD} \tag{3.7}$$

It can be interpreted as the ratio of the inertia forces of the flow to the gravity forces acting on the structure. It is important in wind engineering problems where the contribution of the structure weight to the stiffness is non-negligible as for suspension bridges and transmission lines. The second one is the density ratio:

$$\frac{m}{\rho D^2} \tag{3.8}$$

Table 3.1 Common variables and dimensionless groups

Group	Name	Interpretation	Application
Reynolds number, Re	$\frac{\rho U D}{\mu}$	$\frac{\text{inertia}}{\text{viscous}}$	Generally of importance in all types of fluid dynamics problems
Froude number, Fr	$\frac{U^2}{gD}$	$\frac{\text{inertia}}{\text{gravitational}}$	Problems where both elastic and gravity forces contribute to the stiffness
Strouhal number, St	$\frac{fD}{U}$	$\frac{\text{oscillation}}{\text{mean speed}}$	Oscillating flows (vortex shedding), see also reduced frequency and reduced velocity
Mach number, Ma	$\frac{U}{c}$	$\frac{\text{inertia}}{\text{compressibility}}$	Important only if compressibility of the flow is important ($Ma > 0.3$)
Mass ratio	$\frac{m_u}{\rho D^2}$	$\frac{\text{model mass}}{\text{flow mass}}$	Important in vibrations problems
Scruton number, Sc	$\frac{2\pi m\zeta}{\rho D^2}$	$\frac{\text{dissipative}}{\text{flow inertia}}$	Product of the mass ratio and critical damping ratio
Cauchy number, Ca	$\frac{E}{\rho U^2}$	$\frac{\text{structure internal force}}{\text{flow inertia}}$	Used for scaling structure stiffness
Force coefficient (drag), C_D	$\frac{F_D}{\frac{1}{2}\rho U^2 DL}$	$\frac{\text{drag force}}{\text{dynamic pressure}}$	All aerodynamics problems
Pressure coefficient, C_p	$\frac{p-p_s}{\frac{1}{2}\rho U^2}$	$\frac{\text{surface pressure}}{\text{dynamic pressure}}$	All aerodynamics problems

and finally the structure nondimensional damping ζ that is yet nondimensional. The combination of the last two parameters is often referred as *Scruton Number*:

$$Sc = \frac{2\pi m\zeta}{\rho D^2} \tag{3.9}$$

Table 3.1 summarizes the most important dimensionless groups in wind engineering problems.

Scaling factors

When designing models it is convenient to introduce a parameter scale λ_p of any parameter p:

$$\lambda_p = \frac{p_{model}}{p_{prototype}} \tag{3.10}$$

The first one is generally the length scale:

$$\lambda_L = \frac{D_{model}}{D_{prototype}} \tag{3.11}$$

If Froude Number similarity is required the velocity scale is related to the length scale since the gravity acceleration is fixed on both model and prototype. The Froude requirements results in:

$$\left(\frac{U^2}{gD}\right)_m = \left(\frac{U^2}{gD}\right)_p \Rightarrow \lambda_V = \sqrt{\lambda_L} \tag{3.12}$$

In order to fulfill Froude's model law, the velocity scale λ_V must correspond to the square root of the length scale λ_L. The reduced frequency requirement results in a frequency scale factor of:

$$\lambda_f = \frac{\lambda_V}{\lambda_L} \tag{3.13}$$

The density scale λ_ρ is usually forced to 1 since tests are often performed in air (same fluid for model and prototype). At this point if we require the similitude also on the Reynolds Number we obtain:

$$\left(\frac{\mu}{\rho U D}\right)_p = \left(\frac{\mu}{\rho U D}\right)_m \implies \lambda_V = 1/\lambda_L \tag{3.14}$$

This requirement is in contrast with Froude requirement except for $\lambda_L = 1$ that is testing the full scale prototype. Under Froude scaling Reynolds similarity is distorted and effects must be evaluated carefully. Even if Froude similarity is not required in order to maintain the same Reynolds number, a test on a $1/100$ scaled model would require a model velocity 100 times higher than the prototype, which is in the order of 1 km/s, which is not practicable and causes distortions due to compressibility effects (compressibility is negligible only for Mach Number < 0.3). The alternative is to perform tests in a different fluid (water) or in a pressurized wind tunnels, but these methods are often non feasible. We have to accept that Reynolds similitude cannot be achieved in boundary layer wind tunnel tests. Usually this is not a serious limitation since a Reynolds number independence is found in many practical situations (models with sharp edges). On the contrary, there are particular situations, as when studying structures with curved surfaces, where scale effects can be important and must be estimated. Hence the general requirement for large wind tunnels that can host large models and reduce scale effects. An alternative is performing tests at different wind velocities and/or at different model scales to identify possible Reynolds effects on the force coefficients. To improve local flow similarity model curved surfaces are sometimes roughened as showed in Fig. 3.2 where sand paper has been used to increase the deck roughness Diana et al. (2013).

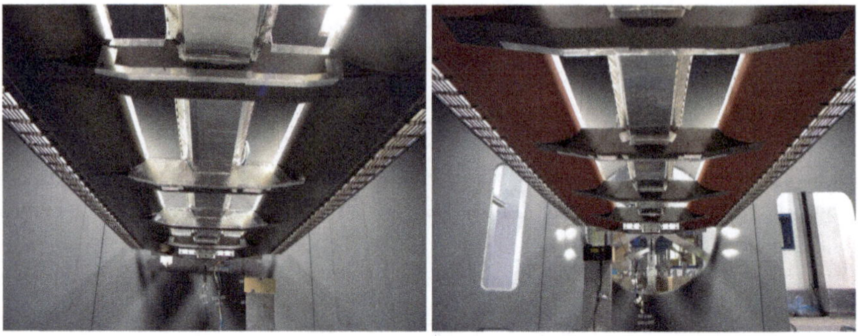

(a) (b)

Fig. 3.2 Deck wind tunnel model: **a** smooth surface and **b** rough surface

3.2 Wind Tunnel Facilities

The wind tunnel is a tool designed to carry out experiments to solve aerodynamic problems involving fluid-structure interaction. In addition, wind tunnels make it possible to use reduced-scale models. Tests on prototype may be expansive (full-scale tests on aircraft, high-speed trains) or not feasible (buildings, bridges). Particular importance is given to wind tunnels capable of simulating natural winds, the boundary layer wind tunnels (BLWT) that are extensively used in civil engineering practice.

Generally, wind tunnels are classified into categories depending on the wind speed range: subsonic (Mach < 0.4), transonic (Mach up to 1.3), supersonic (Mach up to 4–5) and hypersonic (Mach > 5). This section deals only with subsonic wind tunnels that are also called *low speed wind tunnels*. Low speed refers to an air flow velocity lower than 135 m/s, where compressibility is negligible. The description of a wind tunnel and its testing techniques would require a dedicated book, while here a synthesis is given. The topic can be investigated in specific references, such as Barlow et al. (1999) or ASCE (1999).

Early aerodynamic experiments are dated in the eighteenth century, when the aerodynamicist Robins used a rotating arm to reproduce the relative velocity between the air and the model. The "modern" idea of a wind tunnel facility, where the model is stationary and the air is moved past, is placed at the end of the nineteenth century. In that period, we found the first use of a wind tunnel to measure wind forces on buildings, made by Kernot in Melbourne, Australia (1893). At about the same time, Irminger (1894) in Copenhagen studied wind pressures on some basic shapes. In 1902 the Wright brothers built a wind tunnel to obtain data for their No. 3 glider. Important developments in facility layouts have been made by Gustav Eiffel in France (1909) and Ludwig Prandtl in Germany (1908), who built the first closed-circuit wind tunnel at Göttingen University. Wind tunnels for aeronautical applications developed rapidly during the first half of the twentieth century because of the government-funded aeronautical laboratories that were able to achieve technical leadership in aviation. In the second half of the same century, the so-called *boundary layer wind tunnels* were brought to light. In 1962, Cermak constructed the first atmospheric boundary layer wind tunnel at Colorado State University. It has a long test section that allows the development of thick turbulent boundary layers. Few years later, in 1965, Alan G. Davenport built his wind tunnel test facility at the University of Western Ontario, Canada starting the wind engineering science.

3.2.1 Wind Tunnel Parts

Despite the possible different layouts, all wind tunnels have four basic parts that are illustrated in Fig. 3.3 (Barlow et al. 1999):

1. A contoured duct to control the passage of the working fluid through the test section where the model is mounted;
2. A drive system to move the working fluid through the duct;
3. A model of the test object that is either full size or, more often, a reduced-scale model;
4. An instrumentation equipment (e.g. to measure wind force, pressure).

3.2.2 Layouts

There are two basic types of wind tunnel circuits and two basic configurations of test sections. The simplest type of wind tunnel layout is the open-circuit circuit (Fig. 3.4) where the flow follows a straight pattern from the entrance, followed by a contraction cone, a test section, a diffuser and finally an exhaust. The second type is the *closed*

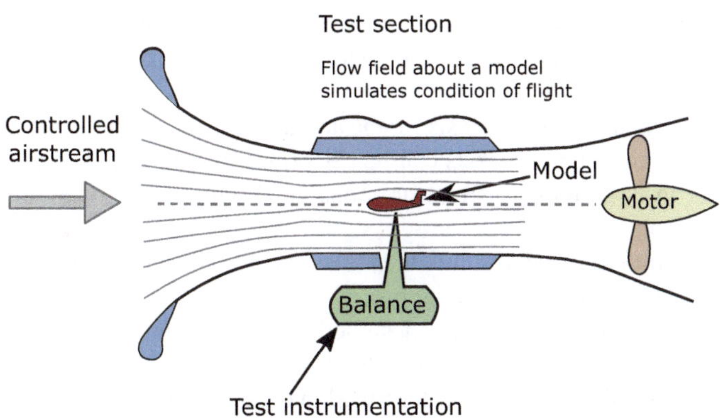

Fig. 3.3 Fundamental parts of a wind tunnel

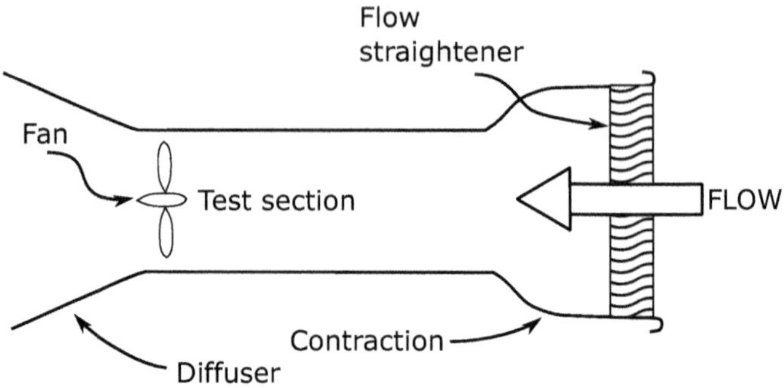

Fig. 3.4 Open circuit wind tunnel

return circuit where the flow has a continuous annular path (Fig. 3.5). Unlike an open circuit, a closed return tunnel has a superior flow quality, is less noisy, and requires less energy to operate, but it has a higher initial cost. An advantage of an open circuit is that there are no purging problems if running combustion engines tests or smoke visualizations. Large wind tunnel facilities has generally a closed circuit layout. In closed circuit tunnels, the flow quality is controlled by corner vanes and screens; a heat exchanger may be necessary to control the flow temperature (Fig. 3.6). Both types of wind tunnel can have a test section with solid walls (closed jet, Fig. 3.7) or without boundaries (open jet, Fig. 3.8). In general, the advantages are for a closed jet test section; in fact, most tunnels are of this type. Some facility is designed to run either with an open or closed test section, based on test requirements, like the Politecnico di Milano wind tunnel (CIRIVE). During the years many tunnels have been built for specific research and testing requirements. To name but a few:

- *Variable density tunnels* (or High Reynolds Number Tunnels) are used to simulate high Reynolds Number on scale models by changing the air density increasing the pressure in the test section (up to 20 atm);
- *Ice tunnels* can be a low speed wind tunnel with the addition of a refrigeration system to reduce the air temperature up to −40 °C to investigate ice formation on aircraft;
- *Automotive tunnels* carry out tests for the aerodynamic parameters that affect car's performance and stability. Tests can be performed on either scale models or full-scale cars in large tunnels. Tests are concerned with drag reduction, car noise etc.;
- *Climatic tunnels* are used to simulate interaction between wind and meteorological conditions (rain, snow, solar radiation);
- *Boundary Layer Wind Tunnels* are designed to simulate the atmospheric boundary layer. They are specifically designed for testing on civil engineering structures. This type of wind tunnel will be described in detail in this chapter.

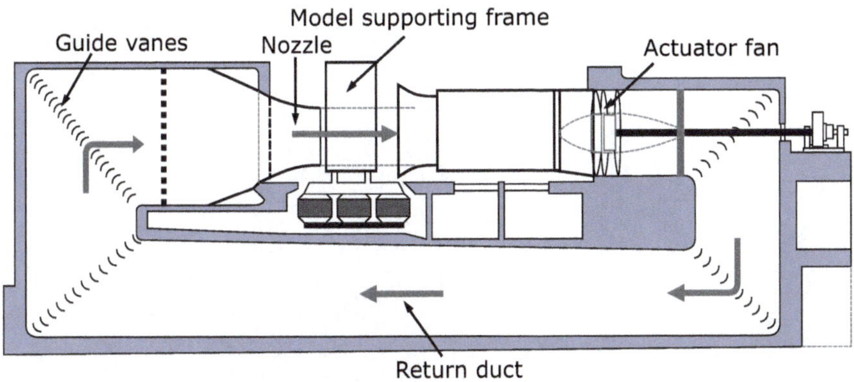

Fig. 3.5 Closed return wind tunnel (authors: Tomia and Liftarn https://commons.wikimedia.org/wiki/File:Windtunnel3-en.svg)

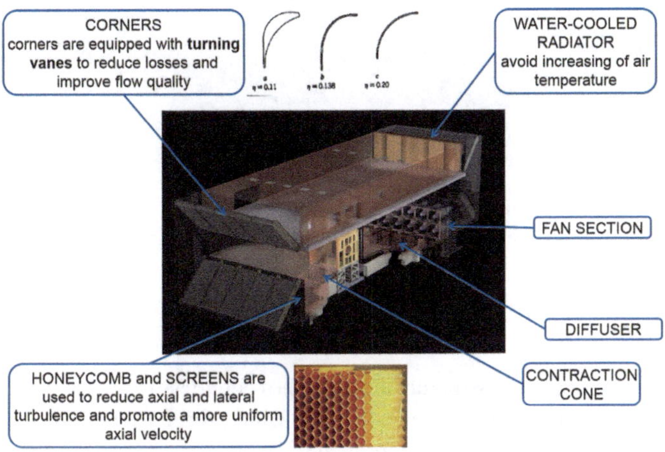

Fig. 3.6 Construction details of the Politecnico di Milano wind tunnel (www.windtunnel.polimi.it)

3.2.3 Simulation of the Atmospheric Boundary Layer

It has been previously stated that for a correct evaluation of the wind effects on civil structures, both in terms of mean and fluctuating values of forces and pressures, the wind tunnel tests should be performed in a correctly scaled atmospheric boundary layer, characteristic of the site and depending on the surrounding territory. Martin Jensen's model law Jensen (1958) states: "The correct model test with phenomena in the wind must be carried out in a turbulent boundary layer and the model law requires that the boundary layer be to scale as regards the velocity profile". It can be expressed as in Eq. (3.15):

Fig. 3.7 Closed Jet test section

$$\left(\frac{h}{z_0}\right)_{model} = \left(\frac{h}{z_0}\right)_{prototipe} \tag{3.15}$$

where h is the height of the structure and z_0 is the roughness length of the surrounding terrain; as a consequence of Jensen's law, the wind profile as well as the turbulence integral length scales should be reproduced in agreement with the geometry scale factor. Jensen reproduced in his experiments different roughness lengths z_0 using a long wind tunnel where the floor was coated by various roughness elements. Then he placed a $1/20$ scaled building model in the tunnel and measured the surface pressure for the different simulated roughness lengths obtaining the different patterns showed in Fig. 3.9 (1958). Comparing the experimental results with full-scale measurements he found that the correct pressure distribution is observed when the value h/z_0 is matched.

In this section, the methods to reproduce in a wind tunnel the characteristics of the atmospheric boundary layer are shown.

Adequate representation of boundary layer flow requires simulation of not only the correct turbulence intensities but also representation of the energy power spectra of the three turbulent velocity components. In particular, we require the representation of the low frequency end of the spectrum, that is, we require that the large energy containing eddies should exist in the simulated flow. Eddies having the size of some

Fig. 3.8 Open Jet test section

tens of meters have been measured in the atmospheric boundary layer, and hence the necessity for their representation.

The Jensen idea of the atmospheric boundary layer simulation was further developed by Davenport at the University of Western Ontario Boundary Layer Wind Tunnel, a facility that was specifically designed with a long floor where various roughness coverages were placed. In this way, the boundary layer naturally develops so that the desired wind profile is obtained at the model location as sketched in Fig. 3.10. The roughness geometry required for the simulation of the desired roughness length z_0 can be estimated according to the experiments by Gartshore and De Croos (1977) even if a "trial and error procedure" is often required for a final set-up. The main limitation of this method is that a long roughness fetch is required, demanding for very long test section wind tunnel facilities. Moreover, many wind tunnels used for aeronautical purposes have been converted for civil engineering application, but having these tunnels a short test section, the presented method cannot be applied as it is. A particular technique has been developed at the Low Speed Aerodynamics Laboratory of the National Research Council of Canada by Irwin (1981). This method is usually named "Spires and Roughness elements" and is shown in Fig. 3.11. Triangular shape obstacles (spires) are designed to be placed at the entrance of the tunnel. Large-scale turbulence is generated by the spires placed upstream of a reduced fetch length covered with roughness elements. The size and spacing of various devices are varied

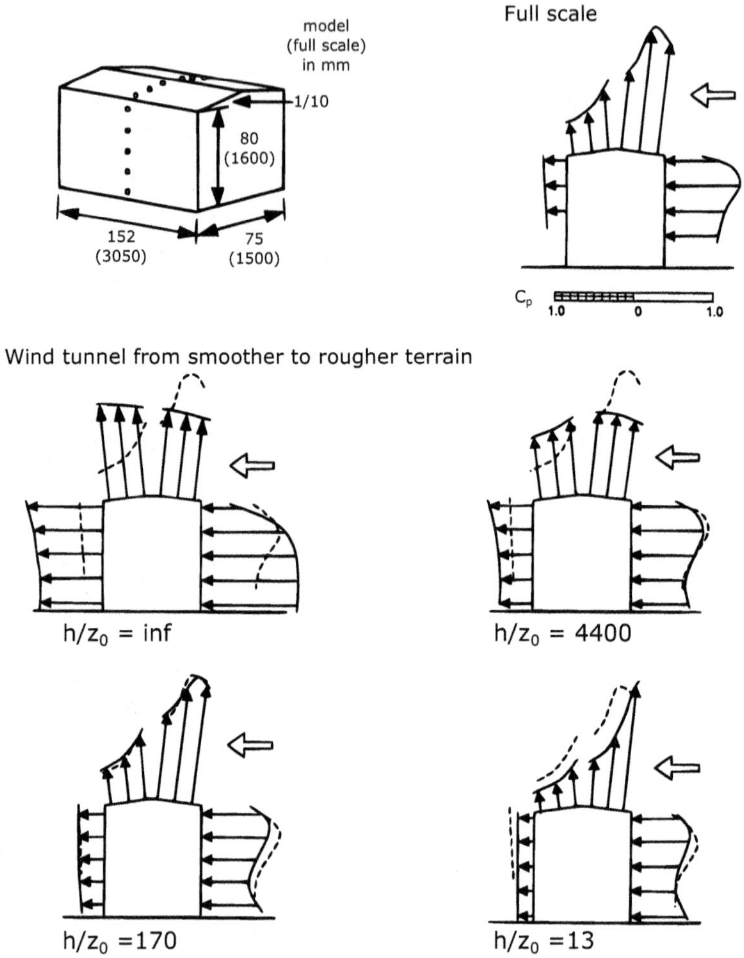

Fig. 3.9 Results from Jensen's experiments (after Jensen (1958))

to generate different wind models. Irwin's experiments confirmed that not only the mean velocity vertical profile but also the longitudinal turbulence intensity and integral length scale L_u^x are properly simulated. Using this approach the boundary layer is established after a distance of about 6 times the spires height.

3.2.3.1 Case Study: The Messina Strait Bridge

The design of a bridge over the Straits of Messina, a structure of outstanding importance and susceptibility to wind, has required wind tunnel tests on a 1 : 300 scaled aeroelastic model in a simulated atmospheric boundary layer. A wind study

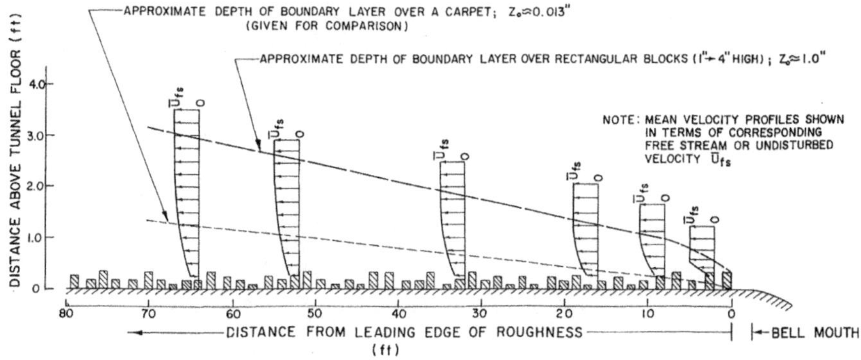

Fig. 3.10 Development of a boundary layer in a long wind tunnel, after Davenport and Isyumov (1967)

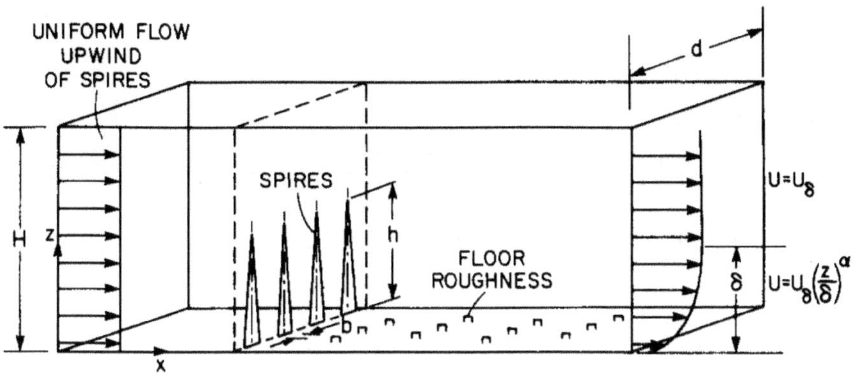

Fig. 3.11 Irwin's "Spires and Roughness elements" method (Irwin 1981). Reprinted with permission from Elsevier

performed by Bocciolone et al. (1993) suggested a roughness length equal to $z_0 = 0.01$m. A "spires and roughness layout" was realized in the boundary layer test section of the Politecnico di Milano wind tunnel to reproduce the correct turbulence properties (Fig. 3.12). Flow measurements using anemometers were performed to describe the characteristics of the simulated boundary layer in terms of mean wind profile and turbulence.

The results obtained are shown in the following figures. Figure 3.13 shows the vertical profile of the mean wind velocity measured at the model location. The velocity is normalized with respect to the velocity at deck level, which is 70 m, corresponding to the scale of the model 0.233 m. The target is calculated using the logarithmic profile equation. The simulated vertical profiles of turbulence intensities are shown in Fig. 3.14 (I_u) and Fig. 3.15 (I_v and I_w). Turbulence is presented as a function of height above ground.

Fig. 3.12 Spires and roughness elements placed in the boundary layer test section of the Politecnico di Milano wind tunnel

Fig. 3.13 Mean wind velocity vertical profile

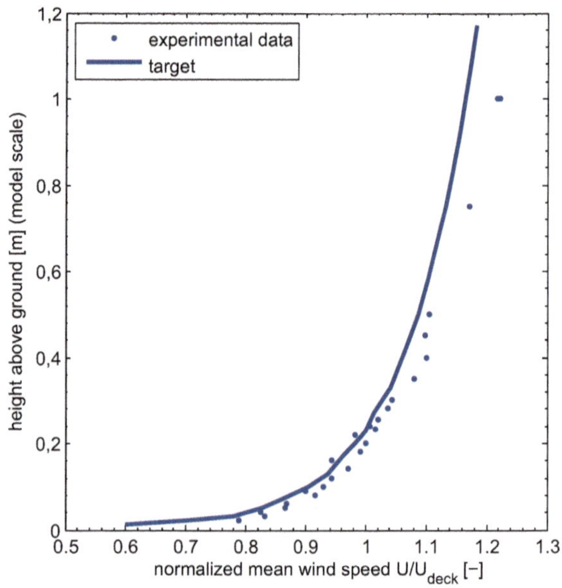

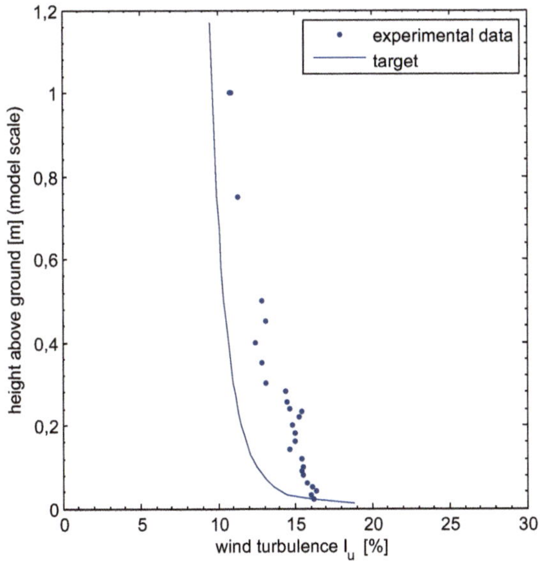

Fig. 3.14 Along wind turbulence intensity vertical profile

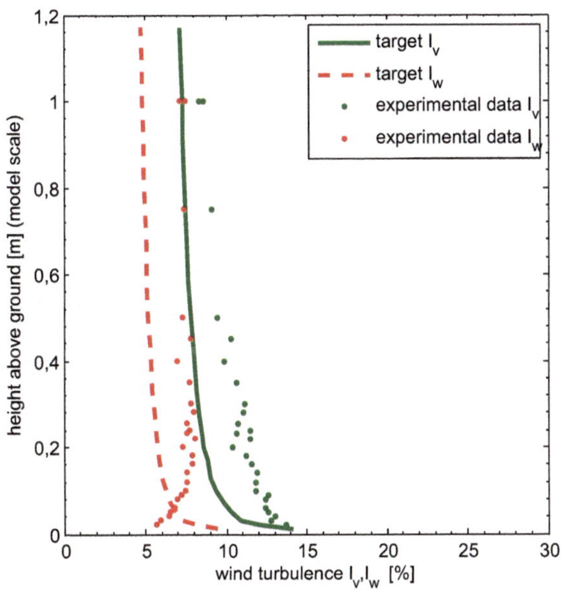

Fig. 3.15 Lateral and vertical turbulence intensities vertical profiles

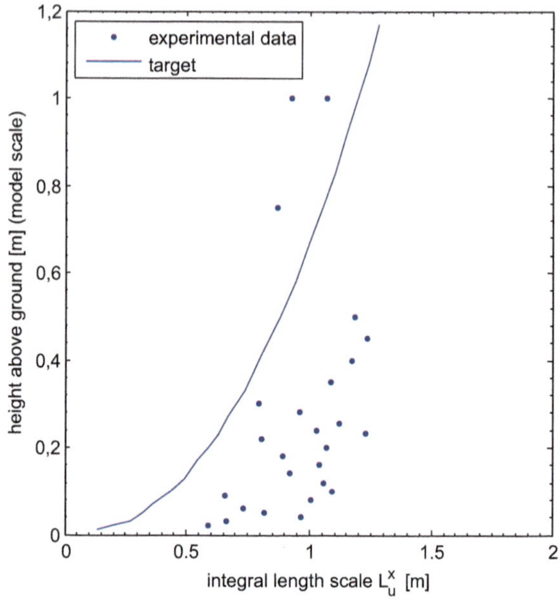

Fig. 3.16 Integral length scale vertical profile

Figure 3.16 shows the vertical profile of the integral length scale L_u^x. This parameter is of particular importance because it is related to the dimensions of typical vortex structures. According to Jensen's model law vortexes dimensions have to be scaled accordingly to the model geometric scale. Figure 3.17 shows the normalized wind spectrum measured in the wind tunnel at deck level. We observe that the energy content at the different scales (frequencies) is correctly simulated reproducing the Von Kármán power spectral density function shape.

The possibility of correctly reproducing the atmospheric boundary layer should be taken into account when deciding on the geometric scale of the model. In fact if a large scale is chosen (big model) it can be difficult to reproduce the low-frequency fluctuations in the wind spectrum since there is a limit to the size of the vortexes that can be generated with passive turbulence generator devices. The result in this case is that the energy content of the wind spectrum is shifted toward higher frequencies with respect to the target, as explained in Fig. 3.18. Using model scales larger than flow scales, the low-frequency velocity fluctuations are underestimated, and the high frequencies are overestimated.

3.2.4 *Active Devices for Turbulence Generation*

Active turbulence generation techniques have been developed with the goal of improving the low-frequency content of the wind spectrum and describing in a deter-

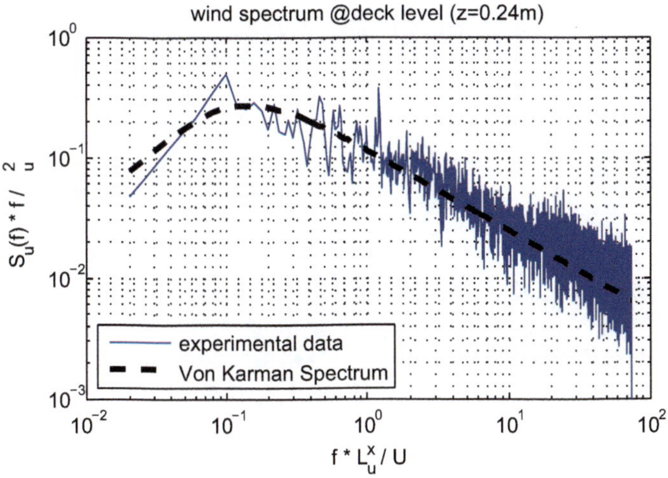

Fig. 3.17 Normalized wind spectrum of the along wind turbulence measured at deck level

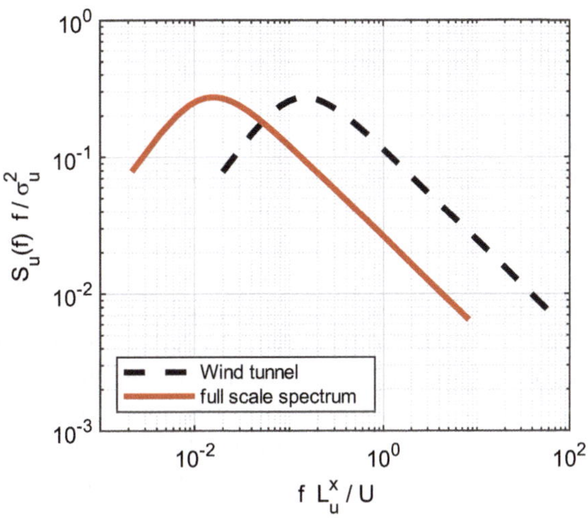

Fig. 3.18 The wind tunnel spectrum of the longitudinal wind turbulence is shifted to the right (Dyrbye and Hansen 1999)

ministic way the incoming flow. An actively controlled device, realized by means of an array of moving airfoils, is proposed by Cigada et al. (2002) and is shown in Fig. 3.19. The device is capable of generating a harmonic flow component in the vertical plane at the desired frequency. This experimental set-up is well suited for the identification of the aerodynamic admittance function on sectional models of long-span bridges.

Fig. 3.19 Device for active turbulence generation

3.2.5 *Wind Tunnel Blockage*

Wind tunnel tests on structures are designed to predict some of the forces, pressures, and other parameters that a full-scale structure will experience in an unconfined or "free-air situation". The walls of the wind tunnel impose constraints on the flow around the model, and the model will partially obstruct the passage of air, causing an acceleration of the flow. This effect is called *blockage* and the consequence is that the force or pressure values measured in the wind tunnel can differ from the free-air ones. The nature of blockage effects is visible in the sketch in Fig. 3.20 where the ideal flow pattern shown by the unconfined-flow streamlines must be modified to be confined within the tunnel walls. A parameter generally used to quantify this problem is named *blockage* and is equal to the ratio of the cross-sectional area of the body S and the cross-sectional area of the wind tunnel A which is the percentage of the frontal area of the tunnel occupied by the model:

$$b = \frac{S}{A} \tag{3.16}$$

For blockage ratios lower than 1% the flow distortion is negligible and no correction is needed. For blockage ratios between 1 and 5% distortions in the streamlines are also negligible, eventually with a correction that takes into account for the speed-up of the flow at model location. In a closed test section the effect of blockage is in a small overestimation of the wind forces, so that frequently no correction on the measured values are applied to be on the safe side. In case of blockage ratios between 5 and 10% distortions are significant and corrections are needed. Higher

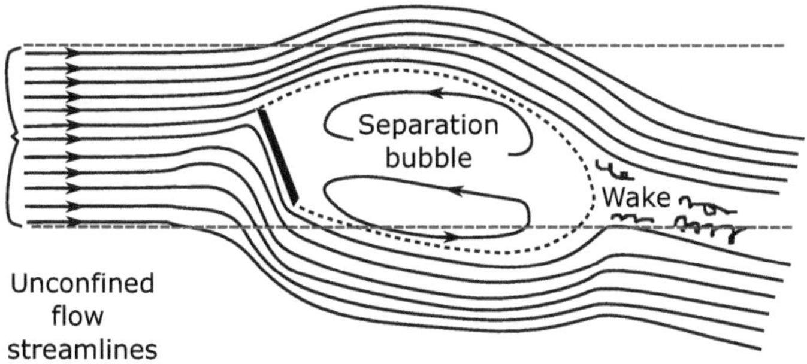

Fig. 3.20 Sketch of the blockage effects in a wind tunnel closed test section (ESDU 80024 1998)

blockage ratios are not generally desirable; in these cases, confined flow can not be representative of the free-air condition.

Most blockage correction methods assume that the blockage effect can be represented as an increase in the free-stream velocity, but it is conventional to perform the correction on the force and pressure coefficients that have been calculated with respect to the uncorrected reference velocity. Several methods have been developed to correct for blockage effects; see, for example, Barlow et al. (1999) or ESDU 80024 (1998). An example of a correction method is (Cowdrey method, reported in ESDU 80024 (1998))

$$\frac{C_{Ff}}{C_F} = 1 - m\frac{S}{A} \tag{3.17}$$

where C_{Ff} stands for free-air force coefficients (or "corrected") and C_F is the measured one. m is a coefficient that has been determined based on the shape of the body. Typically, the values range from 1.3 to 2.2 (ESDU 80024 1998). A very simple method based on the blockage-area ratio leads to the following.

$$\frac{C_{Ff}}{C_F} = (1 - S/A)^2 \tag{3.18}$$

In an open test section blockage effects tend to be smaller and of the opposite sign with respect to a closed tunnel.

3.2.6 Instrumentation for Flow Measurements in Wind Tunnel

The wind flow in the test section and close to the model is described by one or more of the following physical quantities: velocity magnitude and direction, turbulence, dynamic, static or total pressure, temperature. We therefore need specific instrumentation to *measure* these quantities.

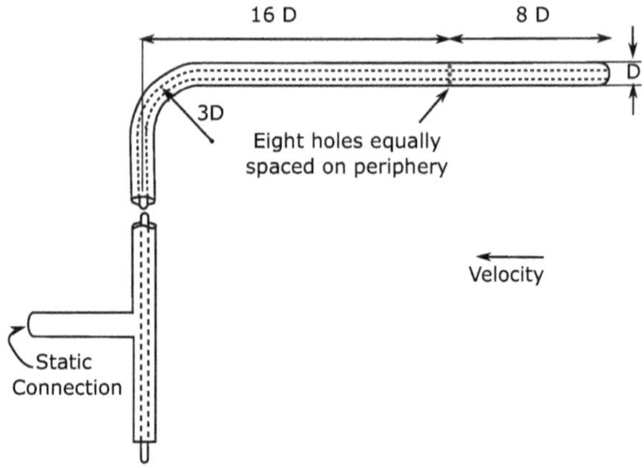

Fig. 3.21 Sketch of a standard Pitot tube (adapted from Barlow et al. (1999))

3.2.6.1 Pitot Tube

The *pitot tube* is the most common device for determining the flow velocity inside the test section. Figure 3.21 shows a standard pitot tube (Arts 1994; Barlow et al. 1999). The orifice at *A* measure the total (or stagnation) pressure that is equal to $p + \frac{1}{2}\rho V^2$ while the orifice at *B* measures the static pressure p. Connecting the two ports to a differential manometer it is possible to measure directly the dynamic pressure q of the wind and then calculate the wind velocity as:

$$\begin{aligned} q &= A - B = \frac{1}{2}\rho V^2 \\ V &= \sqrt{\frac{2q}{\rho}} \end{aligned} \tag{3.19}$$

Instead of a single hole, static pressure is generally measured through a series of circumferential holes. In order to perform accurate measurements, the pitot tube must be aligned with the flow; generally errors in alignment up to $\pm 3°$ are tolerated. Flow turbulence can also influence the readings of pitot tube.

3.2.6.2 Multi-hole Probes

In flow measurements, we often need information not only on the velocity magnitude, but also on direction. Directional probes are capable of identifying the variations in the horizontal plane (yaw) and the vertical plane (pitch) as indicated in Fig. 3.22. A typical flow direction probe is composed of five holes, having two pairs of symmetrically slanted tubes in the horizontal and vertical planes. These probes require a complicated calibration procedure. Once the calibration curves are known, the user can calculate the velocity magnitude, yaw and pitch as a function of the pressure

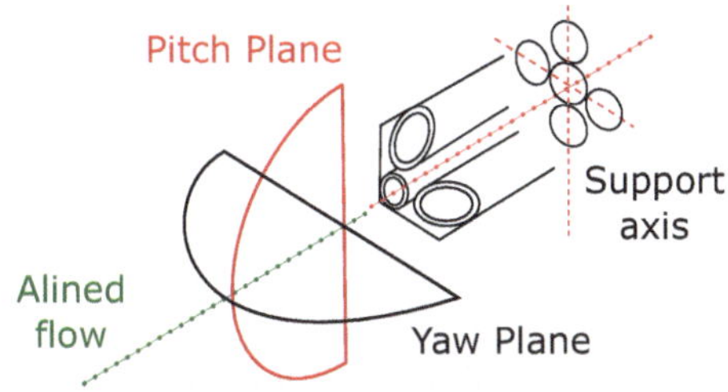

Fig. 3.22 Sketch of a three dimensional flow direction probe (Arts 1994)

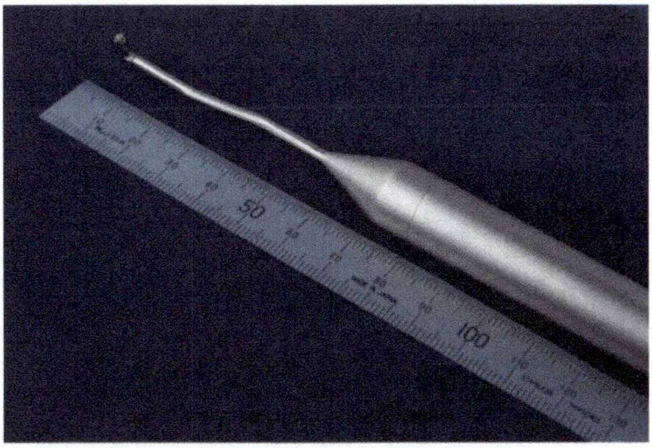

(a)

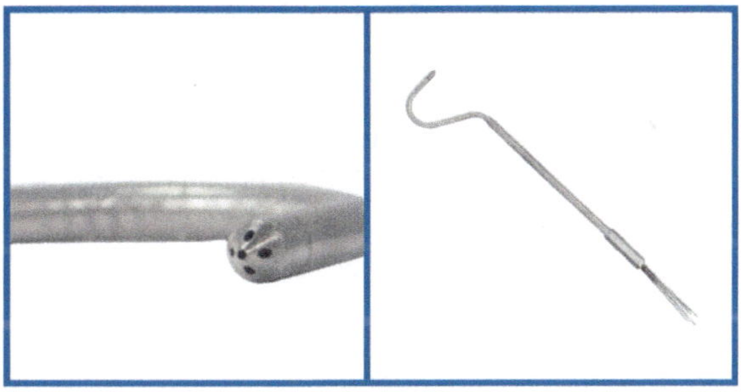

(b)

Fig. 3.23 **a** TFI Cobra Probe with embedded sensors and **b** AEROProbe conventional multi-holes probes

measured at the holes. Probes with a higher number of holes (7 or 12) are used to increase the larger velocity angle measurable. Accurate resolution of velocity vectors as high as 165° from probe axis is available. The frequency response of the probe depends on the pressure sensors and connection tubes used; responses up to 500 Hz are typical of conventional probes. Probes with embedded sensors can increase the frequency response to 5 kHz (Fig. 3.23).

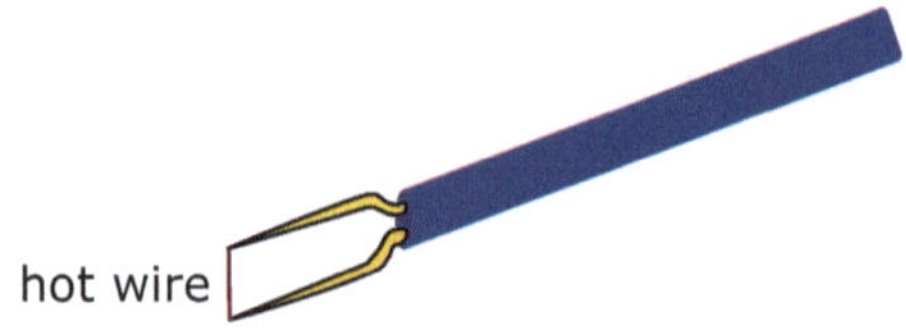

Fig. 3.24 Drawing of an hot-wire anemometer

3.2.6.3 Hot Wire Anemometers

The hot-wire anemometer is widely used in fluid mechanics experiments in particular for the description of turbulence. In fact, the hot-wire typology is indicated for velocity measurements up to several hundred kHz. The probe (Fig. 3.24) consists of a small electrically heated wire exposed to the flow. Typical probes have tungsten wire sensors, 1 to 3 mm long and 5 μm in diameter, mounted on two needle-shaped prongs (Arts 1994). An electronic circuit is used to provide the wire with a controlled amount of electrical current. The working principle is based on the cooling effect of a flow on a heated body: heat transfer is a function of the fluid velocity. In the constant temperature working method (the most commonly used type of operation), the electronic circuit varies the voltage supply to maintain the wire at a fixed temperature (200°C typical). Thus, a relationship between the flow velocity and the electrical output is established. Using probes having 2 or 3 wires, the flow direction can also be measured. These probes have some disadvantages: they are very delicate and sensitive to changes in ambient temperature (a 1°C change gives an error of about 2% in velocity). Moreover, they have a strongly nonlinear relationship between voltage and velocity, and complex calibration is needed, in particular for directional probes.

In fluid velocity measurements, there is also the availability of non-intrusive velocity measurement systems like multicomponent laser Doppler velocimeters and PIV systems. These methods are very expensive and are not widely used in wind tunnel tests related to wind engineering studies.

3.3 Wind Tunnel Models

When designing a model to perform wind tunnel tests, one has to decide which aspects to investigate and the desired information that the tests should provide. We have to state since the beginning that there is generally no "perfect" model that is able to correctly reproduce all the phenomena involved in the fluid-structure interaction problem. The first decision is for a *rigid* (or static) model instead of an *aeroelastic* model. A rigid model reproduces the external features of the prototype but not the structural properties: the tests are used to define the wind forces in terms of static and fluctuating components. The dynamic response of the structure is evaluated using the measured wind forces together with numerical models that simulate the dynamic properties of the structure. One great advantage of these models is that they are simple to make, and the tests can be performed in an early stage of the design of the structure when the external look is given but the structural design is still unknown. Moreover, if, after the tests, changes are due in the structural design, the measured wind forces can still be used to check the changes until the external geometry remains unchanged. This method can be used if the motion of the structure does not influence the fluid-structure interaction, that is, no important motion-induced forces exist. This is the case of very stiff and damped structures, such as low- and medium-rise buildings. In contrast, when studying light, slender and flexible structures (long-span bridges, high-rise buildings), it is often necessary to simulate the complete dynamic behavior of the prototype on the model: these are called *aeroelastic* models. This section describes the main characteristics of the different types of wind tunnel models and the instrumentation needed. Table 3.2 can be used to identify common wind tunnel tests on models and their use (ASCE 1999).

3.3.1 Force, Moment and Pressure Loads

On structures or elements where aeroelastic effects are not important, the overall and local external wind loads can be studied using *static models*. These models have geometrically scaled external features and are as stiff as possible to avoid deflections or vibrations during the tests. Static model tests are typically performed for buildings (Fig. 3.25a) but also vehicles (Fig. 3.25b). These models allow for accurate measurements of the mean and dynamic components of the wind loads. For tests on civil structures, it is also important to have a correct simulation of the natural wind as described in Sect. 3.2.3. The influence of the surrounding buildings must also be considered and the surrounding structures should be included in the model (Fig. 3.25a). Force and pressure coefficients can be measured at a single test wind speed and then full-scale values can be calculated. Typically, tests are carried out at different wind exposures with steps of 10–15° for a complete range of 360°. The total load acting on the model can be directly measured with a force balance: all three perpendicular forces (F_x, F_y, F_z) and the three moments (M_x, M_y, M_z) should be measured. A good wind tunnel balance must accurately separate these forces and moments and distinguish small changes in large force without having appreciable deflections (Fig. 3.26). The measurement is based on strain gauge elements that measure the deformation of balance parts under different loads. A calibration is required to define the relationship between forces and strains. The balance can be placed directly inside the model (internal balance) or outside the model (external balance). In this second case, the balance should be placed outside the airflow or shields should be used (Fig. 3.27). If the force balance is capable of measuring forces and moments up to a high frequency without significant amplification or distortion, the correct spectral densities of the wind forces and moments are hence defined. In case of buildings, these data are used in analytical methods to calculate the dynamic structural response; starting from the base moments the generalized forces for the lowest vibration modes are calculated applying correction factor that takes into account the mode shape. This wind tunnel

Table 3.2 Common wind tunnel model tests and use

Type of test	Information provided
Overall wind loads Direct measurement of the overall wind loads on rigid models, generally using a force balance	• Force and Moment coefficients as a function of the wind exposure • Mean and fluctuating wind force measured to define peak values and spectra • It is possible to evaluate the generalized force for the first fundamental modes (2 sway and 1 torsion) • Wind induced response is calculated analytically
Surface pressure measurements Measurement of the local wind pressure on rigid models, generally at several locations (up to 500 measuring points)	• Mean and peak external pressures for curtain-wall and cladding design • Overall wind loads can be evaluated integrating the external pressures over tributary areas • Simultaneously measured pressures allow the calculation of the generalized force also for higher modes • Wind induced response is calculated analytically
Sectional models Sectional models are a section a structure (bridge deck, leg o a tower, cable...) Static and dynamic tests are performed	• Static force and moment section coefficients as a function of the relative wind angle • Unsteady aerodynamic forces: flutter derivatives, admittance function • Check for the aerodynamic stability and response to vortex shedding • Model data are often integrated with numerical models that predict the wind response of the full scale structure
Aeroelastic models Models designed to simulate the aerodynamic and dynamic properties of the prototype, reproducing the vibration modes that contributes to the wind-structure interaction	• Information on the structure dynamic behaviour due to all the wind forces, inclusive of the motion-induced forces • Direct measurement of the response of the structure to the wind excitation including displacement and acceleration • Evaluation of the aerodynamic stability considering the contribution of all the structure components and mode shapes • Check of the effectiveness of active and passive systems to control dynamic motion

technique is often called the *High Frequency Force Balance Technique* (Tschanz and Davenport 1983; Holmes 1987; ASCE 1999; Holmes 2007). The frequency f_{0p} of the dynamic wind load that corresponds to the full-scale frequency is related to the wind tunnel frequency f_{0m} as to the reduced frequency relationship:

(a) (b)

Fig. 3.25 **a** High rise building 1:200 scaled model (Ibn Battuta Residences 2, Dubai) and **b** passenger train 1:18 scaled model

Fig. 3.26 A Ruag Aerospace wind tunnel force balance

$$f_{0m} = f_{0p} \frac{L_p}{L_m} \frac{U_m}{U_p} \tag{3.20}$$

The analogous wind force frequencies are much higher on the model than the full scale ones and, as a consequence, the model should be stiff enough to avoid significant dynamic amplifications at the highest frequency of interest.

Balances are indicated when the overall loads are needed; if one is interested in the wind force acting on a small area of a structure, surface pressure measurements are the best method. This is typically the case for loads that act on building claddings and curtain walls. In order to identify areas of high local pressure/suction, a large number of pressure taps is often required, up to 500 measuring points. Figure 3.28 shows the set-up of a building model for pressure measurements where the pneumatic internal connection are visible (*pressure taps* or *pressure tapping*).

Generally, a tap is realized through a hole, with a typical diameter of about 2 mm, located on the external surface of the model, as visible in Fig. 3.29. A plastic tube is used to connect the tap to a pressure transducer: Fig. 3.30 shows a multichannel pressure scanner that can sample pressures at 500 Hz per channel. Transducers with

Fig. 3.27 Force balance below the IBN Battuta tower

small dimensions can be placed inside the model to realize a clean setup. Both mean and fluctuating pressures can be measured. Short-duration peak loads are often the most important information for cladding design. Full-scale measurements suggest that peak local pressures acting on a building have an effective duration of about 1–10 s (ASCE 1999). As a consequence, the equivalent full-scale upper frequency response limit should not be less than about 2 Hz. To convert this to the model frequency, assuming a typical velocity ratio of 1 : 3 and a geometric ratio of 1 : 100, according to Eq. 3.20, the frequency ratio is 33 and the desired upper limit is 66 Hz.

In this range, the frequency response of the tubing connection can introduce distortions as visible in Fig. 3.31: the amplitude graph shows a resonant peak with amplification in the amplitude up to 50% for a tube having a length of 1.5 m and an internal diameter of 1.4 mm; the transfer function also presents a phase lag. In the Figure, experimental investigation of the tubing distortion are compared with the theoretical model described by Bergh and Tijdeman (1965): it is possible to observe that the tubing response can be described analytically with high accuracy. The best response can be obtained using a tube as short as possible, but in many situations one is constrained to a minimum allowable tube length.

The transfer function of the tubing can also be used to correct the measured pressures, taking into account the amplification due to the pneumatic connection.

The scaling of surface pressures is performed through the pressure coefficient calculated as:

Fig. 3.28 Pressure taps (about 400) inside IBN Battuta tower. The pneumatic connection inside the model are visible

$$c_{p,i} = \frac{p_i - p_s}{q} \tag{3.21}$$

$$q = \frac{1}{2}\rho U^2 \tag{3.22}$$

where p_i is the pressure measured it the tap i, p_s is the ambient (static) pressure in the test section and q is the wind reference dynamic pressure. If a differential pressure transducer is used and the negative transducer port is connected to the static pressure of the wind tunnel, the numerator of Eq. 3.21 is directly measured. Maxima values c_p are typically found near the corners of high-rise buildings, where values greater than 5 (suction) are often identified; in particular cases peak values up to 10 have been measured.

Surface pressure tests on large roofs (stadia, Fig. 3.32) are performed in a similar way. Roofs are often a membrane structure and the net pressure across the membrane

Fig. 3.29 Model surface showing pressure taps

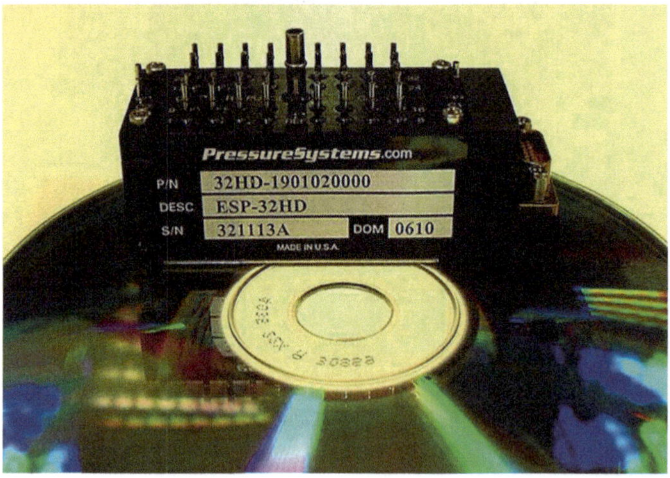

Fig. 3.30 Multi-channel high sample-rate pressure scanner (Pressure Systems ESP32)

is needed. In this case, pressure is measured on both sides of the roof, and the difference between the two measurements is the required net pressure.

Surface pressures can also be used to define the modal loads. This approach consists in the integration of surface pressures over tributary areas to assess modal loads. With this method, it is possible to calculate the generalized force for all mode shapes of interest, resulting in a higher number of information on the wind loads compared to the force balance method. However, proper measurements of the overall wind loads require a large number of simultaneously measured pressure taps and an adequate sample rate to measure the wind excitation at higher modes.

For both force and pressure measurements, if the fluctuating part of wind loads must be identified, the duration of the tests becomes an important issue. If only

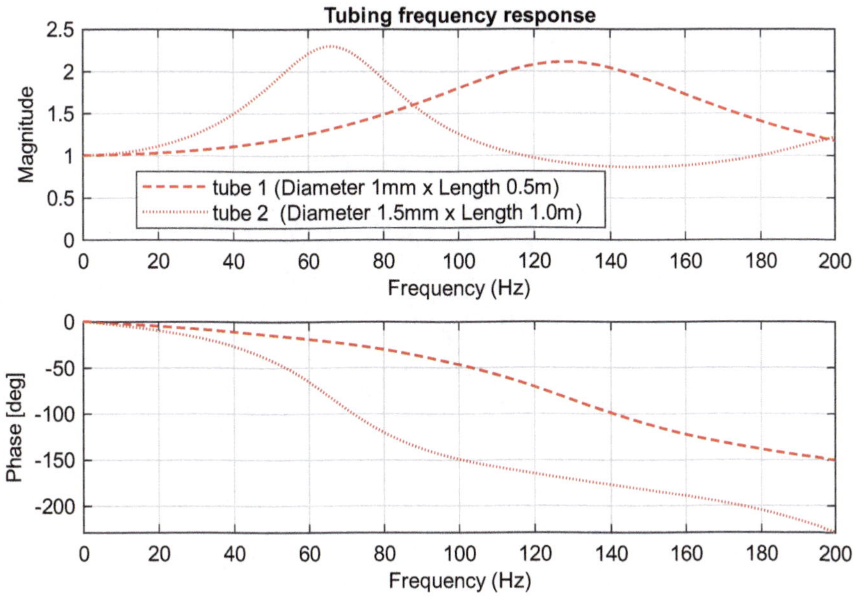

Fig. 3.31 Sketch of a pneumatic connection with a typical transfer function for tubes, connecting the model surface and the transducer

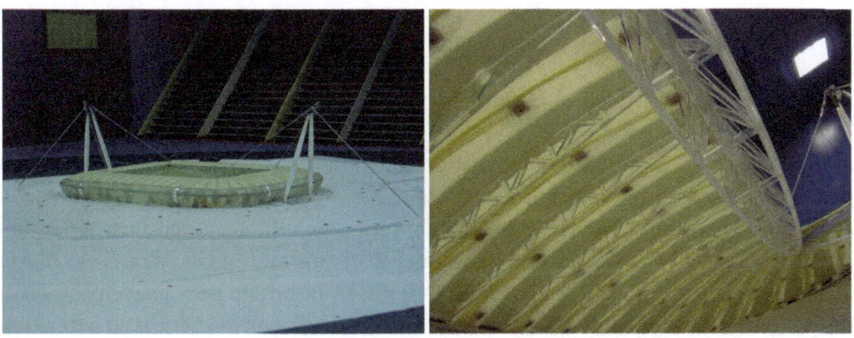

(a) (b)

Fig. 3.32 Stadium model for pressure measurements on the roof (Juventus Stadium, Torino). **a** External side and **b** detail of the internal side showing pressure taps

the mean value has to be measured in boundary layer tests, a measurement period corresponding to 10 min full scale is adequate. In contrast, if standard deviations, peaks and spectra have to be calculated, a time period corresponding to more than 2 h full-scale can be necessary. With the typical scale factors previously used, this corresponds to a duration of about 4 min for each test run.

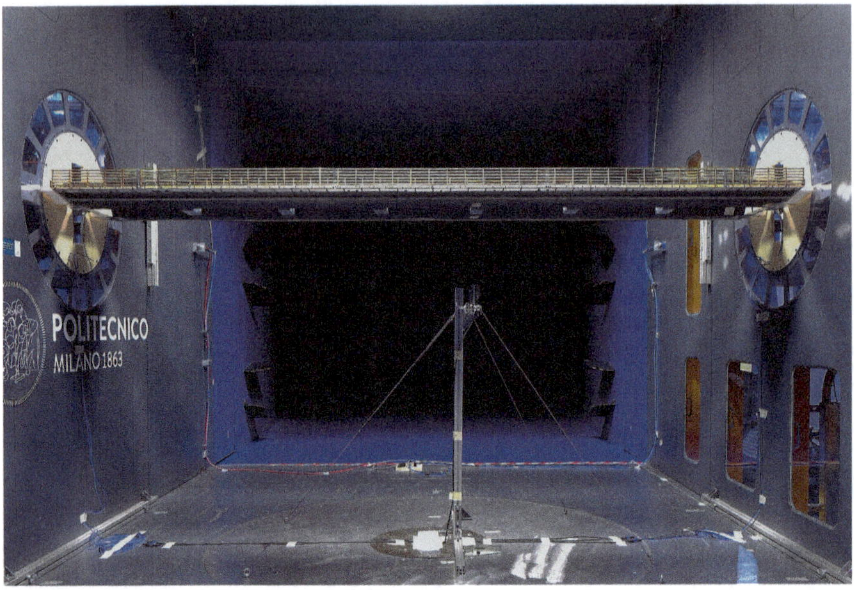

Fig. 3.33 1 : 44 scaled sectional model of the 1915 Canakkale Bridge

3.3.2 Sectional Models

Sectional models are used to study the wind action on slender (line-like) structures like long-span bridges, towers and cables: in these cases motion induced effects are important. The model is a rigid representation of a "section" of the structure and is scaled to partially reproduce the dynamic properties of the prototype (mass, oscillation frequency) (ASCE 1999). To ensure a two-dimensional flow across the model, it must fit between the wind tunnel walls, or end-plates should be placed at the extremities (Fig. 3.33). They are often used in bridge deck aerodynamics to define force and moment coefficients, flutter derivatives, and the admittance function. Aerodynamic stability and vortex shedding can also be investigated (Diana et al. 2006). Experimental data is often integrated with numerical models that predict the wind response of full-scale structures (Diana et al. 2013). Tests can be performed in both smooth and turbulent flow. Models can be mounted on springs to have a scaled mass, moment of inertia, damping, and frequency and to reproduce a free-motion behavior; it is also possible to use an external system to drive the motion (Diana et al. 2004). These models have an important advantage in the scale factor, since large models can be realized: a bridge sectional model typically has a scale factor of 1 : 25 to 1 : 60, while a full bridge model has geometry scale up to 1 : 300.

Fig. 3.34 Replica model of a large roof (Nuovo Polo Fiera Milano). Scale factor 1 : 75

3.3.3 Aeroelastic Simulations

Aeroelastic models are designed to simulate the dynamic properties of the prototype, reproducing the vibration modes that contribute to the wind-structure interaction. The test wind velocity is related to the full-scale velocity through the velocity scale factor: tests at different velocity levels are needed to evaluate the response at different wind conditions. They can be *replica models* or *equivalent models*. The sectional models previously described can also be considered as a sort of aeroelastic model.

3.3.3.1 Replica Models

Replica models are geometrically the exact reproduction of the prototype structure realized with materials that scales the prototype mechanical properties so that mass and stiffness requirements are automatically matched. This method is not diffused due to the difficulties in finding a material to make the model that matches the required scaled mechanical properties: the method is used in structures where the mechanical properties are concentrated in the external skin, such as for cooling towers, chimneys, and large roofs. Figure 3.34 shows a replica model of a large roof.

3.3.3.2 Equivalent Models

In equivalent models, the physical properties that have to be reproduced are divided into two mechanical parts on the model:

- An internal structural part that reproduces the stiffness properties;
- An external skin that reproduces the exterior geometry.

(a) (b)

Fig. 3.35 Aeroelastic model of a suspension bridge tower (1 : 200 scale, Julsundet bridge, Norway). **a** Tower in stand-alone layout and **b** model detail with the internal equivalent spine and an accelerometer placed at tower top

The sum of the mass of the internal structure and of the external skin must match the scaled prototype mass. This method is effective for structures in which bending and torsional effects predominate. This is the case of slender structures such as long span bridges and towers. Figure 3.35a shows a wind tunnel model of a bridge tower: the tower is tested in a standalone layout to simulate the erection stage. In Fig. 3.35b the interior is visible: there is an aluminum spine that reproduces the stiffness, while the external shield is made by machined polyurethane. It is to note that the external shield is not continuous for all the tower height but it is divided into small dimension parts being each one fixed to the internal spine: this to avoid that the external shell introduces additional stiffness. In Fig. 3.35b, an accelerometer placed at tower top is visible: it is used to measure the tower oscillations during the tests.

Case Study: The Izmit Bay Bridge

The design of an aeroelastic model of a full bridge requires attention to reproduce correctly both the aerodynamic and structural properties. Firstly we have to consider that the outstanding length of a cable-suspended but also a cable-stayed bridge often results in requiring large scale factors to fit the model in the wind tunnel; only large

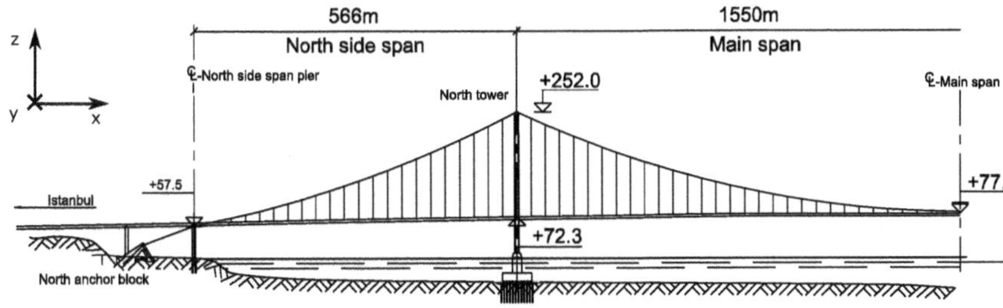

Fig. 3.36 Overview of the Izmit Bay Bridge project (units m)

wind tunnels should be used to perform this investigation to minimize Reynolds number effects on the scaled model. Geometric scales up to 1 : 300 are often used. The case study is the design of the aeroelastic model of the *Izmit Bay Bridge*, located in a bay of the same name in Turkey. The bridge project showing the main dimensions is shown in Fig. 3.36. Taking into account the length of the bridge of 2682 m and the dimensions of the wind tunnel (turntable diameter equal to 13 m) a length scale $\lambda_L = 1:220$ has been decided, resulting in a model length of 12191 mm. Since gravity force contributes significantly to stiffness, the Froude number for the prototype and model must match. As a consequence according to the similarity laws stated, once the length scale is fixed all the other scale factors are calculated as a function of this value. Table 3.3 summarizes the scale factors for all the physical quantities involved in the design of the model.

Since the design of such models is quite complicated, Finite Elements methods are often used in the design. A FEM model of the prototype is required as a starting

Table 3.3 Summary of the scale factors for the bridge aeroelastic model, according to Froude law

Quantity	Symbol	Unit	Scale factor
Length	L	[m]	λ_L
Velocity	V	[m/s]	$\sqrt{\lambda_L}$
Frequency	f	[Hz]	$1/\sqrt{\lambda_L}$
Time	t	[s]	$\sqrt{\lambda_L}$
Acceleration	a	[m/s^2]	1
Mass	M	[kg]	${\lambda_L}^3$
Mass per unit length	m	[kg/m]	${\lambda_L}^2$
Density	ρ	[kg/m^3]	1
Force	F	[N]	${\lambda_L}^3$
Stiffness	K	[N/m]	${\lambda_L}^2$
Young modulus	E	[N/m^2]	${\lambda_L}^4$
Bending stiffness	EJ	[Nm^2]	${\lambda_L}^5$
Axial stiffness	EA	[N]	${\lambda_L}^3$

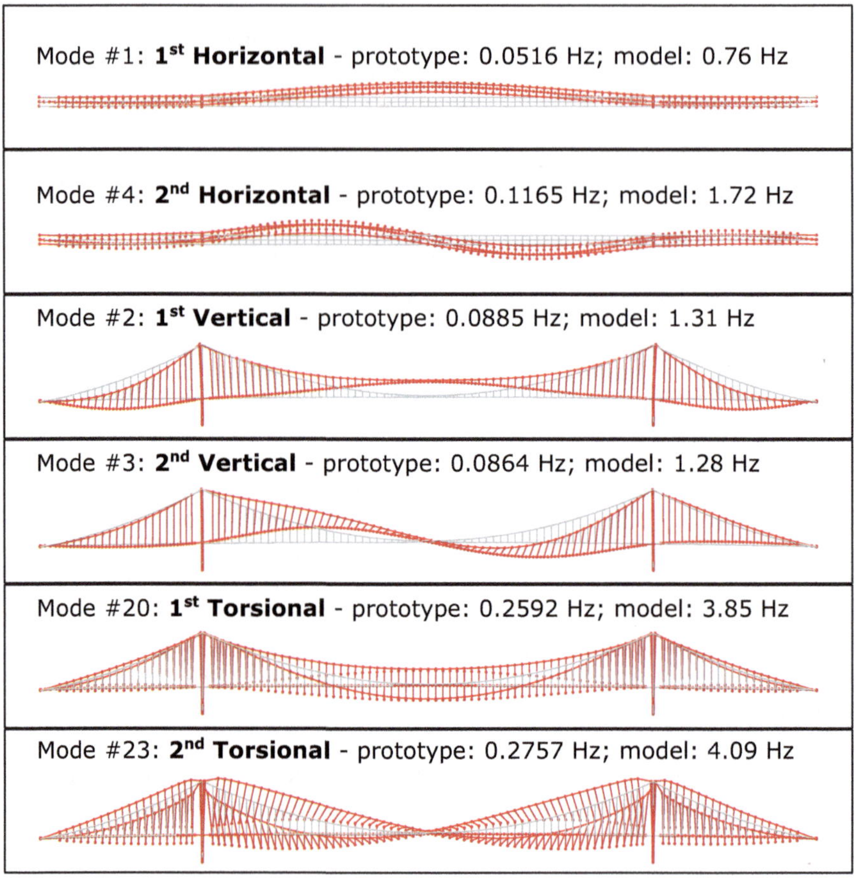

Fig. 3.37 Main vibration modes involved in the wind response

point: the results of the simulation are used to identify the main vibration modes that are involved in the wind response and must be reproduced on the model (see Fig. 3.37). The model design is divided into three main parts that are managed in different ways: the deck, the towers, and the cables. Once the model is completed, a modal identification is performed to verify that the vibration modes are reproduced correctly (Fig. 3.38).

Deck

The deck is responsible for most of the wind forces on the bridge (Fig. 3.39). It is always designed as an equivalent model. The internal structure reproduces the bending stiffness in the two orthogonal directions (EJ_y and EJ_z) and the torsional stiffness GJ_x. The spine is generally made by aluminum ($E_{Al} = 70000\,\text{MPa}$) that is a material easy to machine: then we have to find a section that has the required

Fig. 3.38 Modal identification performed on the bridge model

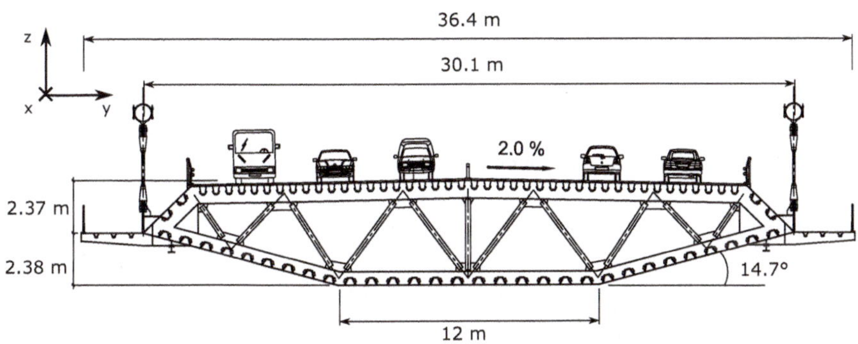

Fig. 3.39 Cross-section of the deck

moments of inertia:

$$J_y = \frac{[EJ_y]_m}{E_{Al}} \tag{3.23}$$

$$J_z = \frac{[EJ_z]_m}{E_{Al}} \tag{3.24}$$

$$J_x = \frac{[GJ_x]_m}{G_{Al}} \tag{3.25}$$

A rectangular section may not be appropriate since we have only two variables (the length of the two sides) and three parameters to match (J_x, J_y and J_z). A different shape may be therefore adopted (e.g."C", "H", etc.).

The external geometry is reproduced using foam, wood, polyurethane: materials that are easy to machine. Fittings (barriers, guardrails, wind screens) may be realized by brass sheets that can be cut using photoengraving or laser techniques. The module is intended as a lumped mass that is fixed on the spine (see Figs. 3.40 and 3.41). The mass has to match the desired mass per unit length m_u and the mass moment of inertia per unit length J_x of the deck. This results is achieved using different density materials. The mass has often to be smaller than the target mass to accommodate possible adjusting of the mass moment of inertia with additional weight placed at a distance from the center of rotation. The values obtained for the IBB deck model are summarized in Table 3.4.

The large-scale factors used and the Froude similarity requirements result in a Reynolds number value on the deck model that is more than 3000 times lower compared to the full-scale one. This difference can produce a different aerodynamic behavior of the deck model and therefore unexpected (and incorrect) wind tunnel results. It is therefore suggested that deck sectional force coefficients have to be measured by performing a sectional model test on the same scale as the full bridge model and then the results compared with those measured on a larger sectional model.

Towers

The towers are made in the same way as the deck (Fig. 3.42). An internal aluminum structure reproduces the stiffness of the tower and machined polyurethane modules are glued on the spine to reproduce the external geometry (Fig. 3.43).

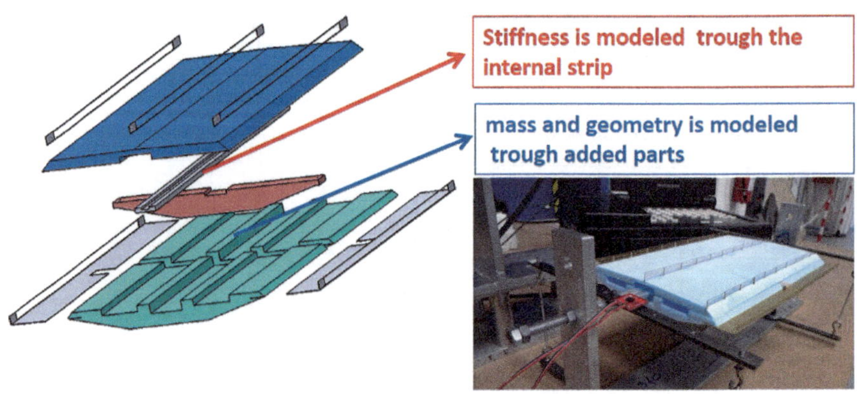

Fig. 3.40 Set-up of a module of the bridge deck

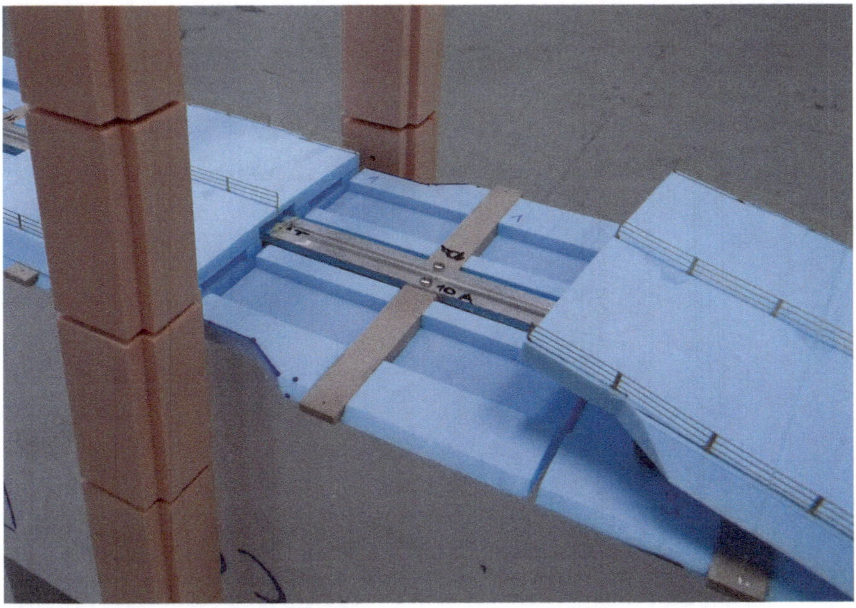

Fig. 3.41 Deck module in the full bridge

Table 3.4 Mass and stiffness properties of the deck section

	Prototype	Model
Mass per unit length	$1.71E + 04$	0.35
Mass moment of inertia per unit length	$1.23E + 06$	$5.23E - 04$
Bending stiffness EJ_y	$3.52E + 02$	1.60
Bending stiffness EJ_z	$9.87E + 03$	44.88
Torsional stiffness GJ_x	$2.97E + 02$	1.35

Cables and Hangers

A completely different approach is used to model the main cables; three properties have to be matched:

- The axial stiffness of the cables EA;
- The normalized drag force per unit length $C_D D$;
- The mass per unit length.

Generally, only steel cables (wire ropes) are used, but it is not possible to match all three properties. One solution is to define the diameter of the cable to match the required axial stiffness and then add lumped cylindrical masses to adjust the drag and the total mass as visible in Fig. 3.44. The parameters used to optimize the cylindrical

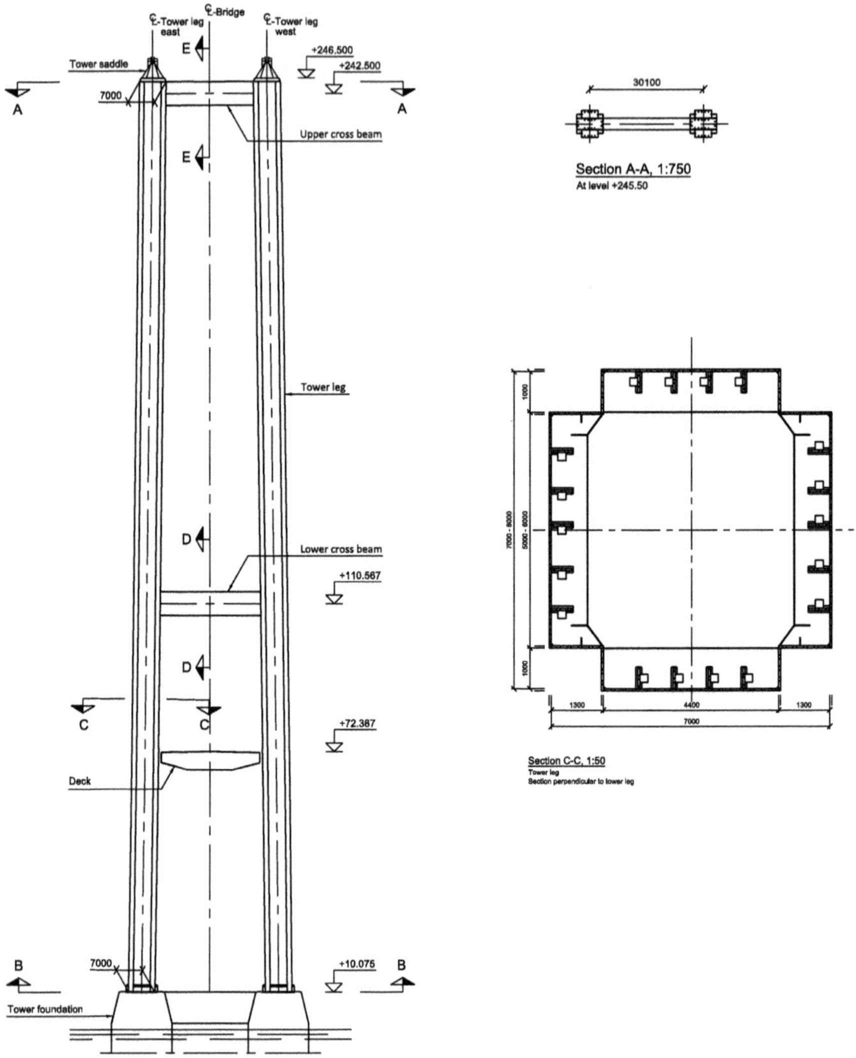

Fig. 3.42 Geometry of the towers (units mm)

masses are the diameter, length, and material density. In the calculations, the effect of Reynolds number on the cylinder drag must also be considered. Eventually ,a spring on the cable anchorages can be added to adjust the longitudinal stiffness. Hangers theoretically has to be considered in the same way, but generally a further simplification is allowed and only a rigid cable is used. The number of hangers can be reduced by lumping two or more cables in one.

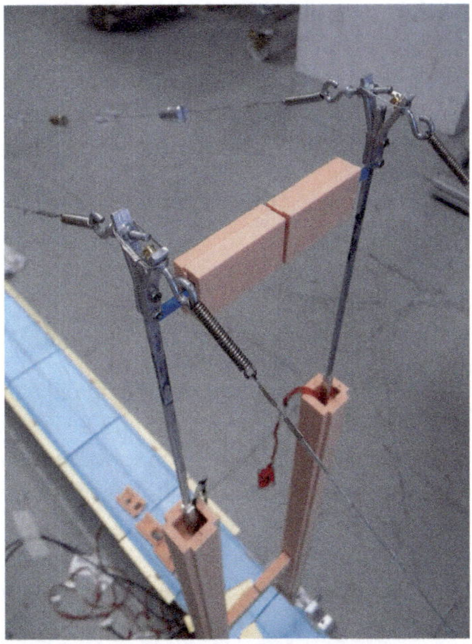

Fig. 3.43 Bridge tower equivalent model

Fig. 3.44 Main cables of the bridge model

3.3.3.3 Aeroelastic Model of a Tall Building

Aeroelastic models of a tall building can be made in a way similar to that described for the towers. A simplified rigid body model with springs at the base is also used if the following conditions are satisfied (Holmes 2007)

- The response of the building to wind loads in torsional (twisting) modes can be neglected;
- The response in sway modes higher than the first in each orthogonal direction can be neglected;
- The mode shapes of the fundamental sway modes can be assumed as linear.

Under these assumptions, the model can be realized through a rigid body pivoted at a level close to the ground and a set of calibrated springs to adjust the stiffness as sketched in Fig. 3.45. In order to reproduce the correct structural frequencies, the model moment of inertia with respect to the ground level must match the value for the properly scaled prototype.

Since body motion has been reproduced, motion-induced aerodynamic forces have also been reproduced in the wind tunnel tests. This is not the case when fixed models are used to measure the fluctuating wind force. In this case, the resonant response of the building is not reproduced.

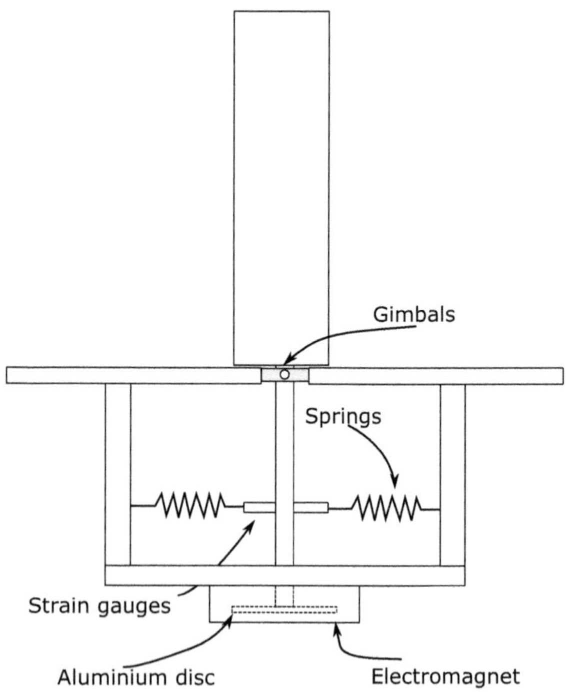

Fig. 3.45 Sketch of an equivalent model for a tall building (adapted from ASCE (1999))

References

Arts T (1994) Measurement techniques in fluid dynamics: an introduction. von Karman Institute for Fluid Dynamics, Rhode Saint Genèse, Belgium

ASCE (1999) ASCE manuals and reports on engineering practice No. 67: wind tunnel studies of buildings and structures. Reston, VA

Barlow JB, Rae WH, Pope A (1999) Low-speed wind tunnel testing. Wiley, New York

Bergh H, Tijdeman H (1965) Theoretical and experimental results for the dynamic response of pressure measuring systems. Report NLR-TRF 238. National Luchtvaartlaboratorium (NLR), Amsterdam

Bocciolone M, Gasparetto M, Lagomarsino S, Piccardo G, Ratto CF, Solari G (1993) Statistical analysis of extreme wind speeds in the straits of Messina. J Wind Eng Ind Aerodyn 48(2–3):359–377

Buckingham E (1914) On physically similar systems; illustrations of the use of dimensional equations. Phys Rev 4:345–376

Cigada A, Diana G, Zappa E (2002) On the response of a bridge deck to turbulent wind: a new approach. J Wind Eng Ind Aerodyn 90(10):1173–1182

Davenport AG, Isyumov N (1967) The application of the boundary layer wind tunnel to the prediction of wind loading. In: Proceedings of the 2nd international conference on wind engineering, pp 201–230

Diana G, Resta F, Zasso A, Belloli M, Rocchi D (2004) Forced motion and free motion aeroelastic tests on a new concept dynamometric section model of the Messina suspension bridge. J Wind Eng Ind Aerodyn 92(6):441–462

Diana G, Resta F, Belloli M, Rocchi D (2006) On the vortex shedding forcing on suspension bridge deck. J Wind Eng Ind Aerodyn 94(5):341–363

Diana G, Fiammenghi G, Belloli M, Rocchi D (2013) Wind tunnel tests and numerical approach for long span bridges: the Messina bridge. J Wind Eng Ind Aerodyn 122:38–49

Dyrbye C, Hansen SO (1999) Wind loads on structures. Wiley, Chichester, New York

ESDU 80024, (1998) Blockage corrections for bluff bodies in confined flows. ESDU International, London

Gartshore IS, De Croos KA (1977) Roughness element geometry required for wind tunnel simulations of the atmospheric wind. J Fluids Eng Trans ASME 99(3):480–485

Holmes J (1987) Mode shape corrections for dynamic response to wind. Eng Struct 9:210–212

Holmes JD (2007) Wind loading of structures. Taylor and Francis, New York

Irwin HPAH (1981) The design of spires for wind simulation. J Wind Eng Ind Aerodyn 7(3):361–366

Jensen M (1958) The model law phenomena in natural wind. Ingeniøren 2:121–128

Munson BR, Young DF, Okiishi T (2002) Fundamentals of fluid mechanics. Wiley, New York

Simiu E, Scanlan RH (1996) Wind effects on structures fundamentals and applications to design. Wiley, New York

Tschanz T, Davenport A (1983) The base balance technique for the determination of dynamic wind loads. J Wind Eng Ind Aerodyn 13:429–439

White FM (1998) Fluid mechanics. WCB/McGraw-Hill, Boston, MA

Chapter 4
Wind Induced Conductor Motion in High Voltage Overhead Transmission Lines

4.1 Introduction

From a structural point of view, wind action on HVTL represents the most important load for the design of towers, conductors, and fittings. A typical HVTL is shown in Fig. 4.1. Moreover, in cold regions the wind action can combine with the presence of ice on cables, resulting in even higher loads on the structure. High wind produces static and dynamic loads on the structural components (cables and towers): both condition the tower design. Tower failures due to high wind are not rare events. For example, recent failures occurred in France and other parts of the world, causing outages and blackouts in entire regions. When this happens, it means that something wrong occurred in the evaluation of the loads acting on the TL. One factor could be a non-correct estimation of the maximum wind typical of the zone where the TL has to be built and its return period (generally 50 years); another factor could be related to the evaluation of the towers, fittings, and cables aerodynamic coefficients.

These problems, mainly related to the static design of towers and fittings, do not differ from those encountered dealing with any other structure subjected to the wind action and should be faced using the same tools, also making reference to standard and codes. Static loads on HVTL is not the subject of this chapter: it will mainly deal with problems related to the conductor wind-induced motion, which represent an important issue for the TL design.

The main problems related to the conductor wind-induced motion are:

1. Aeolian vibrations
2. Wake induced oscillations or subspan oscillations
3. Galloping.

Aeolian vibrations is the name associated to Vortex Induced Vibrations (VIV). They are due to vortex shedding from the conductors and can produce fatigue failures of the conductors themselves.

Subspan oscillations occur on conductor bundles. In this case, each phase of the TL is not represented by a single conductor but by a bundle of conductors, separated

D. Rocchi et al., *Wind Engineering*, Research Topics in Wind Energy 9,
https://doi.org/10.1007/978-3-032-17423-9_4

Fig. 4.1 High voltage transmission line

by spacers (as an example Fig. 4.1 shows a TL equipped with quadruple bundles). When the wind blows on the bundle, some sub-conductors may be in the wake of another (for the quad bundle shown in Fig. 4.1 there are two couples of conductors with one in the wake of the other), so sub-conductors oscillations can be excited by wake effect. The excitation mechanism will be explained in more detail in the following. In this case, conductor failures can also be caused.

Galloping is a kind of instability due to the unstable shape assumed by the conductors when they are covered with ice. This phenomenon can cause failures of the conductor fittings and the tower components and contacts among the phases.

In order to better understand the different mentioned problems, it is useful to briefly describe the main features of an Overhead TL and to recall the structure of the wind blowing across it: this will also allow us to correlate different ranges of wind speed to the different phenomena. The chapter is then organized into the following sections:

Section 4.2: Description of an HVTL
Section 4.3: Wind characteristics, also in relation to the different wind excited vibrations
Section 4.4: Aeolian vibrations
Section 4.5: Subspan oscillations
Section 4.6: Galloping.

4.2 Description of an HVTL

High-Voltage Transmission Lines (HVTL) are used to transfer energy from a point where it is generally produced to the places where it is used. In order to reduce energy losses, the voltage must be as high as possible: typical voltages are between 180 and 1000 kV.

Due to problems related to the electromagnetic field around the conductor and, as a consequence, due to the corona effect, TL equipped with single conductors are used up to a voltage around 200 kV: this is the case of the 180kV line shown in Fig. 4.2, where each of the three electrical phases is represented by a single conductor.

It is possible to have a double circuit TL: in this case three phases are supported at one side of the towers and the other three phases at the other side of the towers, as shown in Fig. 4.3.

The conductors are supported at the tower by suspension clamps, which are connected to the arms of the tower through insulator strings (Fig. 4.4).

The typical distance between the towers is on the order of 300–400 m. The conductors, due to their weight, assume the typical catenary configuration, with a sag-to-span ratio around $3 \div 5\%$.

As the voltage increases, more than one conductor per phase is needed, and then twin, triple, and quadruple conductor bundles are used (Fig. 4.5). The distance among

Fig. 4.2 180 kV TL

Fig. 4.3 Double circuit TL equipped with quadruple conductor bundles

Fig. 4.4 Suspension clamp and insulator string

the sub-conductors of a bundle in the same phase is of the order of 15 conductor diameters. Figure 4.3 presents a typical 480 kV TL equipped with quadruple bundles, one per each phase.

In the case of conductor bundles, the bundle geometry along the span is maintained by the spacers. The spacers can be rigid or flexible and in this last case some damping can be also introduced in the articulations in order to contribute to the control of wind induced vibrations. This point will be discussed in the following.

The conductor of an overhead power line is considered to be the most important component of the overhead line, since its function is to transfer electric power, and

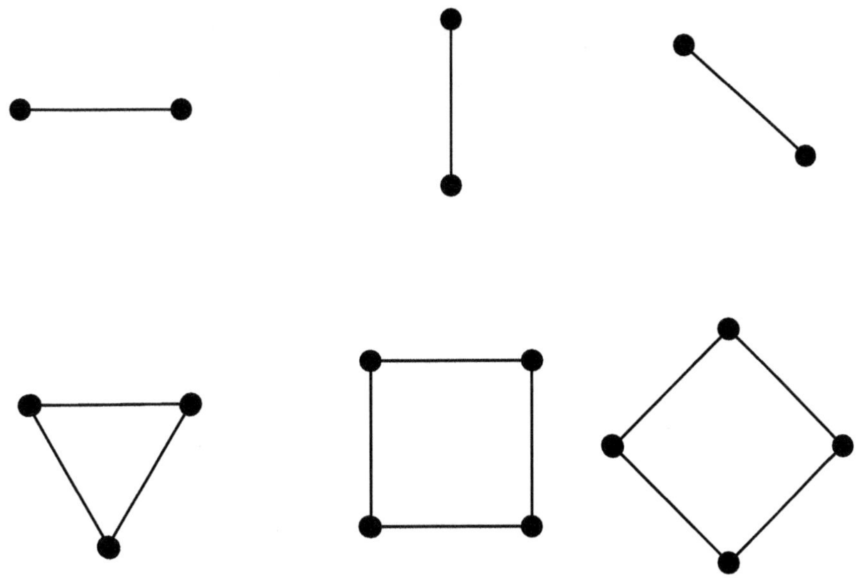

Fig. 4.5 Conductor bundles

its contribution toward the total cost of the line is significant. The conductor selection has moved from simple copper wire or copper-based bare conductors in the early days to more cost-effective solutions, such as aluminium and variations of aluminium alloy conductors.

Conductors are stranded cables. The most widely used form of conductors is that of layers of round wires stranded, first, around a so-called core, which can be of the same material or different, and then around each other. In order to keep the integrity of this construction, the stranding takes place in alternating directions from layer to layer (Fig. 4.6).

Most of the requirements for conductor design come from mechanical constraints. The electrical aspects of conductors are usually limited to current density, electrical resistance, and the associated power loss and voltage gradient, which are solved by adding area and adjusting the outside diameter or using multi-conductor bundles on the line. Some overhead conductors are constructed from commercially pure aluminum, known as AA1350-H19 and referred to as All Aluminum Conductor (AAC) or Aluminum Stranded Conductor (ASC). Because of its relative low strength-to-weight ratio (which is the most important mechanical criterion), these types of conductors are suitable for short spans in distribution networks, and for areas where ice and wind loads are limited, as well as for flexible bus bars in substations.

For added strength, various aluminum alloys have been developed, and these conductors are referred to as All Aluminum Alloy Conductor (AAAC) or Aluminum Alloy Stranded Conductor (AASC). Most alloys used today are of the AA6000 series,

Fig. 4.6 Typical conductor

which are heat-treatable. It should be noted that any improvements in strength are usually to the detriment of conductivity.

When a better strength-to-weight ratio is desired, a strength member has to be added to the conductor. This can be achieved by adding an aluminum alloy core to the AAC to create an ACAR (Aluminum Conductor Alloy Reinforced), but it is usually done with steel wires which offer much higher strength-to-weight ratios than aluminum alloys. Aluminum Conductor Steel Reinforced (ACSR), the most commonly used conductor type, and Aluminum Alloy Conductor Steel Reinforced (AACSR) are variations of the above conductors. In a few special cases—for instance, under extreme corrosive (marine) environments—the use of ACAR eliminates the galvanic reaction that is possible between the steel and the aluminum. Copper conductors may be used because of their superior electrical characteristics, but offer very poor mechanical properties and, therefore, are seldom selected.

Apart from the "standard" conductor designs, there are also a number of special designs, such as conductors with high steel content for very long spans (river crossings—Fig. 4.7), smooth-body conductors, etc. One way to improve and tailor conductors to special situations is to shape the aluminum wires. Over the past 30 years, the development of trapezoidal, and more recently Z-shaped (Fig. 4.7), wires has contributed to the improvement of conductor design. These conductors make better use of their space compared to round-wire conductors, and are, therefore, called "Compact".

High temperature conductors are also becoming popular to increase the transmitted power. This type of conductors has generally a special core (in steel o carbon fiber) to control the sag at high temperatures (up to 200 °C).

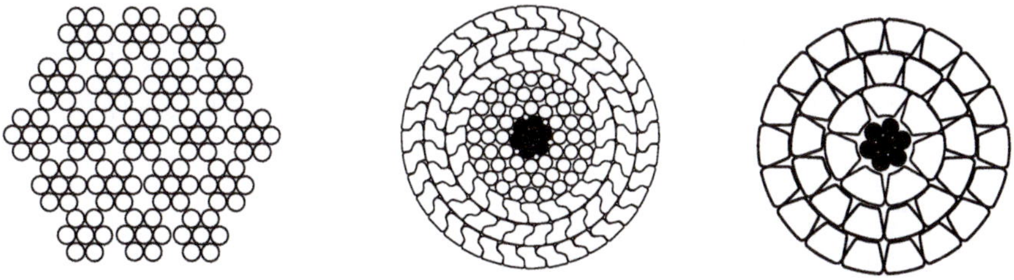

Fig. 4.7 Example of special conductor for river crossings, Z-shaped and trapezoidal conductors with steel core

Fig. 4.8 Types of terrain: **a** flat terrain (low turbulence); **b** flat terrain + small obstacles (low turbulence); **c** undulated terrain (high turbulence)

4.3 Wind Characteristics

The structure of the wind blowing on a line depends on the type of the terrain crossed by the TL. Different types of terrain have been categorized, as reported in Fig. 4.8: different levels of turbulence are associated to them.

In case of flat terrain or river crossing the turbulence level can be of the order of 10%; increasing the terrain roughness, the turbulence level increases and the wind speed vertical profile changes, as reported in the section devoted to the atmospheric boundary layer.

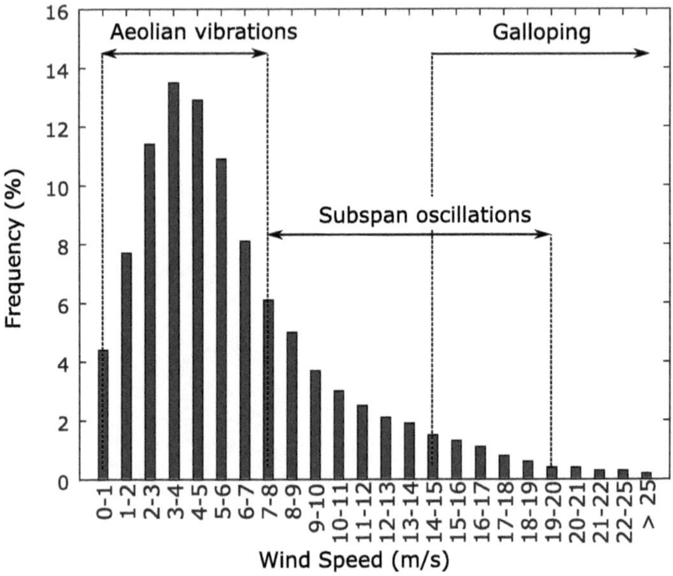

Fig. 4.9 Statistical distribution of wind speed

Dealing with the different types of problems related to the wind excitation, it is important to observe that each one of them (aeolian vibrations, subspan oscillations, galloping) occurs in different wind speed ranges.

Aeolian vibrations occur for low to mean wind speeds—up to about 10 m/s; subspan oscillations are excited by wind speeds in the range $8 \div 20$ m/s and galloping generally occurs for high wind speeds, over 15 m/s. The wind statistics typical of the site where the TL is being built is an important information to evaluate which one of the oscillation phenomena shall be taken into account in the TL design.

A typical wind speed statistical distribution is reported in Fig. 4.9: it can be easily deduced that while aeolian vibrations can be classified as a frequent phenomenon, subspan oscillations and galloping are relatively rare phenomena.

However, if particular situations are considered, like, for instance, the South Arabia desert or other similar sites, the occurrence of wind speeds over 10 m/s can be much higher than in Fig. 4.9 and then the occurrence of subspan oscillations and galloping could not be so rare. Remember that ice galloping not only needs high wind speed, but also low temperatures.

4.4 Aeolian Vibrations

Aeolian vibrations are caused by vortex shedding from the conductors. Due to this fact, before analyzing Aeolian vibrations of HVTL in their main features, it is useful to analyze vortex shedding from a bluff body making more attention to a body with circular section, being this case the most widely studied for vortex shedding.

The analysis is divided in three steps:

1. Vortex shedding
2. Mechanical behaviour of conductors
3. Numerical model to simulate Aeolian vibrations of HVTL.

4.4.1 Vortex Shedding

Vortex shedding occurs in bluff bodies, or, in any case, when there is flow or boundary layer separation. The boundary layer separation produces vorticity and, as a consequence, oscillating vortices arise behind the object (Fig. 4.10).

The vortex shedding from the body exerts a periodic force on the body itself and if the associated frequency is close to of the body natural frequencies, the body starts to oscillate. The vortex shedding frequency is related to the body shape and dimensions, through the following very well known formula:

$$f_S = St\frac{V}{D} \tag{4.1}$$

where V is the flow speed, D is a reference dimension of the body and St is the Strohual number, that is related to the body shape. From (4.1), the Strouhal number can be defined as:

$$St = \frac{f_S D}{V} = \frac{1}{V_r} \tag{4.2}$$

where V_r is the reduced velocity, a fundamental parameter in the fluid-elastic interaction, as it is better detailed in the section devoted to Bridges. Once the bluff body natural frequency f_p is given, the critical velocity V_c, also called Strouhal velocity V_{St}, at which the body will vibrate is given by:

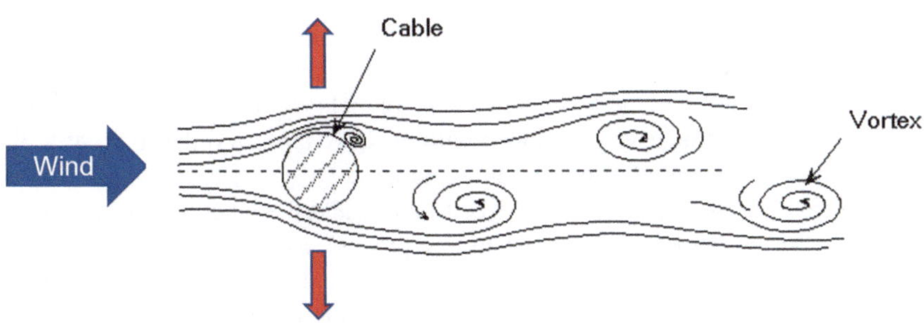

Fig. 4.10 Vortex shedding from a bluff body

Section shape	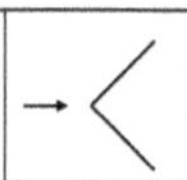	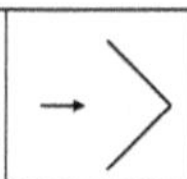	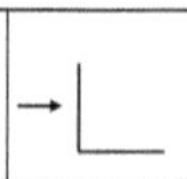	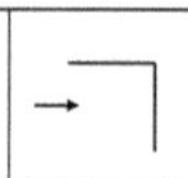	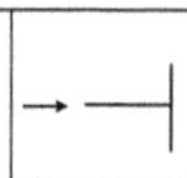	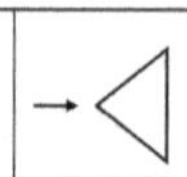	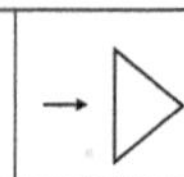	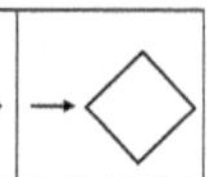
Strouhal number	0.19	0.15	0.14	0.14	0.15	0.19	0.12	0.19

Fig. 4.11 Dependence of the Strohual number on shape and dimensions

$$V_c = \frac{f_p D}{St} = V_S \tag{4.3}$$

For a circular cylinder the Strouhal number St is equal to 0.18; other shapes are considered in Fig. 4.11.

As it will be better clarified in the following, vibrations due to vortex shedding are self-controlled, in that the vibration amplitude is limited: for instance, for the case of a circular cylinder, the maximum vibration amplitude due to vortex shedding equals the cylinder diameter.

Due to the fact that a bluff body can also be unstable (when the slope of the lift or moment coefficients as a function of the wind angle of attack is negative), vortex excitation and one-degree-of-freedom instability can simultaneously occur, increasing the oscillation amplitudes in a very large amount, up to the system collapse, as in the case of the very famous Tacoma Narrows bridge failure. Of course the circular section is not affected by this type of instability, being the average lift equal to zero.

The vortex shedding mechanism is a complex function of the body motion. The vortex shedding street and the forces exerted on the body change in a large amount as a function of the cylinder motion. In order to better understand, it is useful to first consider a non moving body and then to extend the study to a moving body. Reference will be made to the circular cylinder, whose case, as already observed, has been widely treated.

4.4.1.1 Vortex Shedding from a Fixed Circular Cylinder

Good starting references are papers Diana et al. (2006a) and Zasso et al. (2006) where the results of pressure measurements all around the circular section are reported. These measurements were performed in a wind tunnel (Fig. 4.12) and more than one section along the cylinder was equipped with pressure taps, in order to identify the forces applied to the cylinder in different sections.

In the case that the cylinder is not free to move and the incoming flow velocity is constant during the entire vortex shedding process, as can be seen in Fig. 4.13, almost three quarter of the cylinder surface is subjected to a negative pressure, the negative pressure peak position moving from one side to the other as the vortices are formed. This fluctuating pressure results into mean and alternating loads: the

Fig. 4.12 Cylinder sectional model equipped with pressure taps and elastically suspended in the wind tunnel

mean and fluctuating drag and lift forces, acting in the stream-wise and cross-flow directions respectively.

As it can be observed in Fig. 4.13 (and identified from the measurements (Zasso et al. 2008)) the process is not fully periodic, but is of random nature; moreover it does not occur simultaneously along the cylinder axis. In particular there is a random phase lag among the lift forces applied to the different sections along the cylinder axis.

The random nature of the process and the lack of correlation along the cylinder axis make the lift force value small, if compared to the case of a vibrating cylinder, as it will be better explained in the next paragraph.

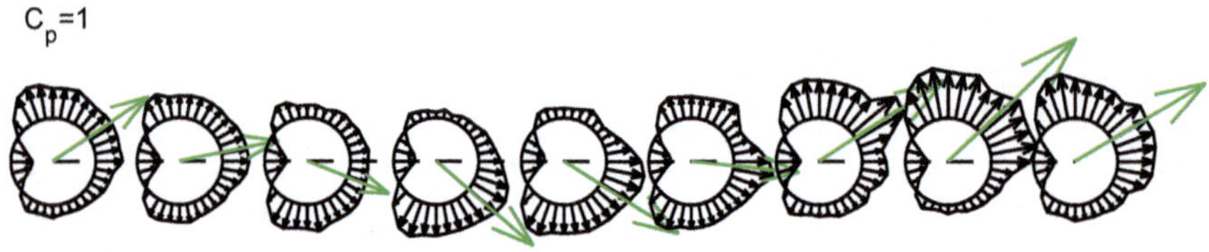

Fig. 4.13 Measured unsteady pressure and force coefficients at nine instants of a period of vortex shedding (case of the stationary cylinder)

4.4.1.2 Vortex Shedding from a Vibrating Circular Cylinder

As already said the vortex shedding mechanism is affected in a big amount by the bluff body vibration amplitudes. A wide research has been done (Zasso et al. 2008; Diana and Falco 1966; Belloli et al. 2006; Zasso et al. 2005; Diana et al. 2006b; Carberry et al. 2004; Govardhan and Williamson 2000; Carberry et al. 2001; Govardhan and Williamson 2004; EPRI 2009) to identify the nature of the phenomenon. Many of them are experimental researches carried out on vibrating circular cylinders, as shown in Fig. 4.12. Recent researches (Diana et al. 2006a; Zasso et al. 2006, 2008), performed at the Politecnico di Milano Wind Tunnel, have been characterized by the simultaneous measurement of amplitudes of vibration and circumferential pressure distribution in different positions along the cylinder axis. The main features of the complex mechanism of vortex shedding will be resumed in the following.

Lock-in and synchronization range

Let us consider an elastically suspended circular cylinder in a WT, as in Fig. 4.12. The arrangement is made in such a way that the cylinder is a single degree of freedom system allowed to vibrate in the vertical direction, i.e. in the direction normal to the approaching flow. The cylinder natural frequency is defined as f_p.

If the flow velocity satisfies Eq. (4.3), i.e. $V = V_c$, the vortex shedding frequency becomes equal to the cylinder natural frequency and the cylinder starts to vibrate.

The vibration amplitude increases and reaches a steady state value, as shown in Fig. 4.14. In this diagram the non dimensional vibration amplitude is reported: u/D is the ratio between the vibration amplitude and the cylinder diameter. As can be observed the vibration amplitude does not exceed one conductor diameter and its value depends on the damping of the vibrating cylinder.

Increasing the damping level, the vibration amplitude reduces down to no vibration, even if Eq. (4.3) is satisfied. The non dimensional parameter identifying the Vortex Induced Vibrations (VIV) severity is the Scruton number, defined as:

$$S_c = 2\pi \frac{hm}{\rho D^2} \tag{4.4}$$

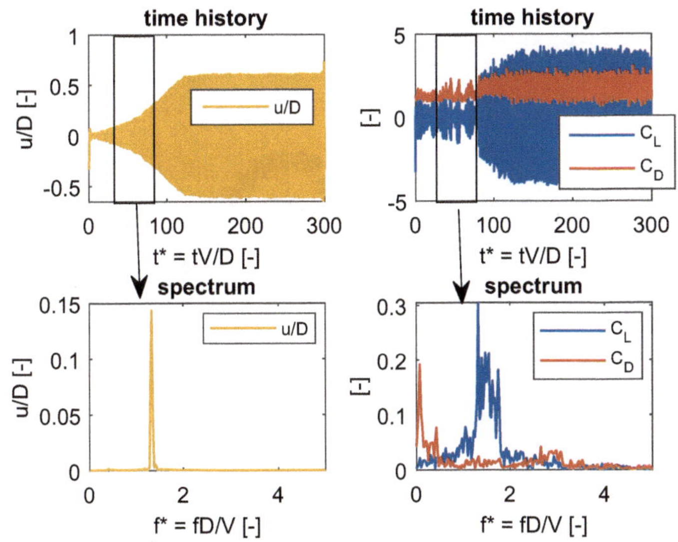

Fig. 4.14 Cross flow vibration amplitudes time histories and drag and lift coefficients (left) with related spectra (right)

where h is the non dimensional damping ($h = r/r_c$), m is the body mass per unit length, D is the body reference dimension (the cylinder diameter) and ρ is the flow density. Coming back to the case represented in Fig. 4.14, where the cylinder reaches a relatively large steady state amplitude of vibration, suppose to increase the wind speed by 10% of its previous value (the one satisfying Eq. (4.3)): the vortex shedding frequency is also increased by 10% with respect to the cylinder natural frequency (Eq. (4.1)) and then the system is now outside the resonance condition; the vibration amplitude is not decreasing, as expected, but, on the contrary, increases. A diagram as the one reported in Fig. 4.15 is obtained: the horizontal axis reports the ratio between the flow velocity and the Strouhal velocity V_S, while the vertical axis reports the steady state vibration amplitude referred to the cylinder diameter. The diagram is taken from Diana et al. (2006a) and Zasso et al. (2006), where more details of this aspect of the phenomenon are described.

As already said, besides vibration amplitudes, the circumferential pressure was also measured in the research work described in Diana et al. (2006a) and Zasso et al. (2006). From the pressure measurements, the lift force produced by vortex shedding has been evaluated.

From this, it clearly appears that in all the flow speed range defined by $0.9 < V/V_S < 1.5$ the vortex shedding frequency becomes equal to the vibration frequency; in other words, vortex shedding is locked to the cylinder natural frequency in all the wind speed range called synchronization range, that is, for very low damping (very low Scruton), between 0.85 and 1.5 V/V_S. This mechanism is known as lock-in mechanism.

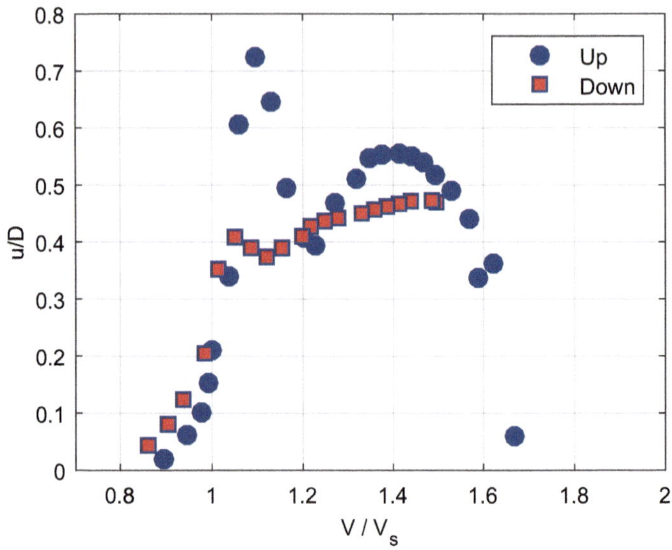

Fig. 4.15 Hysteresis mechanism increasing or reducing the wind speed

The frequency of vortex shedding, given by expression (4.1) is reported in Fig. 4.16, as a function of the wind speed V: of course it is a linear function of V, in case of fixed circular cylinder (left side).

If the cylinder vibrates, in the synchronization range the vortex shedding frequency is triggered by the structure, jumping to the frequency given by (4.1) outside the synchronization range (right side).

Another important feature of the phenomenon is that the lift force, that was observed to have a random nature in case of fixed circular cylinder, in all the synchronization range looses its random nature and becomes fully periodic. Moreover, the lift forces measured along the cylinder axis are all in phase and, as a consequence, the effect is much greater than in the case of fixed cylinder. The mechanism allowing the vortex shedding frequency to jump from the Strouhal formula (4.1) value to the cylinder natural frequency is complex. In order to better understand it is needed to analyse the build up of vibrations when, starting from the condition of cylinder at rest, the wind starts to blow with a wind speed inside the synchronization range.

Going back to Fig. 4.14, the build up of vibrations is reported both in terms of vibration amplitude (u/D) and in terms of measured lift force coefficient (C_L), evaluated from the pressure measurements. Relevant spectra are reported in the bottom of the figure.

In view of analyzing the frequencies involved in the phenomenon during the build up, it is possible to observe that, when the amplitudes are small, less than 0.2 diameters, the vibration amplitude spectrum obviously identifies the cylinder natural frequency, while the lift force shows a non periodic nature, with a spectrum containing two main frequencies: one is the cylinder natural frequency and the other

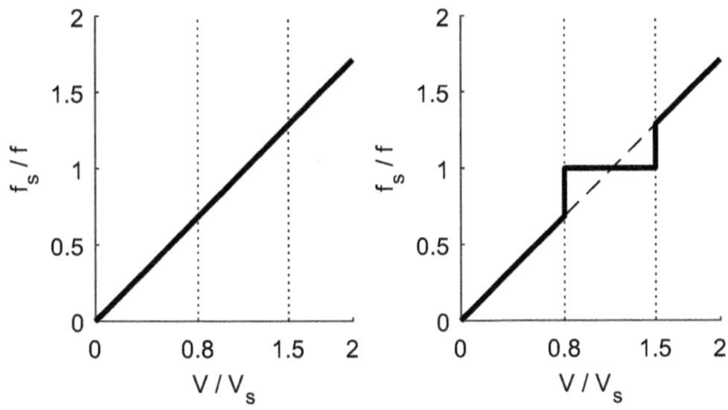

Fig. 4.16 Lock in effect on a vibrating cylinder

is the vortex shedding frequency given by the Strouhal formula. This last one is a little bit higher because in this test the flow velocity is higher than the Strouhal velocity V_S ($V/V_S > 1$).

As the amplitudes are increasing and reach a value around 0.3 diameters, in this case corresponding to a time of 105 s, the lift force suddenly increases and the vortex shedding frequency becomes equal to that of the vibrations, as can be seen better in Fig. 4.17, where the lift is reported as a function of time.

In correspondence to this jump, also the phase of the lift force with respect to the displacement jumps from around 90° to a value close to 0°, as can be seen in the time histories reported in Fig. 4.17. More details can be found in Diana et al. (2006a) and Zasso et al. (2006).

In Fig. 4.18, taken from Zasso et al. (2008) the harmonic component at the vibrating frequency of the lift coefficient C_L is reported as a function of the u/D amplitude of vibration. In the same figure, the phase between the lift force and the vibration (displacement) is also reported. The vibration amplitude effect is of paramount importance on this type of excitation: the different curves reported in Fig. 4.18 refer to different values of the V/V_S parameter, inside the synchronization range. As can be seen, the lift amplitude increases with the vibration amplitude and the phase is close to 90° for low amplitudes while goes close to zero for large amplitudes of vibration: this is a clear indication that vortex induced vibrations have not to be considered a forced motion but a self-excited mechanism. The amplitudes are self contained to one diameter due also to the drag damping effect that drives the phase of the lift force close to zero when the amplitudes are increasing. This is clearly shown in Fig. 4.18.

If, for a given value of the lift force amplitude, its phase with respect to displacement approaches 90°, the energy input reaches it maximum value, while the energy input decreases together with the phase and becomes smaller when the phase tends to 0°.

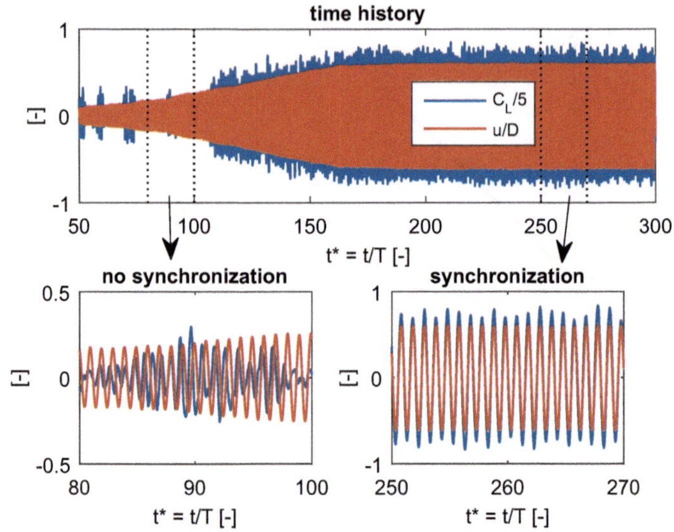

Fig. 4.17 Vortex shedding inside and outside the synchronization range

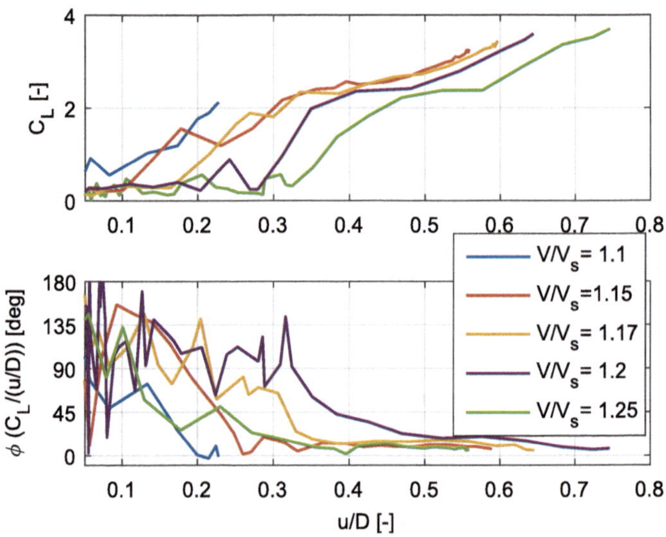

Fig. 4.18 Lift coefficient and phase as a function of the u/D amplitude of vibration Zasso et al. (2008)

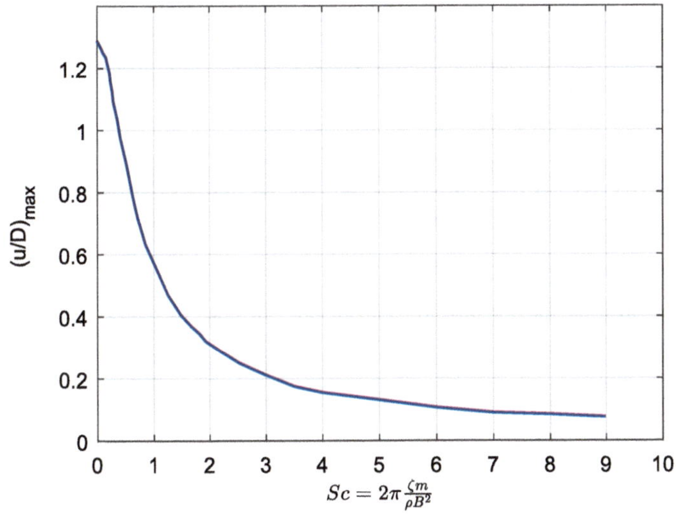

Fig. 4.19 Maximum vibration amplitudes as a function of the Scruton number

Maximum amplitudes of vibration due to vortex shedding

From a practical point of view, it is very important to know what is the maximum amplitude due to VIV for any bluff body exposed to this phenomenon. The maximum amplitude is identified as that one reached in the synchronization range (see Fig. 4.15). As already said, the VIV are increasing as the Scruton number decreases. For any type of bluff body subjected to VIV, it is used to identify a curve reporting the maximum VIV amplitude as a function of Scruton. For instance, in the case of the circular cylinder this relation is shown in Fig. 4.19.

As well as the amplitudes of vibration, also the range of synchronization depends on the Scruton number—see Fig. 4.20. Increasing the Scruton number, the synchronization range width decreases. Comparing Fig. 4.20 to Fig. 4.19, reporting the Scruton number effect on the maximum vibration amplitudes, it is possible to understand that only when large amplitudes of vibration are reached (of the order of one conductor diameter) the synchronization range extends from 0.85 to 1.5 V/V_S, while it is becoming very narrow for $Sc \approx 15$.

Similar figures can be obtained for any other bluff body shape. These curves are very useful because immediately give what is the damping necessary to control the vibration amplitude within a defined value. These curves are non dimensional and allow to extrapolate the figure obtained in a wind tunnel test on a model to a real structure, once the mass per unit length (m), the reference dimension (D) and the non dimensional damping ($h = r/r_c$, with $r_c = 2m\omega$) of the real structure itself are known.

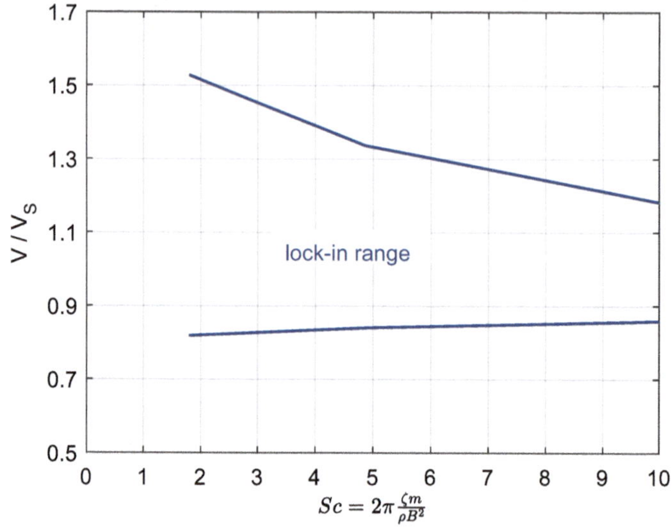

Fig. 4.20 Synchronization range as a function of the Scruton number. Adimensional wind speed is reported on the vertical axis

It will be shown in the following that in the case of cables, more sophisticated approaches are used to identify the damping needed to control VIV.

Reynolds effect

The Reynolds number (Re) is a very important parameter affecting, as already observed, all the fluid-dynamic interaction between the fluid and the body. The Re number is defined as $\text{Re} = \frac{VD}{\nu}$ where V is the fluid velocity, D is a reference dimension and ν is the cinematic viscosity, whose value, in air, can be assumed equal to $1.5 \cdot 10^{-5}$.

The Re number affects vortex shedding in different ways: the boundary layer separation depends on Re, as well as the boundary layer transition from laminar to turbulent. The effect of Re is much more important for a body characterized by a curved shape, than for a body characterized by corners that control the separation point. From this point of view, the circular section is fully affected by Re and, as a consequence, also its VIV response.

Being interested in cables vibrations induced by vortex shedding and due to the fact that cables have a circular section, we will analyze the *Re* effect on a cylinder with circular section subjected to VIV. The Re effect on the drag coefficient of a circular cylinder is well known.

In Fig. 4.21 the drag coefficient as a function of Re is reported: many fluid regimes can be distinguished, as reported in Fig. 4.21; the transition from one to the other depends on Re but the Re value for which each transition occurs is also function of the cylinder roughness and of the incoming flow turbulence. Analyzing Fig. 4.21,

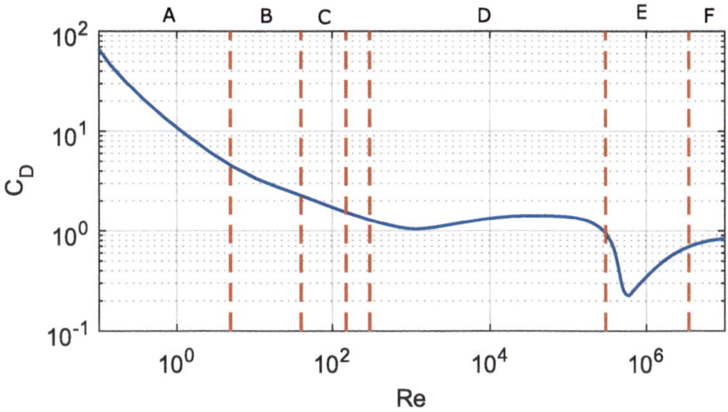

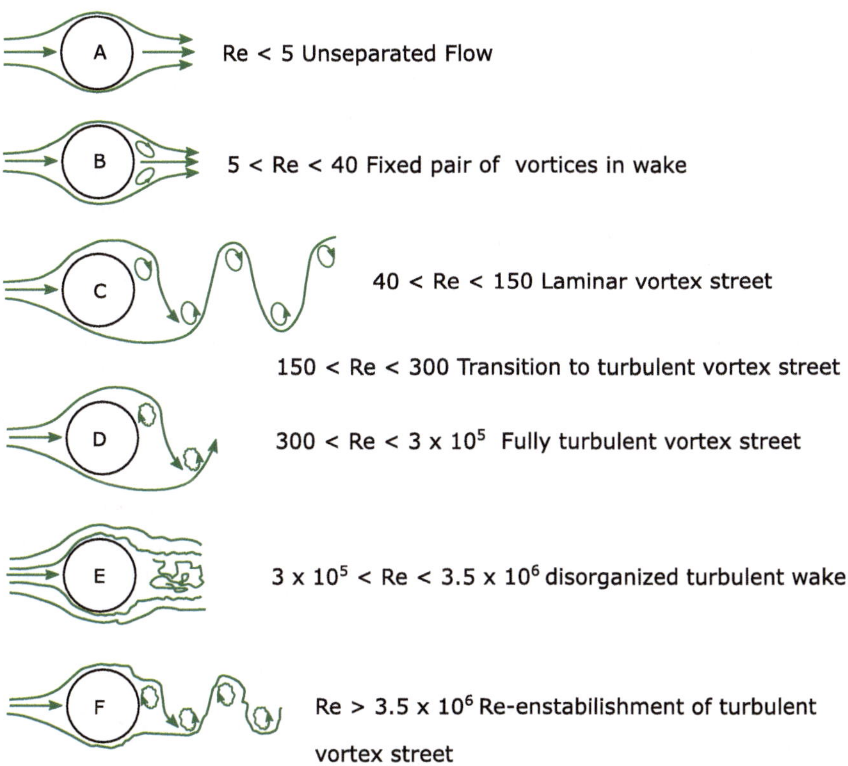

Fig. 4.21 **a** Drag coefficient of a circular cylinder with a smooth surface. **b** Vortex shedding from a circular cylinder—flow regimes, as a function of Re

which is relevant to a smooth surface and to a non turbulent flow, it can be observed that vortex shedding occurs for very low Re numbers, up to $3 \cdot 10^5$, where the drag is more or less constant and equal to 1.2. For Re higher than $3 \cdot 10^5$, the drag starts to decrease because the boundary layer becomes turbulent and the separation point, from a point at 90° with respect to the stagnation point in front of the section, moves downstream, the stream becoming narrower and the drag reaching a minimum: see Fig. 4.21. Vortex shedding disappears in these critical regimes.

However, for Re higher than $3 \cdot 10^6$ (post critical range), even in presence of a turbulent boundary layer, the vortex street reappears and VIV excitation occurs with features that are not far away from those of the subcritical range. The transition from one regime to the other, as already said, is also a function of the surface roughness and of the flow turbulence.

The increase of roughness as well as the increase of turbulence shift the transition from subcritical to critical regime towards lower Re. The shift of the critical range as a function of the turbulence index is reported in Fig. 4.22. Similar figures are available for the effect of the surface roughness: as an example the critical range shift is reported in Fig. 4.23 as a function of a roughness parameter identified as the ratio between the roughness dimension,ϵ, and the cylinder diameter D.

All the data concerning VIV reported in literature are mainly related to the subcritical zone, which corresponds, from a practical point of view, to the range in which OHTL cables are operating. However, if the cable size is increased, like the case of bridge hangers and stays, it is possible to pass to the critical and post-critical ranges. Only a few experimental data concerning VIV and its mechanism in the post critical region are available, due to the difficulties related to performing tests at so high Re numbers.

Some data for a rough circular cylinder in the post-critical range are available (Diana et al. 2008): these show that vortex shedding does not change its basic features.

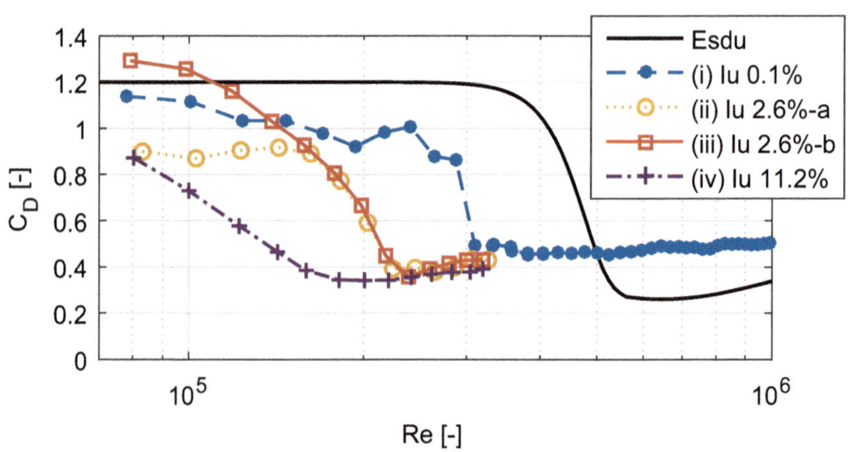

Fig. 4.22 Drag transition from subcritical to critical regime as a function of turbulence

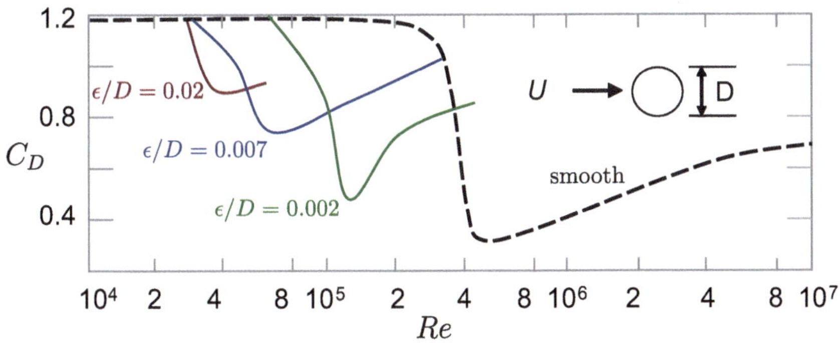

Fig. 4.23 Drag transition from subcritical to critical regime as a function of surface roughness

For this reason, and also to be on the safe side, the approach used for the sub-critical region is maintained also for the post-critical one.

4.4.1.3 Energy Introduced by Vortex Shedding on a Vibrating Cylinder

The energy introduced by the wind to excite VIV is very important because, as it will be seen in the following, one of the methods widely used to predict OHTL conductors amplitude of vibration is based on a balance between the energy introduced by the wind and that dissipated by the conductors plus fittings and damping devices (vibration dampers only in case of single conductors and dampers plus spacer-dampers in case of conductor bundles). Two methods are available to compute the energy input from the wind.

One is based on the measure of the VIV build up on a vibrating cylinder, from which the wind power input is computed, as shown in Fig. 4.24, which is obtained at different V/V_S wind speeds inside the synchronization range, in a wind tunnel.

Knowing the cylinder mass per unit length m and the vibration frequency f_p and defining by u_i and u_{i+1} the vibration amplitudes in two adjacent cycles—as in Fig. 4.25—the cylinder maximum kinetic energy, reported to the cylinder unit length, for $u = u_i$, is equal to: $E_{Ku_i} = 1/2mu_i^2\omega^2$. The kinetic energy increment from cycle i to cycle i+1 is given by:

$$\Delta E_{K_i} = \frac{1}{2}m\omega^2\left(u_{i+1}^2 - u_i^2\right)$$

The energy dissipated by the cylinder in case of no wind, also reported to the cylinder unit length, evaluated at the average amplitude $u_{m_i} = (u_{i+1} + u_i)/2$, is equal to:

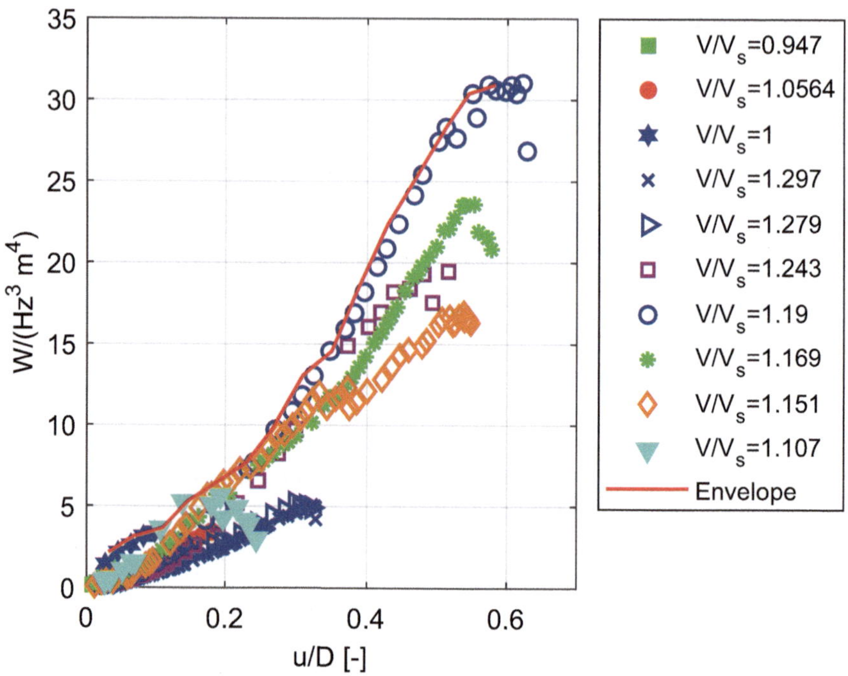

Fig. 4.24 Energy input from the wind due to VIV

$$E_{D_i} = 2\pi h_i m\omega^2 u_{m_i}^2$$

which can be derived by: $h_i = \frac{E_{D_i}}{4\pi E_{K_i}}$ (Diana and Cheli 2010).

The energy input from the wind in one cycle and at the average amplitude u_{m_i} is then given by:

$$E_{W_i} = E_{D_i} + \Delta E_{K_i} = 2\pi h_i m\omega^2 u_{m_i}^2 + \frac{1}{2} m\omega^2 \left(u_{i+1}^2 - u_i^2\right)$$

A decay test in still air can be performed to obtain the non dimensional damping h_i for different amplitudes of vibration. In reality the test should be made in vacuum, but, generally the energy dissipated in still air is small and, in any case, could be evaluated by the available formulations (aerodynamic damping (EPRI 2009)).

From the described tests the energy introduced by the wind as a function of the non dimensional amplitude u_{m_i}/D for different values of V/V_S can be measured—see Fig. 4.24.

Another approach to define the energy input from the wind is based on the measure of the pressure along the cylinder section—as already described in the previous sections of this chapter and also reported in Diana et al. (2007). The force along the motion direction can be deduced from the pressure measurements. This force

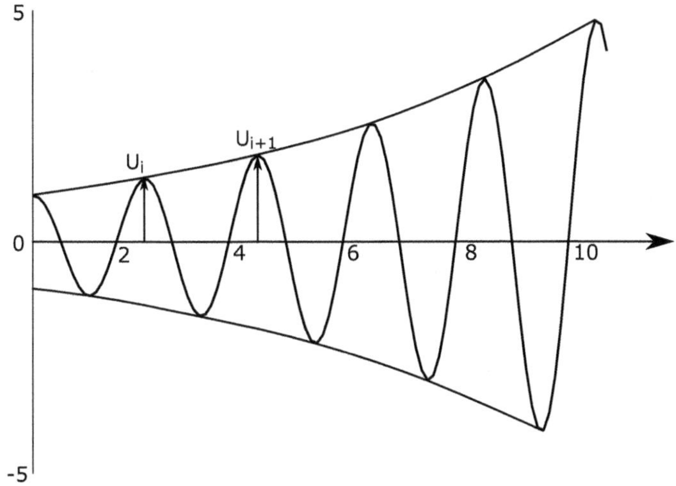

Fig. 4.25 VIV build up on a vibrating cylinder: vibration amplitudes in adjacent cycles

is generally defined as lift force. In reality the lift force should be the one normal to the flow relative velocity (with respect to the cylinder), that is a function of the cylinder vibration velocity too. The energy introduced in one vibration cycle by the lift force F_L at the u_i amplitude of vibration can be computed through the following expression:

$$E_i = \int_0^T F_L v_i \mathrm{d}t \tag{4.5}$$

where $v_i = u_i \cos \omega t$ is the vibration velocity and $T = 1/f_p$ is the oscillation period. This approach has also been used in the tests made at POLIMI (Diana et al. 2006a, 2005; Zasso et al. 2006, 2008; Diana et al. 2007, 2008; Belloli et al. 2006) and its results compared to those of the build up method. As shown in Fig. 4.26, the agreement is fairly good.

The energy input from the wind is generally reported in non-dimensional diagrams, considering the power W introduced by the wind per unit conductor length. Of course $W = E_w f$, where f is the vibration frequency. The power introduced by the wind is reported in these diagrams as a function of the non dimensional amplitude of vibration u/D. To get a non dimensional quantity the power W is divided by $f^3 D^4 \rho$, where ρ is the air density. In practice, generally considering ρ a constant, W is only divided by $f^3 D^4$.

Figure 4.24 reports the function $fnc(U/D) = W/(f^3 D^4)$ as a function of u/D for different values of V/V_S and it is taken from EPRI (2009).

The envelope of all the curves gives the maximum power introduced by the wind due to vortex shedding. Figure 4.27 (EPRI 2009) reports the maximum power input

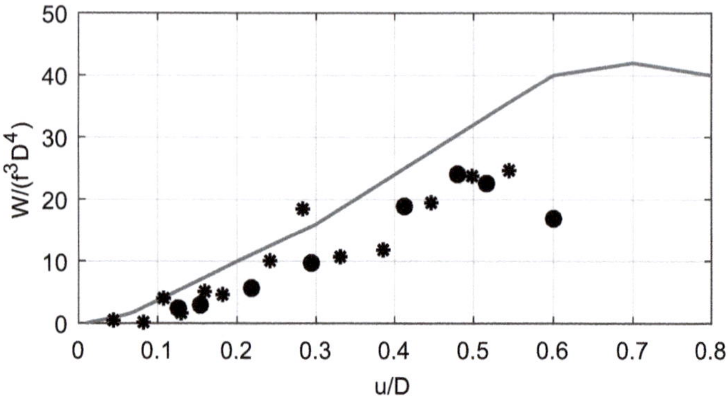

Fig. 4.26 Power input: experimental data measured at V/VS=1.17 with the first (°) and second (*) approach and literature reference data (Diana and Falco 1966) (solid line)

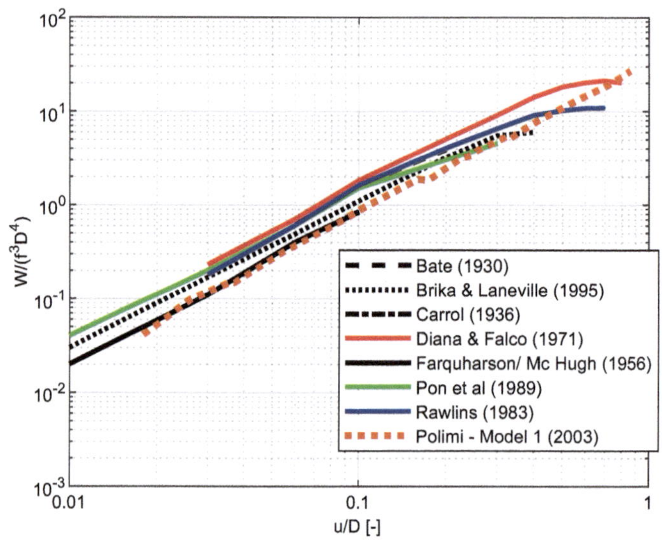

Fig. 4.27 Maximum power input from the wind due to VIV obtained by different researchers

obtained by different researchers. Together with the results of tests made on rigid (translating) cylinders, also those of tests on flexible cylinders vibrating according to a sinusoidal shape are reported. These tests have been performed with the aim of reproducing what happens on a real line, where, as it will be better explained in the following, the conductor vibrates according to one of its modes of vibration, i.e. according to a sinusoidal shape along the span (Fig. 4.28).

For these last tests, the energy introduced by the wind is half of the energy introduced on a rigid cylinder vibrating at the same amplitude u all along its length.

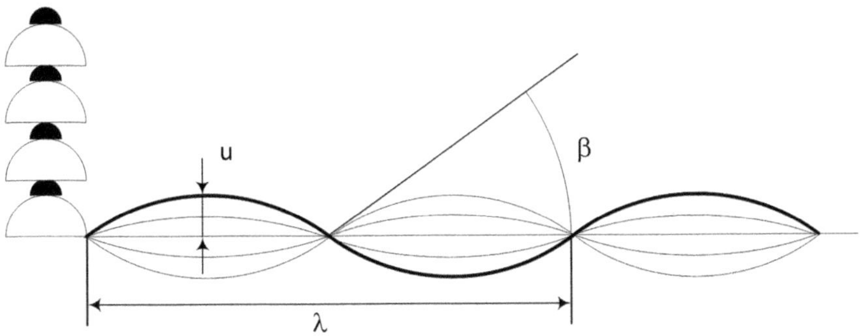

Fig. 4.28 Representation of the antinode amplitude of vibration and wavelength

As a consequence, to reproduce the power input on a flexible cylinder, the power W measured on a rigid cylinder must be divided by 2.

4.4.2 *Mechanical Behavior of Single Conductors*

4.4.2.1 Natural Frequencies and Modes of Vibration

Usually a single conductor is modeled as a taut string, and then its natural frequencies (where n is the mode number, $n = 1, 2, \ldots$), can be evaluated through the following expression (Sturm 1936; Claren and Diana 1969b):

$$f_n = \frac{1}{\lambda_n}\sqrt{\frac{T}{m_L}} = \frac{n}{2L}\sqrt{\frac{T}{m_L}}$$

where L is the span length, λ_n is the wave length, T is the tensile load, and m_L is the cable mass per unit length.

According to this model, the vibration modes are sinusoidal functions of x (Fig. 4.29):

$$y_n = A_n \sin\left(\frac{2\pi}{\lambda_n}x\right)$$

where A_n is the antinode amplitude of vibration. The term $\sqrt{\frac{T}{m_L}}$ is the propagation velocity of a transversal perturbation along the string.

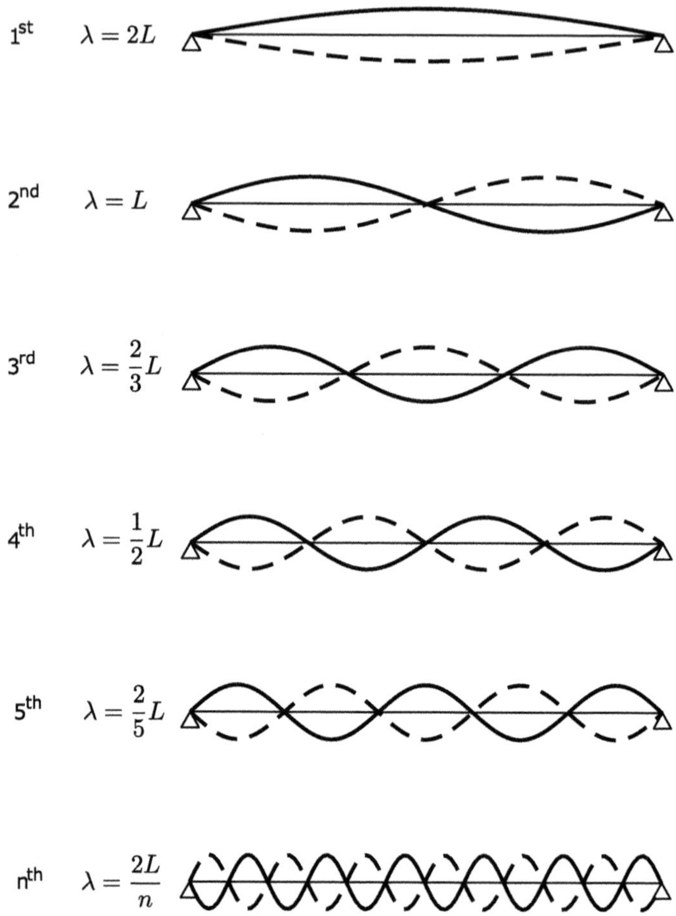

Fig. 4.29 Vibration modes—taut string model

As an example EPRI (2009), for a 366-m span equipped with a Drake conductor tensioned at 28.024 N (about 20% of its Ultimate Tensile Strength), the following values are obtained:

$$T = 28024\,\text{N}$$

$$m_L = 1.628\,\text{kg/m}$$

$$L = 366\,\text{m}$$

$$\sqrt{\frac{T}{m_L}} = 131.2\,\text{m/s}$$

$$f_n = \frac{1}{\lambda_n}\sqrt{\frac{T}{m_L}} = \frac{n}{2L}\sqrt{\frac{T}{m_L}} = 0.179n \quad \text{Hz} \qquad (n = 1, 2, \ldots)$$

and then $f_1 = 0.179$ Hz, $f_2 = 0.358$ Hz, $\ldots f_{10} = 1.79\text{Hz} \ldots f_{28} = 5.012\text{Hz} \ldots f_{168} = 30.07$ Hz.

If a wind velocity (V) of 0.75 m/s is taken, which is usually considered as the lowest wind speed at which aeolian vibration on a conductor occurs, the vortex-shedding frequency, according to the Strouhal formula is:

$$f = 0.185\, V/d = 0.1850.75/0.028 \approx 5\,\text{Hz}$$

which corresponds to the frequency of the 28th mode of vibration. With a wind speed around 3 m/s, we would obtain a frequency around 20 Hz.

So it is clear that aeolian vibration excites the higher modes of vibration of the cable, not the lower ones, which, on the contrary, can be excited by galloping.

In practice a real conductor is not like a string because it is characterized by a certain bending stiffness (EI). In that case, a more realistic model of a conductor is a tensioned beam. So the natural frequencies and vibration modes are represented by the following equations (Claren and Diana 1969b; Morse 1948):

$$f_n = \frac{1}{2\pi}\sqrt{\left(\frac{n\pi}{L}\right)^2 \frac{T}{m_L}\left[1 + \left(\frac{n\pi}{L}\right)^2 \frac{EI}{T}\right]}$$

$$y_n(x) = A_n Sh(z_n x) + B_n Ch(z_n x) + C_n sin(a_n x) + D_n cos(a_n x)$$

The shape of the vibration mode (determined by the values of A_n, B_n, C_n, D_n) depends on the span end conditions, and this is, in turn, fundamental to the correct evaluation of the strains and stresses at the span extremities, as it will be better explained in the following.

If the end conditions are hinges, the shape of the vibration modes is the same as for the taut string, and the maximum bending strain on the conductor is found at the antinode (Claren and Diana 1969a).

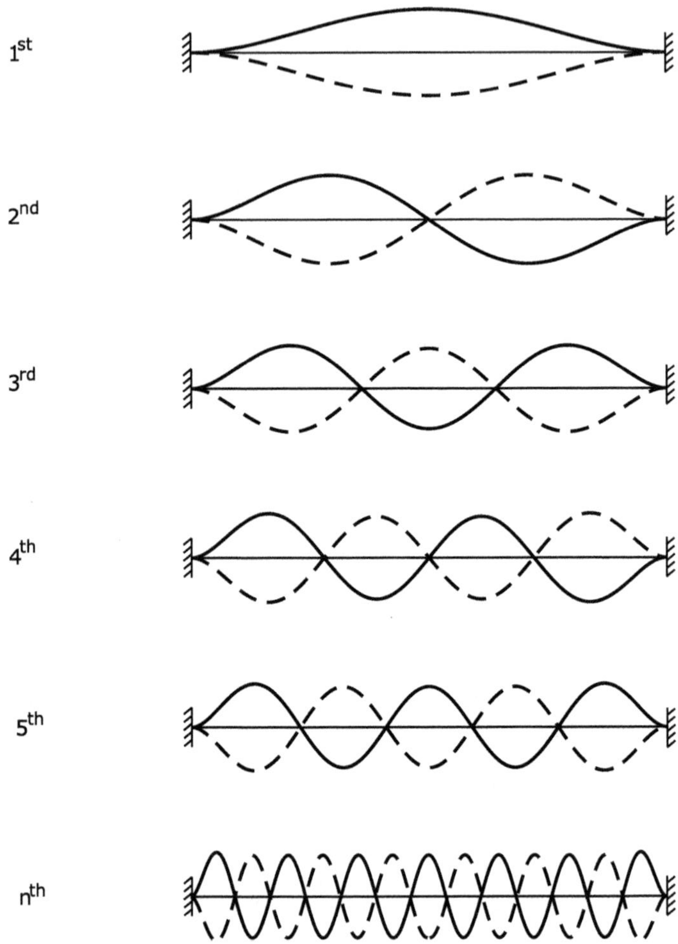

Fig. 4.30 Modes of vibration, beam model, fixed constraints

If the span extremities are fitted with fixed constraints, the vibration mode shape is as in Fig. 4.30, and the maximum bending strain is found at the span extremities (Claren and Diana 1969a).

The determination of the stresses due to the conductor bending requires evaluation of the conductor flexural behavior, which, in turn, depends on its flexural stiffness.

If no slippage between the single layers of an ACSR stranded conductor is considered, the equation to get the cable flexural stiffness $EI_{\max}$ is:

$$EI_{\max} = \frac{E_{St} \sum_i A_{St_i} d_i^2}{2} + \frac{E_{Al} \sum_i A_{Al_i} d_i^2}{2}$$

where A_{St_i} is the single steel wire area, A_{Al_i} is the single aluminum wire area d_i is the distance of the single wire from the conductor centre.

The conductor curvature is defined as the second derivative of the conductor displacement, and then in the no-slippage hypothesis, the maximum bending strain (on the cable outer layer) is obtained through:

$$(\varepsilon_{\max})_n = \frac{\partial^2 y_n(x)}{\partial x^2}\frac{D}{2}$$

where D is the cable diameter.

In practice, however, a slippage mechanism among the individual wires is present, and the maximum bending strain value is lower than that defined above, the cable stiffness is not equal to the $EI_{\max}$ value and it is not constant along the span.

By experimental tests reported in CIGRE (2005a), Claren and Diana (1969b), and Diana et al. (1973), the $EI_{\max}$ value is corrected through a reduction coefficient, whose value is generally around 0.5.

Moreover, the static configuration of a conductor is represented by a catenary, and the tensile load is not constant along the span. This fact makes the frequency and related shape of the first vibration mode different from that reported in Fig. 4.29. However, as already observed, aeolian vibration does not excite the first conductor modes, and therefore, this is not significant.

In both cases, whether the cable bending stiffness is accounted for or not, in the free span, the cable mode of vibration is represented by a sinusoidal function of the type:

$$y_n = A_n \sin\left(\frac{2\pi}{\lambda_n}x\right)$$

where λ_n is the wave length. As a consequence, in case of hinged extremities, the maximum bending strain is at the antinode (ϵ_a) and is given by:

$$(\varepsilon_a)_n = \frac{\partial^2 y_n(x)}{\partial x^2}\frac{D}{2} = \frac{4\pi^2}{\lambda_n^2}A_n\frac{D}{2}$$

where A_n represents the conductor antinode amplitude of vibration.

In case of clamped extremities (no rotation), the strain $(\epsilon_c)_n$ at the span extremities is higher than the antinode strain, and the following relation applies:

$$\left(\frac{\varepsilon_c}{\varepsilon_a}\right)_n = K\lambda_n\sqrt{\frac{T}{EI_{\max}}} \tag{4.6}$$

$$(\varepsilon_c)_n = 4\pi^2 K\sqrt{\frac{T}{EI_{\max}}}\frac{A_n}{\lambda_n}\frac{D}{2} \tag{4.7}$$

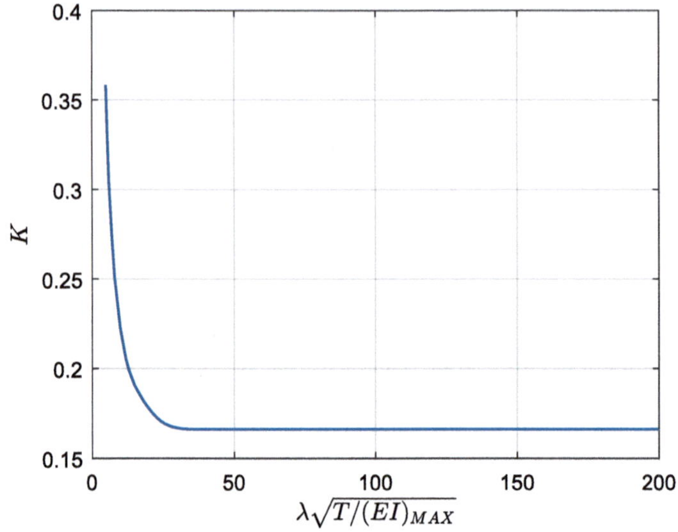

Fig. 4.31 *K* coefficient as a function of $\lambda_n\sqrt{\frac{T}{EI_{max}}}$ (adapted from Claren and Diana (1969a))

where the K coefficient depends on the $\lambda_n\sqrt{\frac{T}{EI_{max}}}$ value Claren and Diana (1969a) according to Fig. 4.31. Also, in this case, the slippage between the single wires of a stranded conductor modifies the above relationships through the EI reduction coefficient previously referred to.

A general observation can be made that λ_n is related to the mode natural frequency f_n through a relation with or without the flexural stiffness EI. As the cable stiffness increases, considering the same vibration frequency, the loop length increases, as shown in the following example, always related to the case of the Drake conductor, already used.

Example

For a Drake conductor:

$EI_{max} \approx 1600\text{Nm}^2$

$EI = 0.5 \times 1600 \approx 800\text{Nm}^2$

Considering a 366-m span (L), the 168th mode has a wave length of 2L/n = 4.36 m.

Without EI:

$$f_n = \frac{1}{\lambda_n}\sqrt{\frac{T}{m_L}} = 30.07\ \text{Hz}$$

With EI:

$$
\begin{aligned}
f_n &= \frac{1}{2\pi}\sqrt{\left(\frac{n\pi}{L}\right)^2 \frac{T}{m_L}\left[1+\left(\frac{n\pi}{L}\right)^2 \frac{EI}{T}\right]} \\
&= \frac{1}{2\pi}\left(\frac{2\pi}{\lambda_n}\right)\sqrt{\frac{T}{m_L}\left[1+\left(\frac{2\pi}{\lambda_n}\right)^2 \frac{EI}{T}\right]} \qquad (4.8)\\
&= \frac{1}{\lambda_n}\sqrt{\frac{T}{m_L}}\sqrt{1+\left(\frac{2\pi}{\lambda_n}\right)^2 \frac{EI}{T}} = 30.95\ \text{Hz}
\end{aligned}
$$

with an increment of about 2.9%.

Considering a higher stiffness, say $EI = 1200\text{Nm}^2$, the frequency corresponding to a wave length of 4.36 m (i.e., to the 168th mode) would be 31.38 Hz, with an increment of about 4.3%.

As it can be seen, the effect of stiffness is not important for the evaluation of the natural frequencies, while, of course, it is of primary importance for the evaluation of the bending strains.

4.4.2.2 Dissipation Mechanism

Energy in a vibrating conductor is dissipated through friction due to the relative motion of the wires. A simple formula to estimate the bending amplitude for which this so-called macroslip starts is given in the discussion by Papailiou (2000) relating to the paper Diana et al. (2000). It is less evident, however, that energy would still be dissipated at the wire-core interface without any gross slipping over any segment taking place. In practice, examination of conductors from the field and from laboratory fatigue tests does not show the fretting that testifies to gross sliding, except near clamps.

This phenomenon is explained by considering that the—points—of contact between two wires are actually elliptically shaped areas (Fig. 4.32) (Hardy et al. 1999). The pressure grows from zero on each side of the strip to a maximum at the strip centre-line. This means that some slip, called microslip, occurs on each side of the contact strip as soon as some friction forces appear at the contact interface, which causes the energy dissipation leading to self-damping. As the friction forces grow, the—slip—region also grows, while the inner—stick—region narrows down. It is worth noting that almost all self-damping in a vibrating conductor is associated with the energy dissipation mechanism, as described above, between the wires of the outer layer and the so-called penultimate layer just below it.

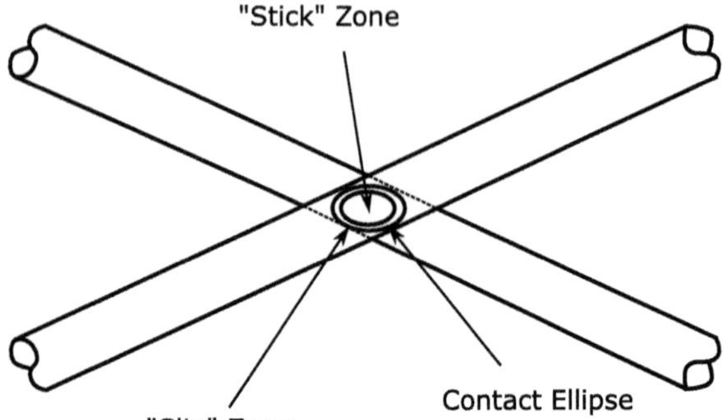

Fig. 4.32 Elliptical interface between adjacent layers

4.4.2.3 Measurement of Conductor Self Damping

Conductor self-damping is generally measured in a laboratory test span, as sketched in Fig. 4.33. The experimental methods described below are also used for the laboratory testing of damping hardware.

The test span comprises two massive blocks, 30 to 90 m apart, onto which the conductor to be tested is strung to the required tension and held rigidly. The conductor is then excited at a sequence of resonance frequencies at controlled antinode amplitude by means of an electromagnetic shaker.

Several problems exist in dealing with a test program intended to provide data on the self-damping of a particular conductor. Self-damping is influenced by conductor tensile load, so that tests are necessary at various conductor tensile loads. It is necessary to be certain that losses assigned to the conductor were not due to other sources. Loss of energy through end fixtures or support hardware must be either reduced to a minimum or properly accounted for in the measurement system.

In the artificial excitation of the indoor test span by means of an attached vibration drive system, the vibration exciter becomes a part of the system being measured. If the mass of the moving system within the shaker is high, conductor distortion is induced in that portion of the span where the shaker is attached. Springs or soft couplings are sometimes used to overcome this effect or to allow greater motion at the drive point than the shaker is capable of generating.

Apart from the main energy loss due to the cable transversal motion, some other phenomena take place in the experimental span, also giving energy losses; these extraneous effects must be carefully evaluated and or eliminated. Energy dissipation is mainly due to:

- Cable clamping system at the span ends
- Local deformation induced by the device used to force it to vibrate
- Cable motion in the air (aerodynamic drag)
- Torsional and other transversal motion; the mode of cable vibration should be examined to ensure the absence of this kind of motion.

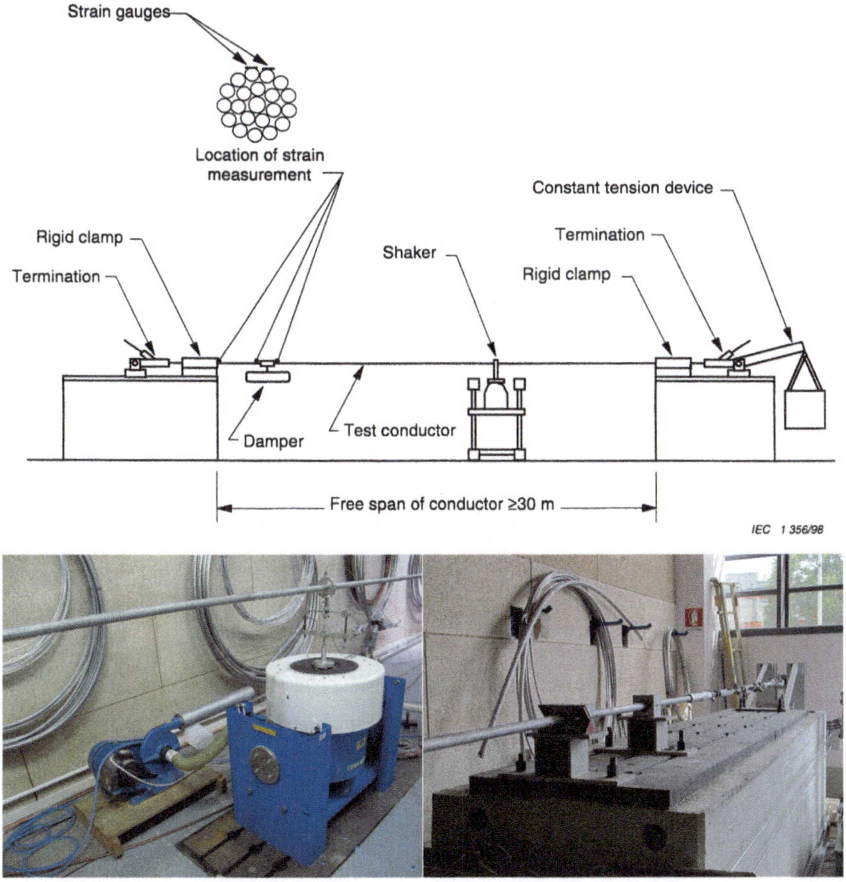

Fig. 4.33 Test span arrangement for self-damping measurements

In an actual span, the contribution to the overall energy dissipation given by the span ends is less significant than in a laboratory testing span due to the reduced length of an experimental span with respect to actual spans.

Also proper conductor conditioning is an important prerequisite for repeatable test results, itself a formidable task. Any excessive looseness in the aluminum layers should be eliminated from the conductor by artificial aging—i.e., by pre-stressing it at the highest tension at which the tests are to be made for a minimum of 2 h and, preferably, overnight.

The methods to measure the self-damping of cables are essentially two: the Power method (PM) and the Inverse Standing Wave Ratio method (ISWR). As these methods are widely described in IEEE (1978) and CIGRE (1979), only a brief summary is given here.

Power Method (PM)

The cable is tensioned on the experimental span and is forced to vibrate at one of its resonant frequencies, with both amplitude and frequency being controlled by means of an electrodynamic shaker.

When a stationary condition is reached, the energy introduced by the shaker to the span is equal to that dissipated by the span over one cycle of vibration. The energy introduced in the cable—and largely dissipated by its self-damping mechanism—is calculated by measuring the force F developed between the cable and the shaker and the displacement of the forcing point μ_F. The result is then given by the formula:

$$E_{diss} = \pi F \mu_F \sin(\varphi)$$

where φ is the phase angle between force and displacement

The non-dimensional damping coefficient, h, can be calculated by dividing the energy introduced in the cable E_{diss} by the total kinetic energy of the cable E_k (EPRI 2009):

$$h = \frac{1}{4\pi}\frac{E_{diss}}{E_k}$$

being the total kinetic energy of the cable given by the formula:

$$E_k = \frac{1}{4} m l \omega^2 u^2$$

where l = span length, ω = circular frequency, u = antinode amplitude, and m = mass per unit length.

While the application of this method is quite simple since it requires a limited number of measurements, all the external dissipation is part of the total calculation of the cable self-damping, and special care must, therefore, be devoted to reducing all these external loss sources.

For instance, in a laboratory span, it is comparatively easy to determine the total amount of vibration energy dissipation in the span, because it is equal to the total amount of energy introduced into the system. This would be quite sufficient for determining the self-damping of the conductor if all the loops of the span had equal energy dissipation. Unfortunately, the loops at the ends of the span and at both sides of the shaker behave quite differently from the rest of the span, having an energy dissipation that can be much higher than that of all of the rest of the span. As the energy dissipation of the conductor is, to a first approximation, proportional to the square of the curvature, it is easy to explain the large dissipation of energy near the end of the span.

Therefore, in order to provide correct self-damping data, the end constraints of the experimental span must be hinges in order to allow the rotation of conductors at the span extremities.

Inverse Standing Wave Ratio Method (ISWR)

Another measurement method, resulting from the work of Tompkins et al. (1956), is based on the measurement of nodal and antinodal amplitudes along the test span. To understand the principle involved, it is necessary to trace the waves leaving the vibration shaker as they are reflected at the span ends. The shaker is assumed to be attached near one of the span terminations. Harmonic forces induced by the shaker travel to the far end of the span to return as reflected waves. If no losses are present in the system, the incident and reflected waves are equal. Perfect nodes are formed where the two waves meet and pass. That is, zero motion exists at the nodes. The antinodes have an amplitude equal to the sum of the incident and reflected waves. If losses are present in the system, however, motion appears at the nodes. The amplitude of this motion is the difference between the incident and the reflected waves. The ratio between nodal amplitude and antinodal amplitude is indicative of the dissipation within the system. Where low span losses are present, the very fine measurements necessary for determining nodal amplitude can pose a problem.

From an electromechanical analogy—but also a mechanical reformulation of the problem is possible (Tompkins et al. 1956)—if the conductor is modelled by a taut string, the mechanical power P_i flowing in one section of the cable is given by:

$$P_i = \frac{V^2}{2} S_i \sqrt{Tm}$$

with:

$$V = \omega\, u$$

$$S_i = \frac{a_i}{u} \text{ (inverse standing wave ratio-ISWR)}$$

where ai is the amplitude of vibration in a node and u that of an antinode.

The power dissipated between the node j and the node k will be:

$$P = P_k - P_j$$

And the power dissipated per unit length will be:

$$P_{diss} = \frac{P_k - P_j}{n_v \frac{\lambda}{2}}$$

where nv is the number of antinodes between k and j, and λ is the wave length. Considering the kinetic energy of the portion of cable between the two nodes:

$$E_{k(\max)} = \frac{1}{2} u^2 \omega^2 \frac{m}{4} \frac{1}{f} \sqrt{\frac{T}{m}} n_v$$

the value of the nondimensional self-damping coefficient is given by:

$$h = \frac{S_k - S_j}{\pi n_v}$$

The advantage of this method is that the measured dissipation relates to the considered portion of cable only; therefore, the estimated self-damping value is not affected by the above-mentioned factors (that is, span ends and shaker-cable link).

The main problems that the method presents are the correct estimation of the node positions and the measurement of the node amplitude of vibration, which is a very small value on the order of a few micrometers, since, as happens with small quantities, an error in the antinode vibration amplitude significantly changes the self-damping estimation. A method to simplify the use of ISWR and reduce errors is reported in Diana et al. (2000).

Decay Method

Application of the vibration decay test to transmission line conductors provides a simple method of evaluating conductor self-damping in laboratory spans Hard and Holben (1967). This method, if correctly employed, can give a first approximation of the value of the self-damping at all amplitudes in one trial—i.e., it is very quick and easy, requiring in its simplest form just one vibration transducer measuring the decay after the exciter Diana et al. (1986). The cable is forced to vibrate at one of its natural frequencies, and then the exciter is stopped. The rate of decay is a function of the system losses.

If a lightly-damped system ($h << 1$) is left free to vibrate from a forced resonance condition, it undergoes a transient decay of motion that looks like Fig. 4.34.

In highly damped spans, the decay is reported by some authors in the form of a step curve Slethei and Huse (1965). This is probably due, as observed by the authors themselves, to the transient induced by the exciting force disconnection. A method to avoid this problem is to provide an elastic link between the cable and the exciter. This should be soft enough to dynamically uncouple the cable from the forcing device.

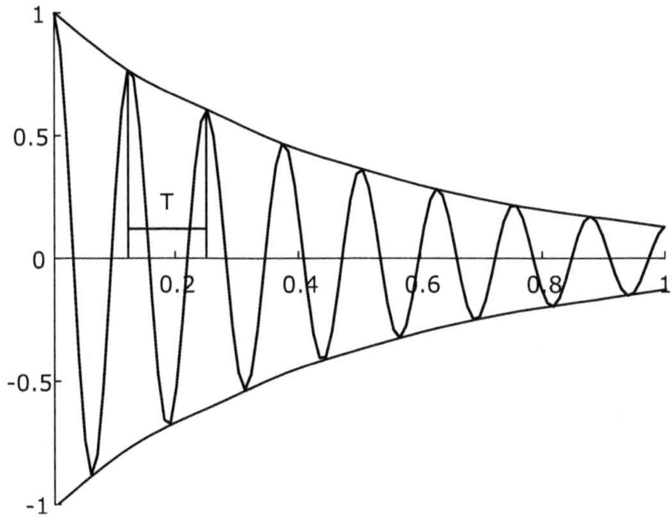

Fig. 4.34 A decay trace

Considering two successive peaks, X_i and X_{i+1}, and defining δ as:

$$\delta = \ln\left(\frac{X_i}{X_{i+1}}\right) = \ln\left(\frac{e^{-h\omega_0 t}\,|X|\cos(\omega t + \varphi)}{e^{-h\omega_0 t + T}\,|X|\cos(\omega(t+T) + \varphi)}\right)$$

simplifying:

$$\delta = \ln\left(e^{h\omega_0 T}\right) = h\omega_0 T$$

if $h << 1$, as in the case of cables, we can consider T as a function of ω_0—i.e. $T = 2\pi/\omega_0$—giving:

$$h = \frac{\delta}{2\pi}$$

To improve the results, it is possible to calculate the energy transferred from the cable to the shaker during the decay with the same setup already described for the power method. However this shaker loss is usually one order of magnitude less than that of the cable. It is, therefore, possible to remove the shaker loss from the total dissipation, but the end losses are still included in the measurement.

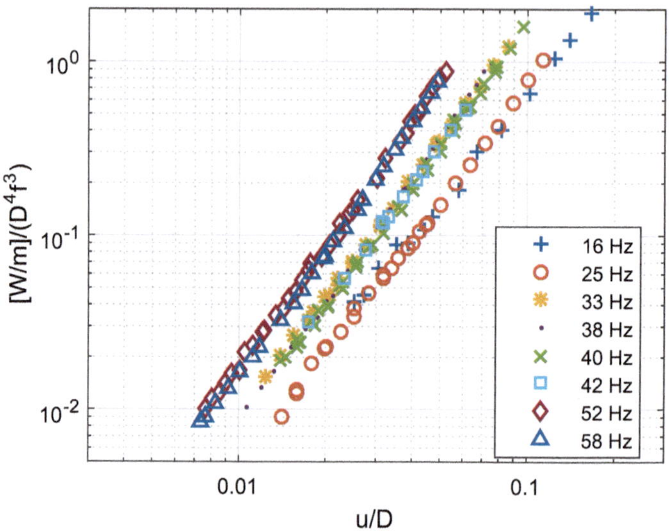

Fig. 4.35 Power dissipation characteristics of a Drake conductor tensioned at 28.5 kN (20% RTS) versus (antinode amplitude/conductor diameter)

Measurement Results

Data measured in the laboratory span are generally expressed empirically through a power law:

$$\frac{P}{L} = k\frac{y_0^l f^m}{T^n} \tag{4.9}$$

in which P/L describes the power per unit length dissipated by the conductor, k is a factor of proportionality, y_0 is the antinode displacement of vibration, f is the frequency of vibration, while l, m, and n are the amplitude, frequency, and tension exponent, respectively. Using the above empirical rule, self-damping determined in short laboratory spans could be extrapolated to actual much longer spans.

IEEE (1978) and CIGRE (1979) standards recommend that the measurement results be presented in diagrams as illustrated in Fig. 4.35, showing the power dissipated per unit conductor length, as a function of the ratio of the antinode displacement amplitude to conductor diameter for each loop length and corresponding frequency and tensile load T.

Table 4.1 from CIGRE (1998) summarizes the exponents obtained by a number of investigators for Eq. (4.9), together with the method of measurement used, the test span length, span end conditions, and number of conductors and tensions tested.

Table 4.1 Comparison of conductor self-damping empirical parameters

Investigations	l	m	n	Method	End condition	Span length [m]	n.cables x tensions
Tompkins et al. (1956)	2.3–2.6	5.0–6.0	1.9 (1)	ISWR	N.A.	36	1 × 2
Claren and Diana (1969b)	2	4	2.5; 3.0; 1.5	PT	M.B.	46	3 × 3
Seppä (1971) and Noiseux (1991)	2.5	5.75	2.8	ISWR	N.A.	36	1 × 8
Rawlins (1958)	2.2	5.4		ISWR	N.A.	36	1 × 1
Lab. A (CIGRE 22.01 1989)	2	4		PT	M.B.	46	1 × 1
Lab. B (CIGRE 22.01 1989)	2.3	5.2		PT	P.E.	30	1 × 1
Lab. C (CIGRE 22.01 1989)	2.44	5.5		ISWR	N.A.	36	1 × 1
Kraus and Hagedorn (1991)	2.47	5.38	2.8	PT	P.E.	30	1 x ?
Noiseux (1991) (2)	2.44	5.63	2.76	ISWR	N.A.	63	7 × 4
Tavano (1988)	1.9–2.3	3.8–4.2		PT	M.B.	92	4 × 1
Möcks and Schmidt (1989)	2.45	5.38	2.4	PT	P.E.	30	16 × 3
Mech. Lab Politecnico di	2.43	5.5	2	ISWR	P.E.	46	4 × 2
Milano (2000)							

ISWR: Inverse standing wave method
PT: Power method
N.A.: Non applicable
M.B.: Massive block
P.E.: Pivoted extremity
(1): Extrapolated
(2): Data corrected for aerodynamic damping

4.4.3 The Energy Balance Principle

Natural frequencies and related vibration modes can be excited by the wind when the vortex-shedding frequency approaches one of them. However, as already observed, the wind velocity variation in space and time (Fig. 4.36) is such that more vibration modes can be simultaneously excited, and so the vibration amplitude along the span exhibits the well-known beat pattern.

Once the aeolian vibration phenomenon is initiated, it is self-sustained due to the lock-in effect.

The presence of more than one vibration mode causes the disappearance of the vibration nodes, and the phenomenon can become unsteady and very complex; this

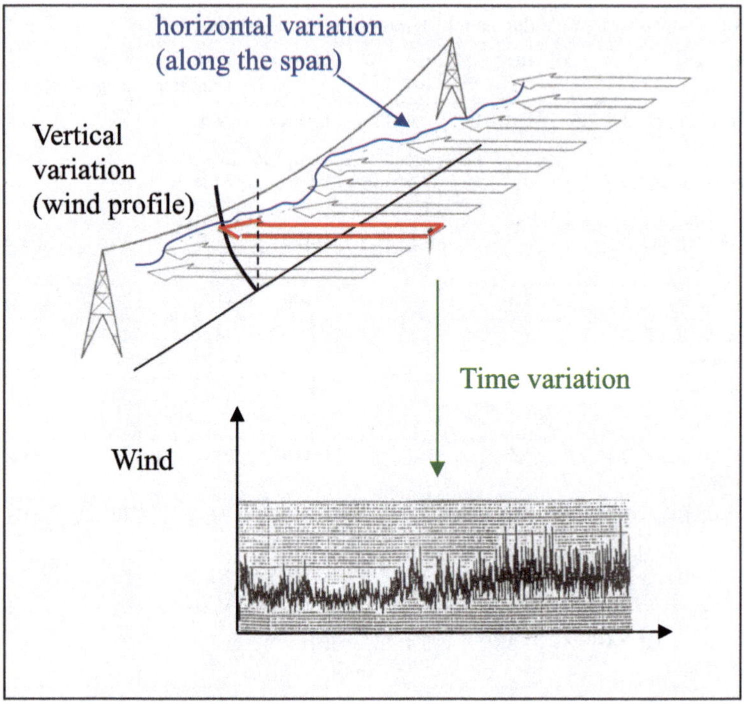

Fig. 4.36 Real wind

makes it difficult to use the maximum energy input from the wind as measured in wind tunnels on rigid or flexible cylinders.

The phenomenon is not easy to reproduce analytically, and what is generally and cautiously done is to apply an EBP approach; the EBP works in the frequency domain. In its simplest form, one mode of vibration at a time is considered, and the steady-state solutions computed correspond to the maximum vibration amplitude that could be excited on that conductor at that frequency.

The reliability of results of these analytical computations is no better than the background data used in them, particularly data on the power supplied by wind during aeolian vibration (see Sect. 4.4.1.3) and data of self-damping in stranded conductors (see Sect. 4.4.2.3). More sophisticated approaches are available in order to reproduce the complexity of the real phenomenon, as reported in Belloli et al. (2006), that can take into account the turbulence of the wind and the characteristics of the VIV as reported in Sect. 4.4.1.1.

The balancing of conductor self-dissipation against wind power input to obtain predicted amplitudes of natural vibration basically requires determining the intersection of wind input and conductor dissipation curves. A graphical solution is shown in Fig. 4.37.

Figure 4.38 presents the same data in an alternate form.

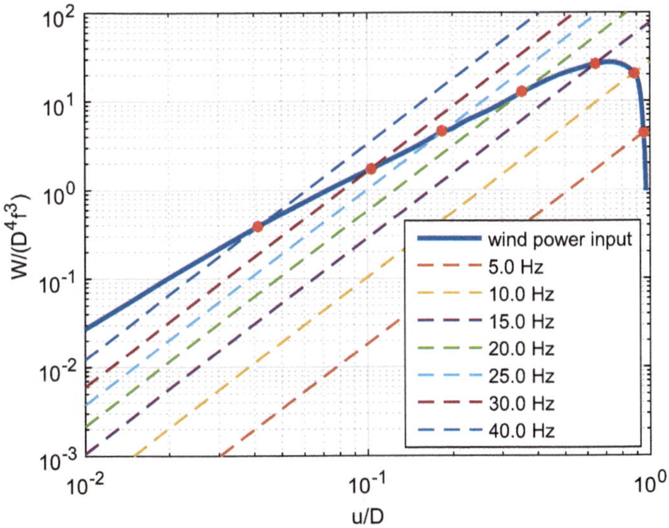

Fig. 4.37 Prediction of conductor vibration amplitudes at various frequencies, through cross-points of wind input and conductor self-dissipation (values at 10 Hz are extrapolated)

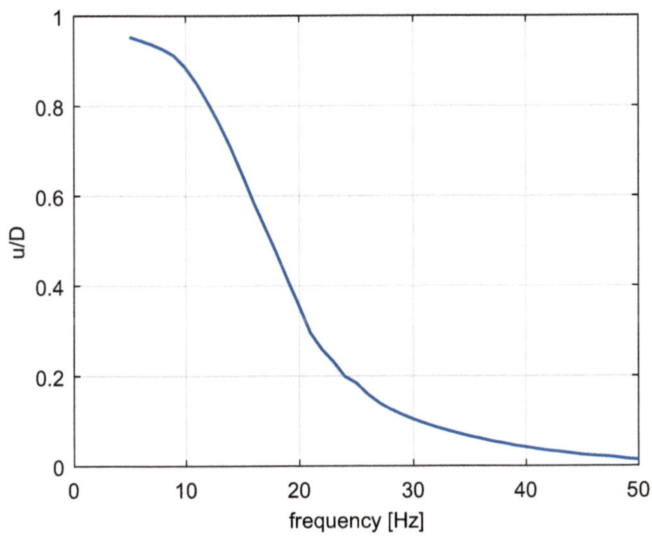

Fig. 4.38 Predicted vibration and amplitudes

The graphical procedure shown in Figs. 4.37 and 4.38, relevant to the Diana and Falco (rigid cylinder) Diana and Falco (1966) and to the Rawlins (1958) (rigid cylinder) Rawlins (1958) wind power input curves of Fig. 4.27 and to conductor self damping measured in laboratory, can be substituted by an automatic evaluation performed by a software based on the EBP.

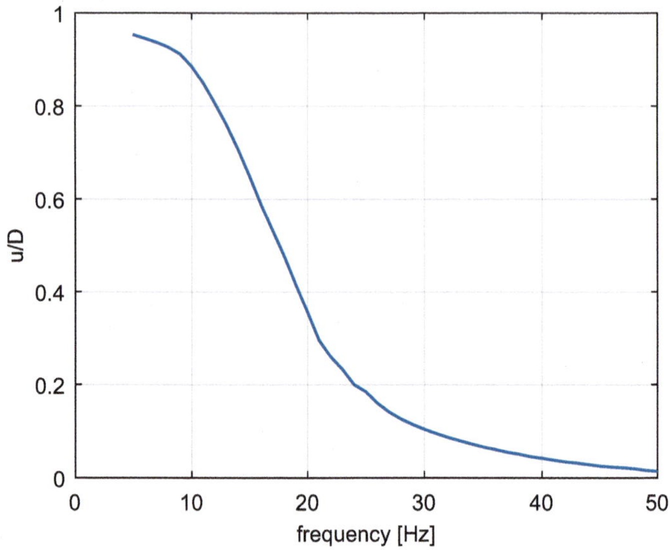

Fig. 4.39 Drake conductor response to aeolian vibrations: maximum antinode amplitudes of vibration (0-peak values) and bending strains at the clamped extremities (0-peak values) as a function of frequency

Each rendition of the EBP technology contains an energy input curve among those in Fig. 4.27 and an empirical relation for conductor self-damping (Eq. (4.9)) with parameter values shown in Table 4.1. As an example, the case of the Drake conductor already worked out is solved. The maximum antinode amplitudes of vibration (0-peak values) and relevant bending strains at the clamped extremities are shown in Fig. 4.39 as a function of frequency.

Drake conductor	
Diameter	28 [mm]
Mass	1.628 [kg/m]
Stiffness	800 [Nm^2]
Tension	28024 [N]
Damping constant k	249740
Constraints	Fixed clamp
Type of wind	No turbulence

As with all analytical tools, the software based on the EBP can also be used to perform a sensitivity analysis—i.e., to evaluate the influence of the different line parameters on the aeolian vibration level. The other way to the sensitivity analysis is to collect data from field measurements and/or experimental spans, with an obvious limitation of the number of conditions that can be considered.

In Fig. 4.40, which considers a Drake conductor, the influence of the conductor tension is shown. As can be seen in the empirical function used to reproduce the

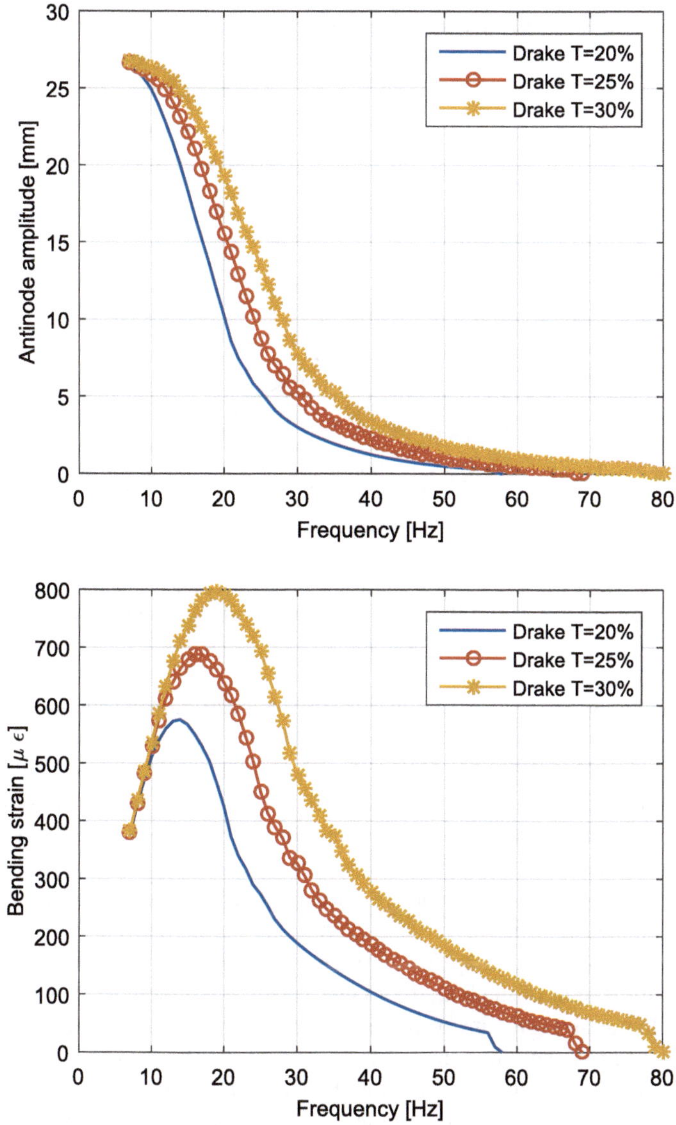

Fig. 4.40 Conductor tensile load effect

dissipated energy (Eq. (4.9)), a change of tension reflects in a change of dissipated energy and then in a variation of the vibration amplitude.

Physically, it happens that, at a given frequency, if the tension increases, the wave length increases, and then the strain at the antinode, given by $(\varepsilon_a)_n = \frac{4\pi^2}{\lambda_n^2} A_n \frac{D}{2}$, decreases and, as a consequence, the slippage between the single wires of the conductor and the dissipated energy decrease. If the dissipated energy decreases, the vibration amplitude and related strains increase.

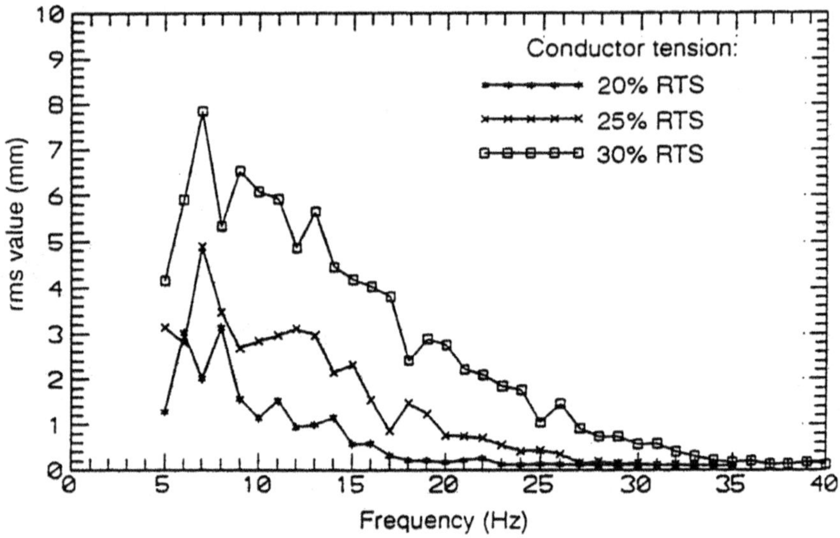

Fig. 4.41 Conductor tension effect, experimental findings, rms aeolian vibration amplitude versus frequency (Hardy and Van Dyke 1995). Reprinted with permission from Elsevier

Approximately the same trend of variation of the vibration amplitude with respect to the tensile load variation could be found experimentally on the Isles de la Madeleine test line Hardy and Van Dyke (1995) (see Fig. 4.41). The tested conductor is the ACSR Bersfort. Both experimental and analytical results confirm that the effect of tensile load on the vibration level is very important: an increment of tensile load causes an enlargement of the frequency spectrum and an increase of vibration amplitude.

In Fig. 4.42, which also considers a Drake conductor, the influence of the wind turbulence level is shown.

As already observed in Chap. 2, the roughness of the earth surface plays an important role in the height of the boundary layer (the gradient height), as well as on the mean and fluctuating (gusting) velocities. Accordingly, the relevant characteristics of the flow at the location of the conductor must be determined in order to evaluate the static and dynamic interactions between the wind and a conductor: they are the mean wind speed and the turbulence. These characteristics, as expected, are functions of the topology of the local terrain and meteorological data. In particular, different turbulence levels—qualified by the turbulence intensity—can be associated with different terrain categories.

What happens physically is that important wind velocity fluctuations cause the loss of synchronization between conductor vibration and vortex shedding: the wind continuously changes, and the phenomenon is always in transient conditions. This does not allow the vibration amplitude to increase up to the maximum values.

The phenomenon is reproduced in the EBP-based software by a simplified approach, just reducing the energy input with respect to that relevant to low tur-

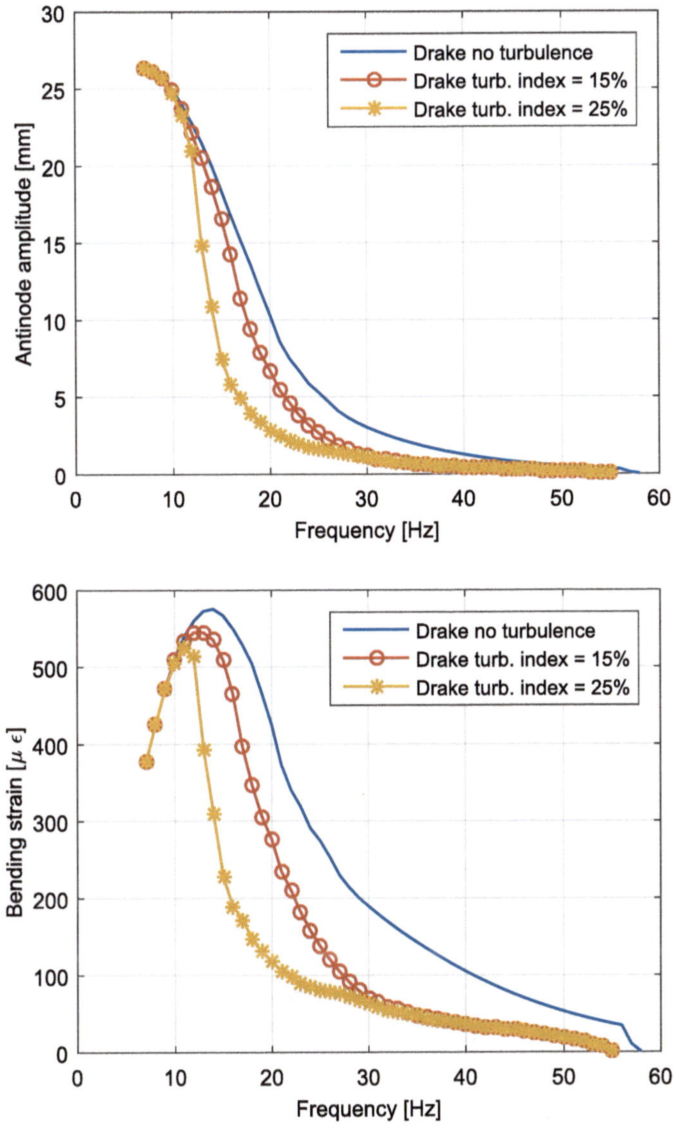

Fig. 4.42 Wind turbulence effect

bulence (see Fig. 4.43 for reduced wind power curves Diana et al. (1979) and also the data in Rawlins (1998)).

At low frequencies, the cable self-damping is so low that, even if the energy input is reduced due to turbulence, the vibration amplitude exhibits a small variation with respect to the low turbulence condition. Moreover, the estimated normalized wind input curves (Fig. 4.27) converge at the highest vibration amplitudes. Also, in the

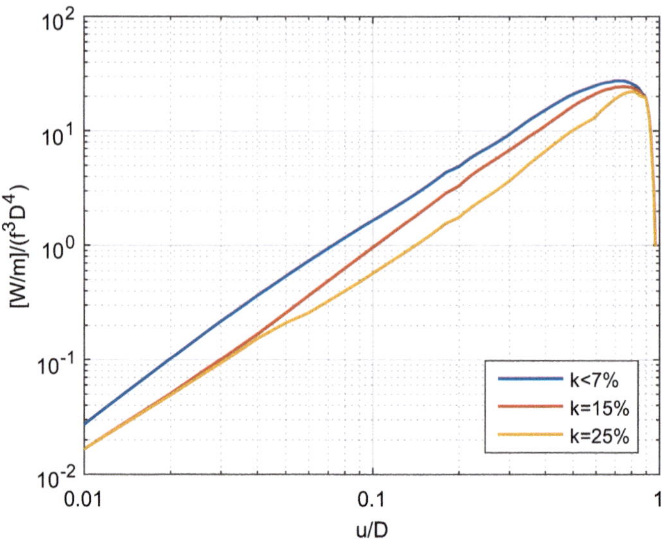

Fig. 4.43 Wind turbulence effect on power input curves Diana and Falco (1966) and Diana et al. (1979)

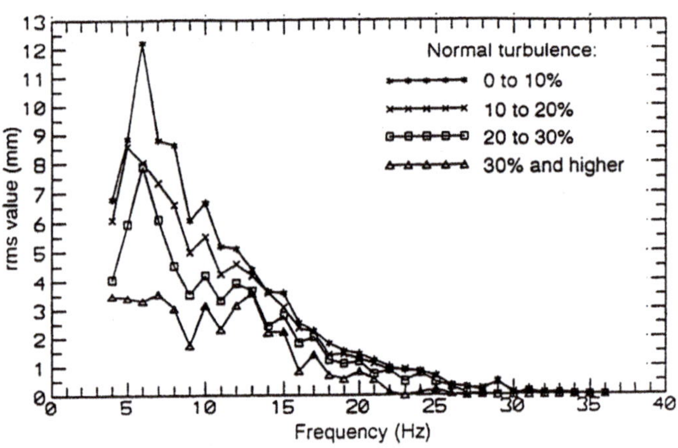

Fig. 4.44 Wind turbulence effect, experimental findings, rms aeolian vibration amplitude versus frequency (Hardy and Van Dyke 1995). Reprinted with permission from Elsevier

case of the turbulence effect evaluation, a comparison between experimental and EBP-predicted findings is possible Hardy and Van Dyke (1995): Fig. 4.44 shows the results found on the Isles de la Madeleine test line on a Bersfort conductor.

For the experimental data, a 10% increase in normal turbulence corresponds roughly to a decrease in amplitude of 20%; however, bigger reductions are predicted by the EBP analysis. However, this aspect of the problem has not yet been fully resolved, and research is continuing in this area.

The last worked example compares the aeolian vibration behavior of the ACSR Drake conductor to that of an ACS OPGW (around 15 mm diameter) and that of an ACAR 1300 cable. For the three different conductors, the T/w parameter (ratio between conductor tension and weight per unit length) is kept constant and equal to 1720 m. The T/w parameter expresses the cable sensitivity to aeolian vibrations.

What can be observed in Fig. 4.45a, b, c is the significant enlargement of the aeolian vibration range of frequencies in the case of the ACS cable.

4.4.4 Maximum Allowed Strain on the Conductors

Conductor vibration causes localized bending which, depending on its level, may cause, sometimes in a short period of time, fatigue failures of the conductor strands at the suspension clamps or at the clamps of spacers, spacer dampers, dampers, and other devices installed on the conductor, as shown in Fig. 4.46. The conductor vibration may also cause fatigue damage of items associated with the support and protection of the conductor itself—i.e., tower arms, spacers, dampers, and warning spheres, etc.

Aeolian vibrations can be successfully controlled in most cases using dampers and/or spacer-dampers. Reliable transmission-line design requires that aeolian vibration of the conductors is controlled below critical levels to avoid fatigue damage.

The value of the maximum allowed strain—to avoid fatigue failures—on an aluminum conductor is around 150 microstrains (0-peak). Taking into account that the aluminum modulus of elasticity E is $7\ 10^{10} \mathrm{N/m^2}$, this corresponds to a maximum stress of 10.5 MPa or 10.5 $\mathrm{N/mm^2}$. Figure 4.47 (EPRI 2009) reports the Wölher curve (experimental results)—i.e. the fatigue limit as a function of the number of fatigue cycles - for a conductor with aluminum wires in the external layer. The value of 10.5 MPa seems to be a reasonable value to be considered as limit value to avoid fatigue.

The strain computed by the EBP for different types of single conductors and at different tensile loads generally exceeds the fatigue limit, and then fatigue failures would be produced on the conductors by aeolian vibrations.

Due to this reason the conductors usually need to be protected by aeolian vibration dampers which have the task of introducing additional damping and then limiting the aeolian vibration level within safe values.

4.4.5 Single Conductor Plus Dampers

It is well known that if the conductor tensile load (or, more precisely, the catenary parameter H/w defined as the ratio between tension H—initial value, before creep—and cable unit weight w) exceeds certain limit values (i.e., 1000 m) CIGRE (2005b), aeolian vibration may cause serious damage to both conductor and fittings. The limit

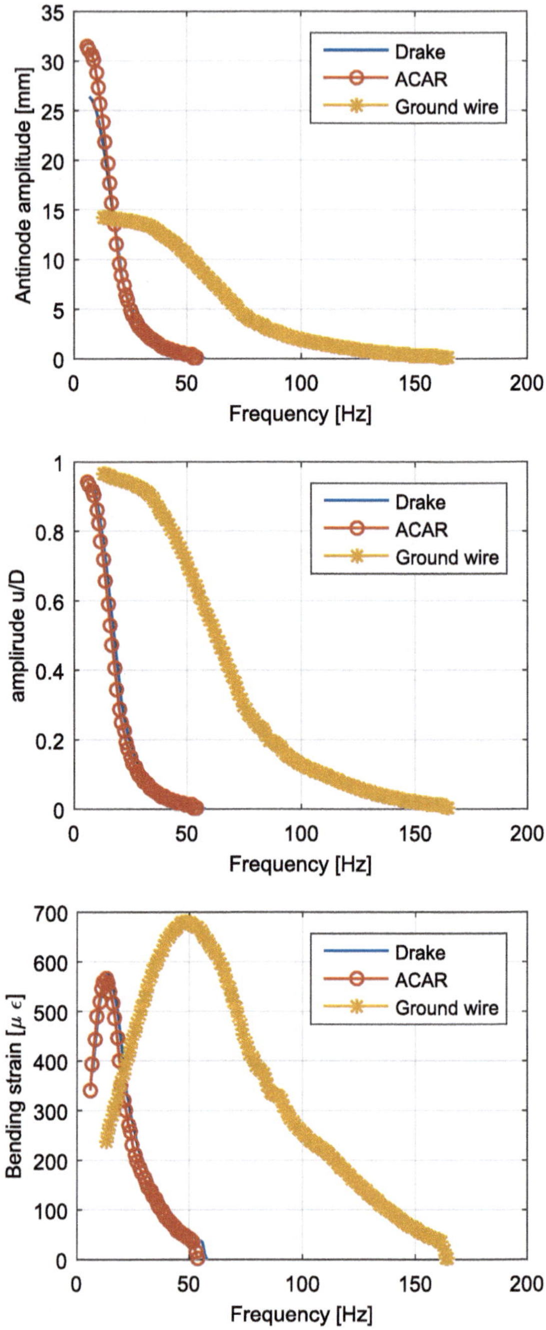

Fig. 4.45 Aeolian vibration behaviour of an ACSR Drake conductor, an ACS OPGW (around 15 mm diameter) and an ACAR 1300 cable, with T/w = 1720 m.

Fig. 4.46 Fatigue failure of conductor strands at the suspension clamp

value of H/w is generally exceeded on transmission lines; therefore, it is common practice to protect conductors with suitable dampers.

For transmission line design, it is important to know how much additional damping is needed to control aeolian vibration within safe levels. To this end, various researchers have developed calculation methods, based on the energy balance principle (EBP), to predict the aeolian vibration level of a cable plus damper and then to allow for the selection of the suitable damping.

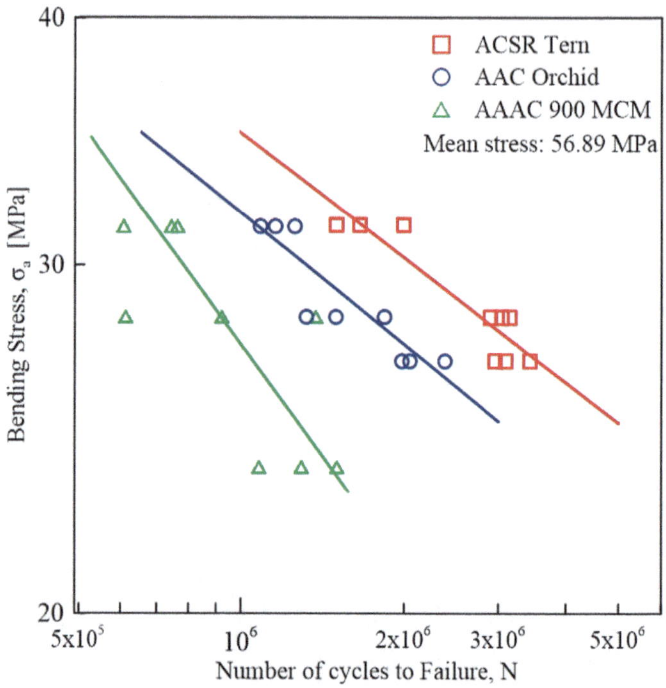

Fig. 4.47 Comparison between Wöhler curves for ACSR, AAC and AAAC conductor Kalombo et al. (2018). Reprinted with permission from Elsevier

The most commonly used damping devices on single conductors are the Stockbridge type dampers. This type of damper is mounted locally on the conductor and is forced to vibrate due to the conductor motion. When the damper vibrates at a frequency close to one of its natural frequencies, a resonance condition occurs. The damper masses vibrate at high amplitude and dissipate power, which corresponds to the transmission of a force component in phase with the conductor vibration velocity. This force produces a damping effect and also modifies the conductor vibration mode—i.e., it creates a distortion (Fig. 4.48). Consequently, the vibration modes of a cable plus damper differ from those of the cable alone.

Figure 4.48 shows a vibration mode amplitude (the frequency is around 35 Hz) for a cable plus damper system; the damper is positioned 0.5 m from the suspension clamp, and only the part of the span close to the clamp is shown. The antinode vibration amplitude in the free span is normalized to the value 1. The curve "damper A" represents the mode obtained using a damper suitable for the cable in question; the curve "damper B" represents the case of too heavy a damper. Figure 4.48 clearly demonstrates that the bigger the damper force, the higher the cable distortion and the smaller the vibration amplitude at the damper clamp, and as a consequence, the lower the power dissipated by the damper. In fact, the ratio between antinode and node vibration amplitude is higher for curve "damper B" than for curve "damper A",

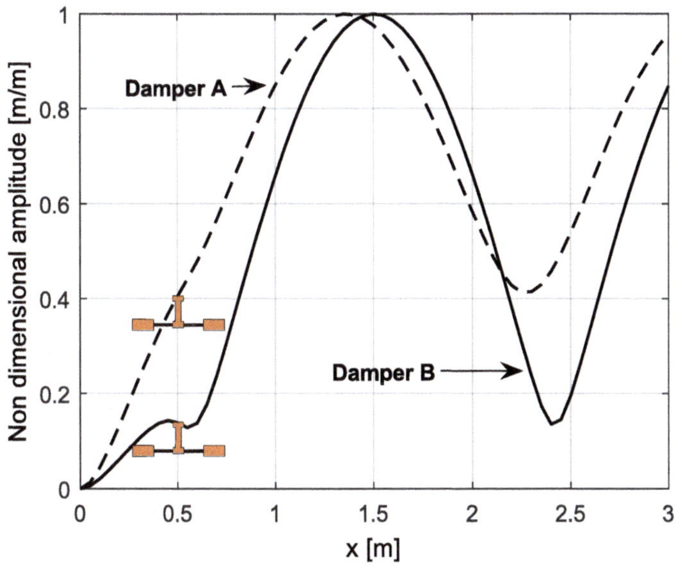

Fig. 4.48 Vibration mode amplitude for a cable plus damper system

thus indicating a lower damping efficiency for the system with damper B than for the system with the "correct" damper A Tompkins et al. (1956).

By contrast, if the damper force is too small, the cable vibration mode is only distorted by a small amount but the dissipated power could be too small. It is clear that for a given cable at a particular tensile load, there will be, as a function of frequency, an optimum damper force, i.e. the force giving the maximum dissipated power.

Another important fact in determining the damper dissipated power is the position of the damper itself on the cable. If the damper is situated at a node for a particular vibration frequency, no dissipation takes place and no distortion occurs. As will be shown below, these analyses lead to the choice of the optimum damper for a certain application and to the optimization of its position.

As already seen, the actual aeolian vibration behavior of a cable plus damper system depends on the energy introduced by the wind and the energy dissipated by the cable plus damper system. With reference to the evaluation of the energy introduced by the wind, all the considerations described in Sects. 4.4.1, 4.4.2 and 4.4.3 hold true. However, for the evaluation of the energy dissipated by the cable plus damper system, different possibilities are available as follows:

1. If a wholly analytical procedure is to be used, a mechanical model of the damper must be prepared. Then the modes of vibration of the cable (modeled as reported in Sect. 4.4.2) plus damper system should be calculated.
2. The need for a mechanical model of the damper can be avoided if the damper response corresponding to a harmonic excitation imparted to the damper clamp

is directly measured by mounting the damper on a shaker in a laboratory. The measured response can be introduced into the cable model as a force transmitted by the damper (impedance matching analysis). This procedure is generally referred to as the "direct method" IEEE (1993). The cable model must be able to correctly reproduce the cable-damper interaction, i.e., the mode of vibration distortion due to the damper presence.

3. The power dissipated by the damper is directly measured on a laboratory span as a function of the cable antinode vibration amplitude in the free span at all the vibration modes of interest for the aeolian vibration phenomenon. The span must be equipped with the same cable as that to be used for the transmission line to be studied. This procedure is generally referred to as the "basic method" IEEE (1993) and Rawlins (1988). It has the advantage that it avoids the difficulties associated with the cable-damper interaction simulation. It has the disadvantage of a significantly greater cost of testing. The measurement of the power dissipated depends on the damper position on the cable; a change of position requires the measurements to be repeated. The measurement of the power dissipated by a combination of many dampers, as in a crossing, could become impractical, because too many tests would be needed in order to optimize the relative position of the dampers.

This situation is summarized in the sketch in Fig. 4.49 CIGRE (2005c), where the various elements present in this technology are represented.

Each step in the analytical chain is affected by errors caused by assumptions and approximations required by the analytical procedures and by inaccuracies in input data. Thus, accuracy deteriorates with progress down the chain. However, accuracy of the final predictions can be improved by entering the chain with independent data at a lower point. For example, vibration amplitudes are more accurately determined from field recordings than from power balance analysis based on laboratory span testing of dampers. Damping efficiency on the conductor is more accurately determined by direct measurement on a laboratory span than by using impedance matching analysis.

Analytical simulations based on impedance matching analysis or on laboratory span testing may be considered conservative with respect to the actual situation if the maximum wind energy input is used and if a suitable safety factor for the damping system is introduced.

4.4.5.1 Natural Frequencies and Modes of Vibration of the Cable Plus Damper System

The natural frequencies and modes of vibration of the cable plus dampers system are computed, assuming for the different sections of the conductor between the dampers a taut beam model. The equation of motion of each section admits a steady-state harmonic solution containing four unknown parameters. End conditions for the first and last section of the conductor, and equilibrium and congruence equations for each

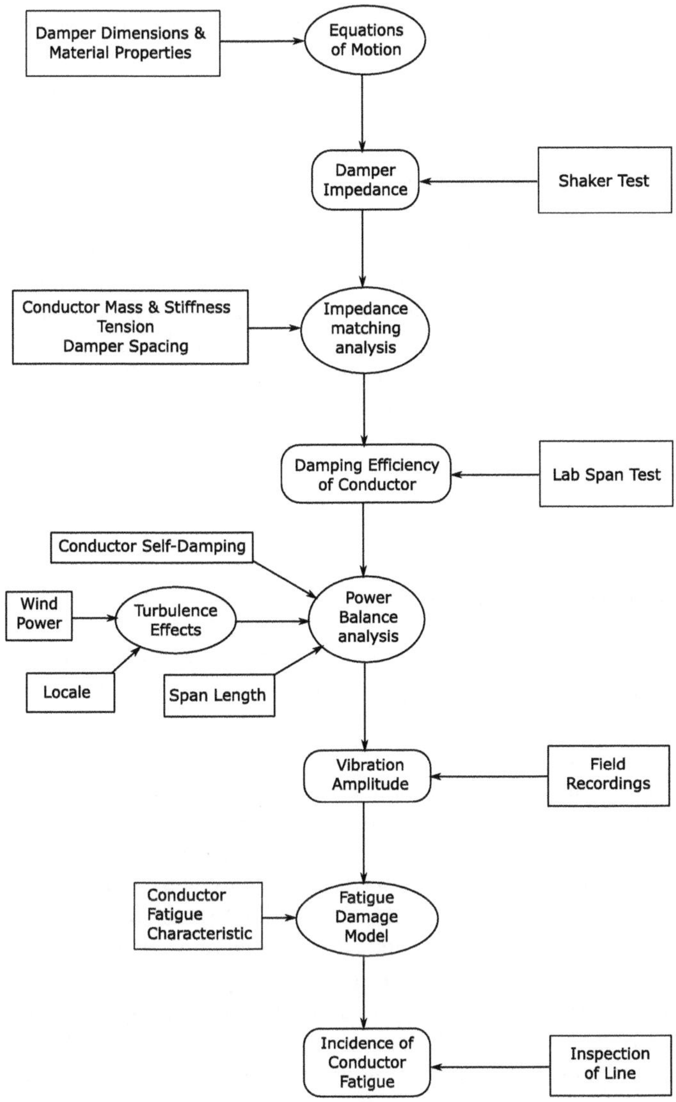

Fig. 4.49 Chain of data analysis for dissipation of power by dampers (adapted from CIGRE (2005c))

of the dampers present along the span, can be written, and an homogenous system in all the unknown parameters can be obtained. By zeroing the determinant of this system, the natural frequencies and modes of vibration are found.

With this approach, the obtained vibration modes account for the damper presence. Then, in general, with respect to the vibration mode of an undamped cable, a distortion, due to the damper interaction with the cable, can be observed. Far from the dampers, in the free span, the vibration mode continues to be represented by an harmonic function of space.

Another point to be put in evidence is that the homogenous system, from which natural frequencies and modes of vibration are obtained, contains the force transmitted by the damper, which is a complex quantity, whose real and imaginary parts—or its modulus and phase with respect to the clamp displacement—are measured, as a function of frequency, through laboratory tests.

It is well known that Stockbridge dampers have a significant nonlinear behavior as a function of the clamp amplitude of vibration due to the variation in the relative sliding of the messenger cable wires. This modifies their flexural stiffness and damping and, therefore, the damper dynamic behavior. For example, in Fig. 4.50, a two-resonance damper response, in terms of dynamic stiffness—amplitude and phase—is reported for different constant velocities of the damper clamp.

In order to take these nonlinear effects into account, the damper impedance can be defined at a number of clamp vibration velocities, or clamp vibration amplitudes, and the intermediate velocity, or displacement, values are obtained through interpolation.

4.4.5.2 Energy Balance Principle

Once the conductor plus damper (and/or other devices) system natural frequencies and associated vibration modes have been defined, it is possible to compute the steady-state amplitudes of vibration using the EBP.

For each natural frequency, the conductor vibration amplitudes can be defined along the span as a function of a reference amplitude, which then enables the wind energy input and the energy dissipated by the conductor to be calculated. With the mode of vibration known, the motion of the dampers is determined as a function of the reference amplitude, and the energy dissipated can then be calculated. According to the EBP, the steady-state amplitudes of vibration for each of the excited vibration modes (i.e., for each of the excited natural frequencies) are obtained through a balance between the wind energy input and the energy dissipated by the system.

Considerations of the wind energy input and the energy dissipated by the conductor already described in Sects. 4.4.1, 4.4.2, and 4.4.3 also hold true in this case.

Regarding the energy dissipated by the damper system, the following relationship is used for each one of the dampers present on the cable, and for each natural frequency Claren and Diana (1969b),

$$E_{damper} = \pi F_i u_i^2 \sin \varphi_i$$

where F_i is the damper force per unit displacement of the damper clamp, as measured through a shaker test at different vibration velocities. u_i is the conductor vibration amplitude at the clamp of damper i. φ_i is the phase between force and displacement, as measured through a shaker test (see Fig. 4.50).

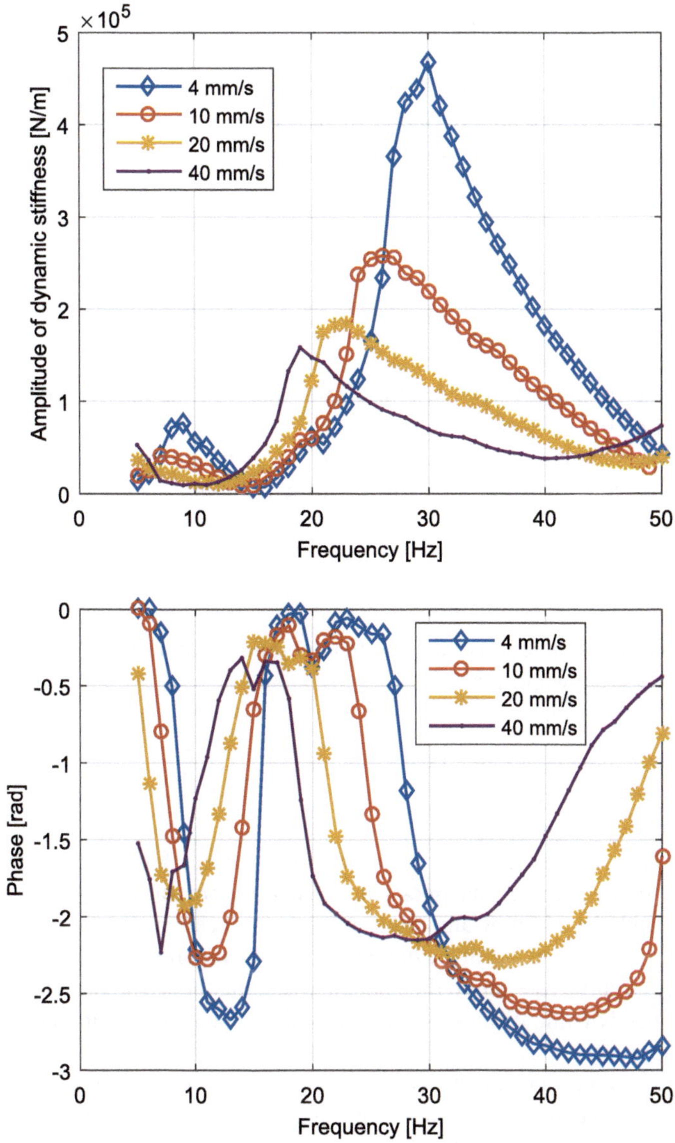

Fig. 4.50 Nonlinearity of the damper response (adapted from CIGRE (2005c))

4.4.5.3 Optimum Damper

Changing the damper force, the maximum damping that can be introduced in the system can be evaluated and, from this, the formulation of the "optimum damper" is derived Claren and Diana (1969b) and CIGRE (2005a).

$$F_{opt} = \omega\sqrt{Tm}$$

$$\varphi_{opt} = \frac{\pi}{2}$$

F_{opt} is the optimum force per unit displacement of the damper clamp and is a linear function of the vibration frequency f.

$\omega = 2\pi f$ is the circular frequency [rad/s]. T is the conductor tensile load m is the conductor mass per unit length. φ_{opt} is the optimum phase between force and displacement.

Example

The results of the simulation of the Aeolian vibration response of a single conductor plus dampers is shown in Fig. 4.51.

Data of the simulation are:

- 366-m span equipped with a Drake conductor.
- Conductor tensile load: 28024 N.
- Low turbulence wind.
- The results are given in terms of vibration amplitude and bending strains as a function of frequency. The free-span antinode vibration amplitude is reported together with the vibration amplitude at the damper clamp (mm 0-peak). Strains at the suspension clamp and at the damper clamp are reported (microstrains 0-peak).
- A reference curve relevant to the undamped conductor is also reported.
- The damper dynamic stiffness is not supposed to depend on the vibration amplitude (linear approach).
- Two dampers have been installed on the conductor, one at each side of the span, at the optimum damper position for this application, which has been evaluated as 1.2 m from the suspension clamp.

As can be observed, a suitable damper placed in an optimized position practically suppresses the conductor aeolian vibrations.

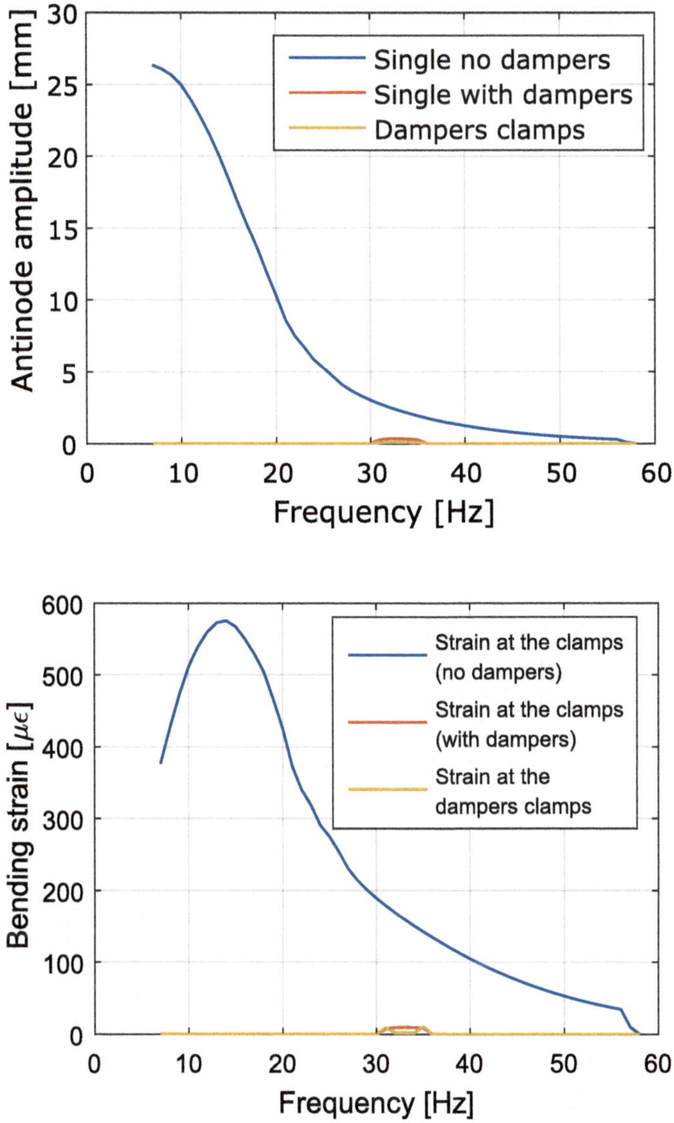

Fig. 4.51 Aeolian vibration amplitudes and bending strains of the damped and undamped cable

4.5 Subspan Oscillations

Subspan oscillations is a well-known phenomenon in High-Voltage and Ultra-High-Voltage Overhead Transmission Lines (HV and UHV OHTL) (EPRI 2009). It occurs on conductor bundles and it is due to the effect of the wake produced by the windward conductor on the leeward one. For this reason the phenomenon is also classified as wake induced oscillations. The oscillations are characterized by a frequency around 1 Hz, which corresponds to the mode of vibration of a section of conductor between two spacers, with an antinode at mid-subspan. A typical oscillation shape for a quad bundle is shown in Fig. 4.52: the single sub-conductor motion is elliptical and is the result of the combination of an essentially horizontal mode of vibration of the bundle with an essentially vertical one.

The field of aerodynamic forces which originates on the leeward conductor is non-conservative and may cause a flutter type form of instability: the horizontal and vertical mode frequencies, initially different, become equal due to the field of forces, as it will be better described in the following. The instability phenomenon occurs for wind speeds in the range between approximately 8 m/s and 20 m/s, as a function of the bundle parameters. Amplitudes of vibration may be as high as to cause sub-conductors clashing and, in any case, the oscillations are associated to strains on the subconductors at the spacer clamps and forces on the spacer arms that may lead to subconductors wires damage and spacer clamp loosening.

The phenomenon has been widely studied from the analytical and experimental point of view. Experimental studies have been carried out both in wind tunnel and on experimental lines. Details of the state-of-the-art-of-the research on the subject are given in Subsect. 4.5.1. As clearly found by experimental observations, among

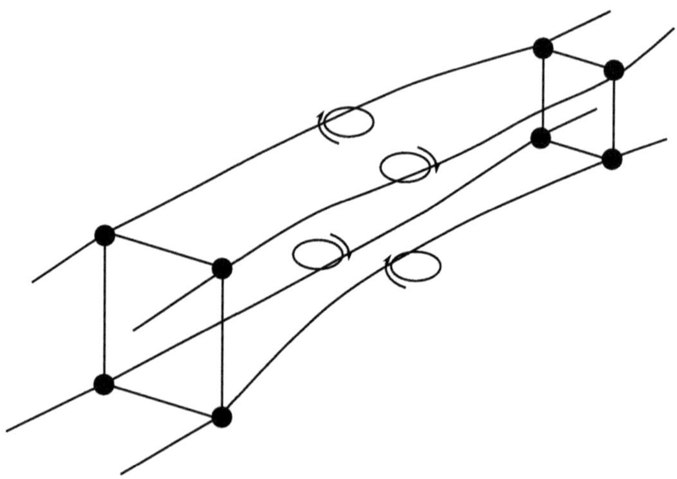

Fig. 4.52 Typical subspan oscillation shape for a quad bundle (adapted from EPRI (2009))

the various types of bundles, quad bundles seem to be the more sensitive to subspan oscillations EPRI (2009).

Subspan oscillations recently raised the transmission lines operators attention because many quad OHTL have been built in Saudi Arabia and, due to the terrain configuration (desert) and the high probability of medium-high wind speeds, several occurrences of subspan oscillations have been observed, causing subconductors and spacer-dampers damage. This fact gave new impulse to the studies and activities on the subject and gave us the opportunity to initiate a research aimed at clarifying some still not completely solved aspect of the phenomenon. In particular, the computation methodologies developed to simulate this problem Simpson (1971), Wardlaw and Cooper (1974), Diana and Gasparetto (1972) are based on the quasi-steady theory, i.e., the aerodynamic forces acting on the leeward conductor are identified through static measurements in the wind tunnel, as a function of the relative position between the windward and the leeward conductor. The quasi-steady theory (QST) accounts for the motion effect introducing the relative velocity of the two conductors with respect to the approaching flow and using the static aerodynamic coefficients measured in the wind tunnel. This approach generally holds for very high reduced velocities V_r:

$$V_r = \frac{V}{fI}$$

where f is the oscillation frequency, V the wind speed and I a characteristic dimension which in this problem is hardly defined: it could be the bundle separation. If we make the hypothesis of V = 10 m/s, f = 1 Hz and bundle separation I = 0.45 m, the reduced velocity is about Vr = 20: one of the research topics is to verify if the QST is adequate to reproduce the phenomenon for these values of reduced velocity Vr. Another important aspect of the problem is the Reynolds number effect: for the range of wind speeds (V) interested by the subspan oscillations phenomenon and the conductor diameters (D) the Reynolds number (Re $= VD/\nu$) value is in the range $2 \cdot 10^4 \div 1 \cdot 10^5$. For stranded conductors, i.e. rough cylinders, these values are close to the critical zone and then the Re number significantly affects the phenomenon because in this zone the drag coefficient sensibly varies with Re Zdravkovich (1997).

The above mentioned problems have been studied through an experimental research in wind tunnel on smooth and rough cylinders allowed to move one in the wake of the other with different frequencies and amplitudes, in such a way to operate at different reduced velocities and Re numbers. The experimental set-up allows for the measurement of the aerodynamic forces while the cylinders are moving, as described in the following sections. The measured aerodynamic forces are finally compared with those introduced in the analytical models used to simulate the phenomenon.

Section 4.5 is organized in different sub-sections:

- 4.5.1 state of the art of the research on the subject
- 4.5.2 experimental set-up used for the wind tunnel tests

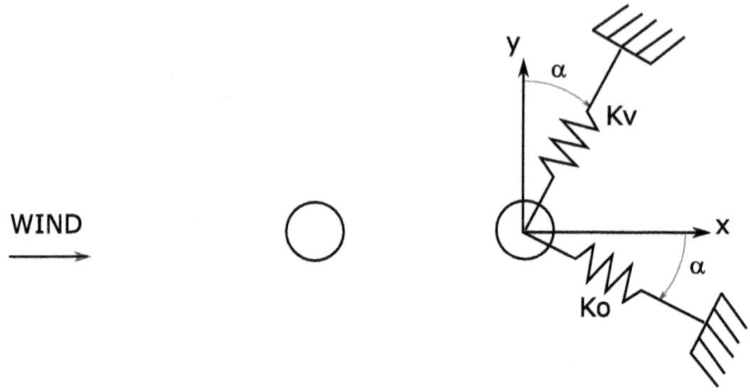

Fig. 4.53 2 d.o.f. analytical models

- 4.5.3 measurement results and the comparison with the analytical ones, improvements introduced in the analytical model, as a function of the results obtained through the wind tunnel experimentation.

4.5.1 State of the Art

The first analytical models developed to understand the cause of subspan oscillations are those of Simpson (1970), Ikegami (1971), Diana and Giordana (1972), Ko (1973) and Tsui (1975). These are two degrees of freedom models (Fig. 4.53): the motion of the leeward cylinder has two modes of vibration inclined of α with respect to the x and y directions, and k_o and k_v represent the modal stiffness; the windward cylinder is still.

The linearized QST is employed and the drag and lift coefficients on the leeward cylinder are deduced from static measurements in wind tunnel, as a function of the relative position of the leeward cylinder with respect to the windward (see Fig. 4.54 taken from Diana et al. (1974)).

These first studies show how the subspan oscillations are caused by an instability phenomenon due to the non conservative field of forces which is generated on the leeward cylinder by the wake of the windward cylinder. It is a flutter type instability with two different frequencies, being K_v different and lower than K_o,which become equal due to the effect of the field of forces. The field of forces is non conservative and then can introduce energy into the system, justifying the observed instability.

These models are linear and greatly simplify the structural behavior of the bundle sub-conductors, taking it back to a two degrees of freedom system in which the leeward conductor is the only one moving (see Fig. 4.53): hence they are not in condition to compute the oscillation amplitudes in a real span, also because these last greatly depend on the non linear effects of the field of forces. Many measurements have been done on experimental spans in order to identify the nature of the problem.

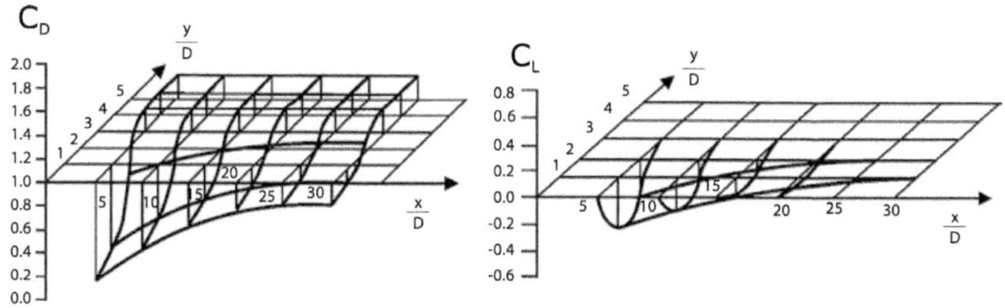

Fig. 4.54 WT static measurements of drag and lift coefficients on the leeward cylinder

More significant results come from the IREQ test spans at the Magdalen Island and in Varennes and from the Enel test span in Pradarena EPRI (2009). Figure 4.55 (Diana et al. 1974, 1973) reports the subspan oscillation amplitudes for a three bundle measured on the Pradarena test span: as can be observed the instability field, in this case, ranges between 10 and 20 m/s.

Figure 4.56 (EPRI 2009) reports the instability field observed at the Magdalen Island test span for a quad bundle equipped with different types of spacer systems.

Experimental evidence well proved that providing the span with different subspan lengths is an useful measure to control the subspan oscillation amplitudes. Another very important factor is the wind angle of attack with respect to the bundle: one of the reasons for which the bundle becomes stable for high wind speeds (>20 m/s) may be connected to the fact that the bundle rotation increases with the wind speed and then, for high bundle rotation angles the leeward conductors are no longer in the wake of the windward ones (see Fig. 4.57).

As already shown, a typical vibration mode of a quad bundle is represented in Fig. 4.52. Many attempts have been made to faithfully reproduce the above described mechanism with analytical methods: they will be briefly described in the following.

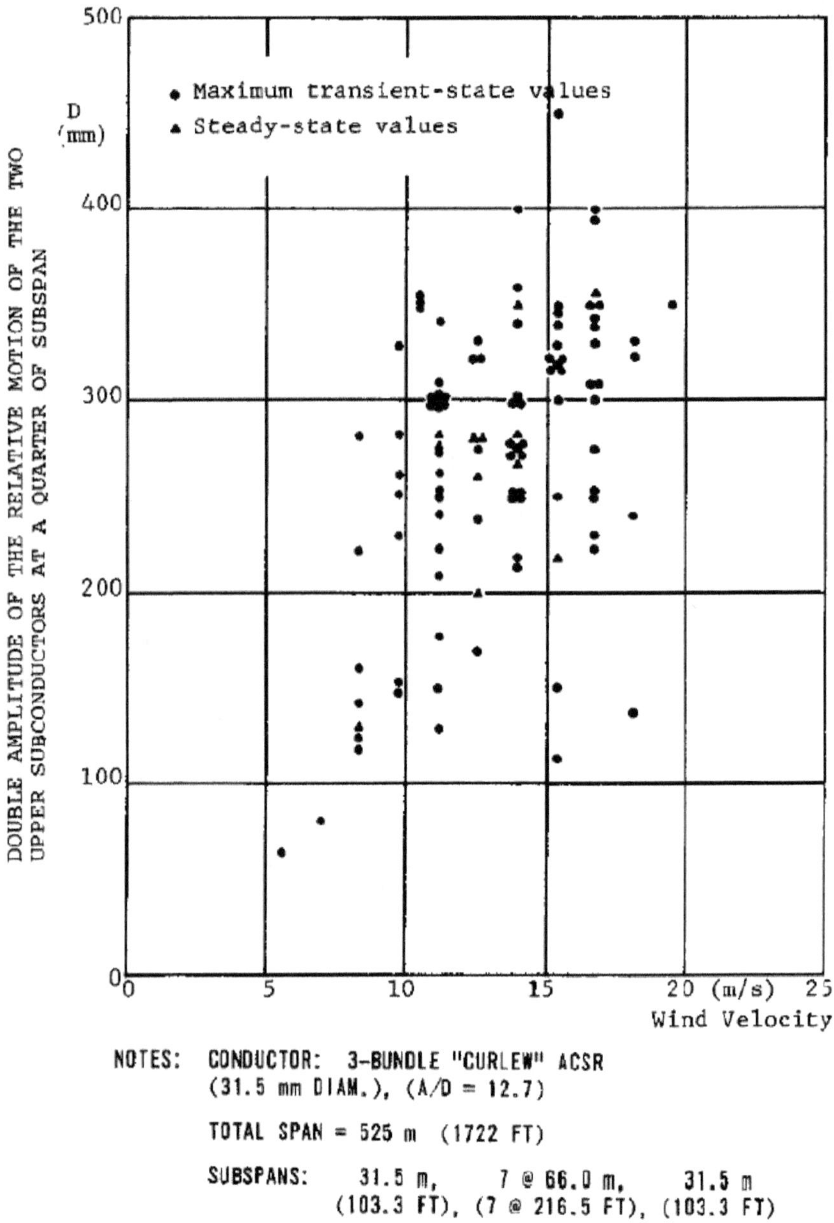

Fig. 4.55 Subspan oscillation amplitudes as a function of the wind speed observed at the Pradarena Pass test span for a three bundle. Courtesy of Giorgio Diana

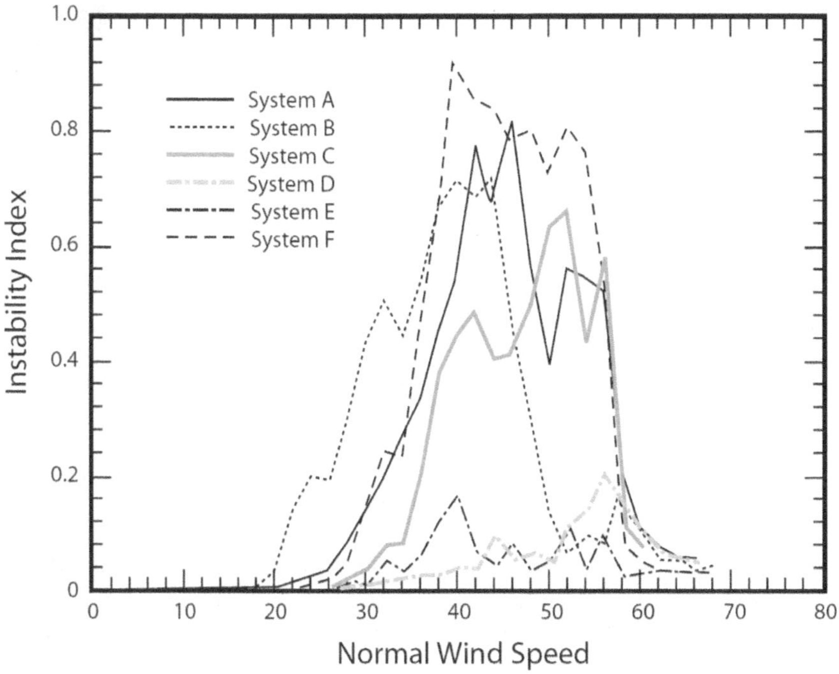

(a)

	Absolute (m)						
Subspan n.	1	2	3	4	5	6	7
System A	40.5	65	78	58	74.5	50	
System B	37	83.1	83.5	68.1	59.6	34.7	
System C	40	60	83	60	83	40	
System D	41	55	63	58	51	55	43
System E	34	56	64	57	65	56	34
System F	38.1	59.5	54.9	62.5	59.4	53.4	38.1

(b)

Fig. 4.56 Instability field (wind speed in km/h) observed at the Magdalen Island test span for a quad bundle equipped with different types of spacer systems, as in the table (EPRI 2009)

4.5.1.1 FEM Approach

Nowadays the FEM analysis allows for the reproduction of the bundle dynamics and for the application of the aerodynamic forces to each finite element using the QST. The first papers showing the application of this approach date back to 1977 Curami et al. (1977) while recently, other results are exposed in a PhD Thesis Snegovskiy (2010). Another researcher using this approach is prof A. Vinogradov.

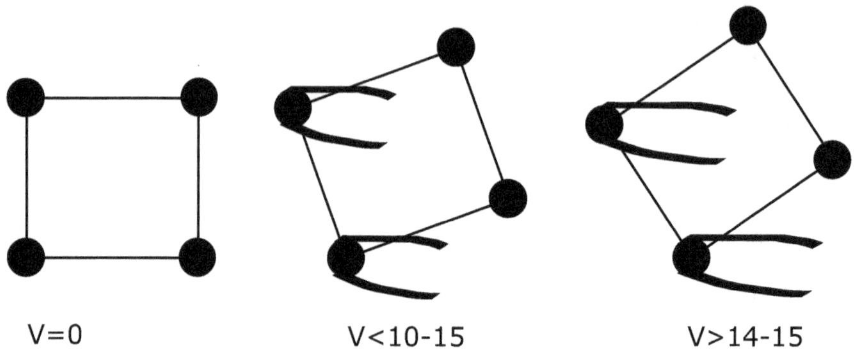

Fig. 4.57 Angle of attack of the wind with respect to the bundle as a function of wind speed

The FEM approach, however, does not succeed in adequately reproducing what is observed in the field and, moreover, operating in the time domain, is time consuming and not as much user-friendly as to be easily used as an engineering method. It still remains, generally, a researcher approach. For the expanded bundles[1] FEM approaches are successfully used to reproduce the bundle sub-conductors motion, which is, in this case, more affected by turbulence than by the wake effect Diana et al. (1999) and Belloli et al. (2010).

4.5.1.2 Energy Approach

To overcome these problems our research group developed an energy based method Diana and Gasparetto (1972) and Diana et al. (1974). The method substantially relies on the following approach:

1. Natural frequencies and vibration modes of the bundle in the field of frequencies between 0.5 and 2 Hz, interested by subspan oscillations, are computed. It has to be pointed out that the conductors vibration mode in a section of the bundle essentially depends on the spacer characteristics. For a quad bundle, as an example, the type of modes in a cross section of the bundle, for this range of frequency, are those reported in Fig. 4.58, together with the related spacer stiffness matrix eigenvalues (the mode eigenvalue decreases going from mode 1 to mode 8).
2. For each one of the vibration modes in Fig. 4.58, a number of natural frequencies and related spatial vibration modes can be computed. The analysis of the type of modes in Fig. 4.58 allows for the identification of those modes having a predominant horizontal component, which are named horizontal modes, and those modes with a prevalent vertical component, named vertical modes. As already observed, the subspan oscillation phenomenon is caused by the combination of

[1] Expanded bundles are the bundle with large spacing between the sub-conductors of the range of 1 m or more.

horizontal modes with vertical modes whose frequencies are initially different and become equal due to the field of forces, giving rise to the unstable elliptical motion.

3. To define which ones of the horizontal and vertical modes can be coupled by the wind to give subspan oscillations, the mathematical model initially analyses the stability of the generic pair of modes, linearizing the equations of motion obtained using the QST—see Diana and Gasparetto (1972) and Diana et al. (1974) for further details. For each possible pair of one horizontal and one vertical mode, the possibility of having instability is evaluated and the range of wind speeds for which subspan oscillation may be excited, together with an instability index, are computed. This preliminary analysis also gives the maximum difference between the vertical mode frequency and the horizontal mode frequency allowing the coupling between the two. The vertical frequency must always be lower than the horizontal one, because the aerodynamic forces tend to increase the vertical frequency Simpson 1971.
4. Once the types of modes that can be coupled are defined, as an example, those indicated in Fig. 4.59, two independent modal coordinates $q_o(t)$ for the horizontal mode and $q_v(t)$ for the vertical mode are defined. Due to the fact that the linear analysis produces in the instability field an elliptical expansive harmonic motion, harmonic laws are imposed to the two coordinates $q_o(t)$ and $q_v(t)$, in such a way to reproduce the elliptical motion of the conductors. The maximum instability index is found for a phase between the two modes equal to $\pm\pi/2$. Hence, imposing

$$q_o(t) = S_{am}\, e^{i\omega t} \tag{4.10}$$

$$q_v(t) = S_{im}\, e^{i(\omega t \pm \frac{\pi}{2})} \tag{4.11}$$

the conductors motion all along the span is obtained multiplying $q_o(t)$ and $q_v(t)$ by the modal shapes. Given that the modes are normalized in such a way that the maximum amplitude along the span is equal to unit, Sam and Sim also represent the maximum amplitudes for the horizontal and vertical modes. As an example, the orbits obtained for a bundle rotation angle α of 3° with respect to the wind direction are reported in Fig. 4.60. The obtained motion—in one section of the bundle—is for the modes indicated in Fig. 4.59.
5. Using the QST it is easy to define the forces acting on the conductors in the different positions occupied along the elliptical trajectories. The windward conductors are only subjected to the drag force, while the leeward conductors are subjected to lift and drag forces depending of the position, as measured through static tests in wind tunnel (Fig. 4.54). The absolute oscillation velocity of the conductors in the different sections of the bundle can be easily computed multiplying

$$\dot{q}_o(t) = i\omega\, S_{am}\, e^{i\omega t} \tag{4.12}$$

$$\dot{q}_v(t) = i\omega\, S_{im}\, e^{i(\omega t \pm \frac{\pi}{2})} \tag{4.13}$$

by the modal shapes.

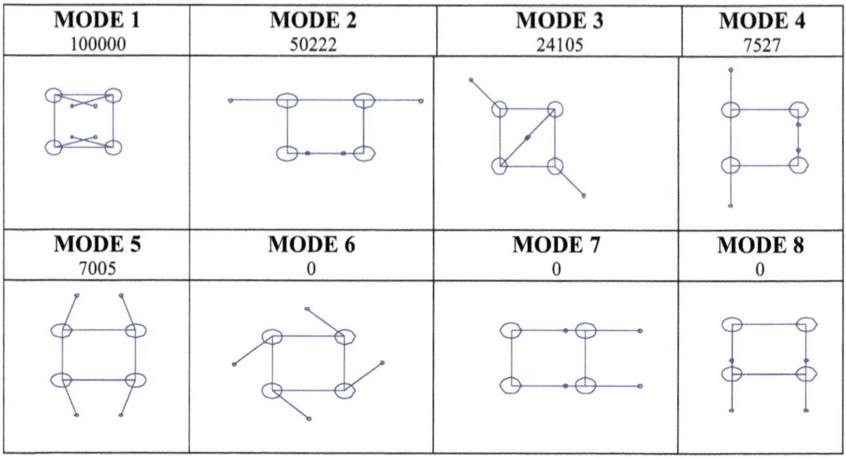

Fig. 4.58 Eigenvalues and eigenvectors of a quad spacer stiffness matrix

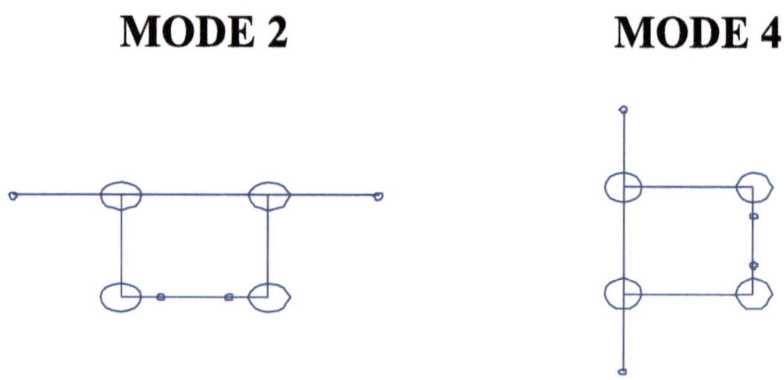

Fig. 4.59 Type of modes that can be coupled by the wind to give sub-span oscillations

Once the aerodynamic forces on the conductors are known, it is possible to compute, in each section of the bundle, the energy introduced by the forces themselves in one complete elliptical cycle. This is a function of the imposed horizontal and vertical orbit amplitudes S_{am} and S_{im} in Eq.(4.13).

The computation method gives the energy introduced by the wind in the bundle sub-conductors all along the span, as the sum of the energies computed in each section: the evaluation is performed both for $+\pi/2$ and $-\pi/2$, obviously choosing the situation for which the aerodynamic forces can introduce energy into the bundle system. The steady state amplitudes of oscillation are defined through the balance between the energy introduced by the aerodynamic forces and the energy dissipated by the bundle, equipped with the spacer-dampers. Reference can be made to Diana and Gasparetto (1972) and Diana et al. (1974) for the computation of the energy dissipated by the bundle.

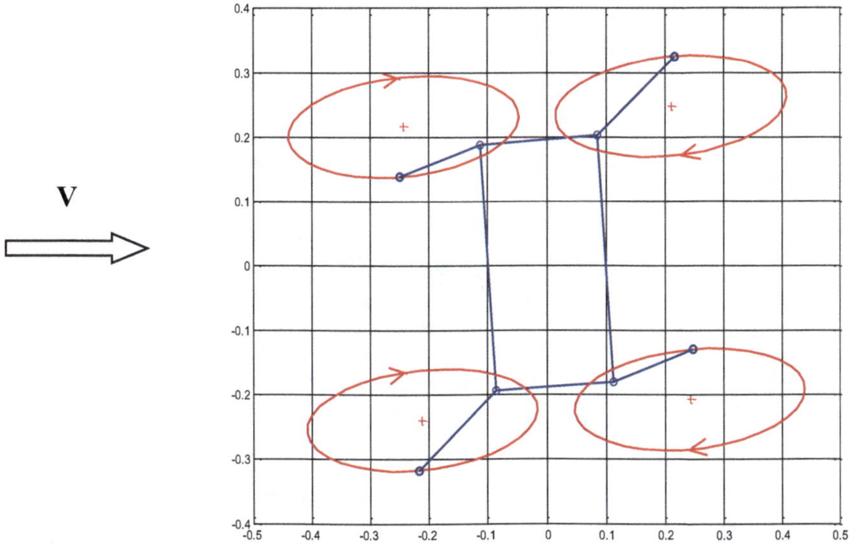

Fig. 4.60 Sub-span oscillation motion in one section of the bundle for the modes indicated in Fig. 4.59 and for a bundle rotation angle α of 3° with respect to the wind direction

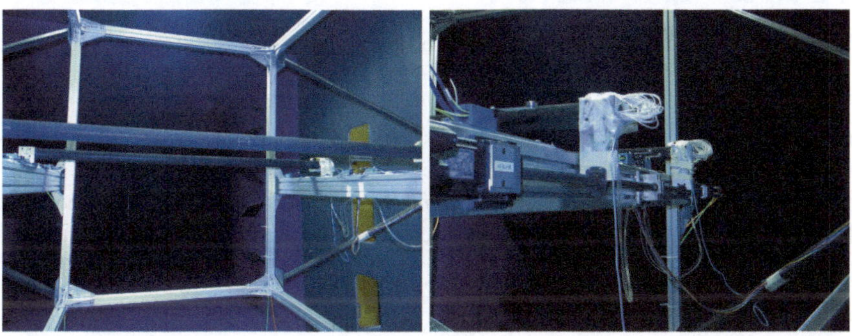

Fig. 4.61 Experimental set-up. **a** Cylinders with the support frame **b** equipment to control the cylinder motion

4.5.2 Measurement Set up

The experimental set up used for this research consists of a frame, shown in Fig. 4.61a, allowed to rotate around an horizontal axis. Four ball screws are mounted on this frame, each of them actuated by a permanent magnets motor. Two cylinders are coupled to the four ball screws, as shown in Fig. 4.61b.

Figure 4.62 shows the frame installed in the 4 × 4 section of the Politecnico di Milano Wind Tunnel, where a 55 m/s wind speed can be reached. The two cylinders diameter is 0.06 m, i.e. more or less double with respect to standard conductors used

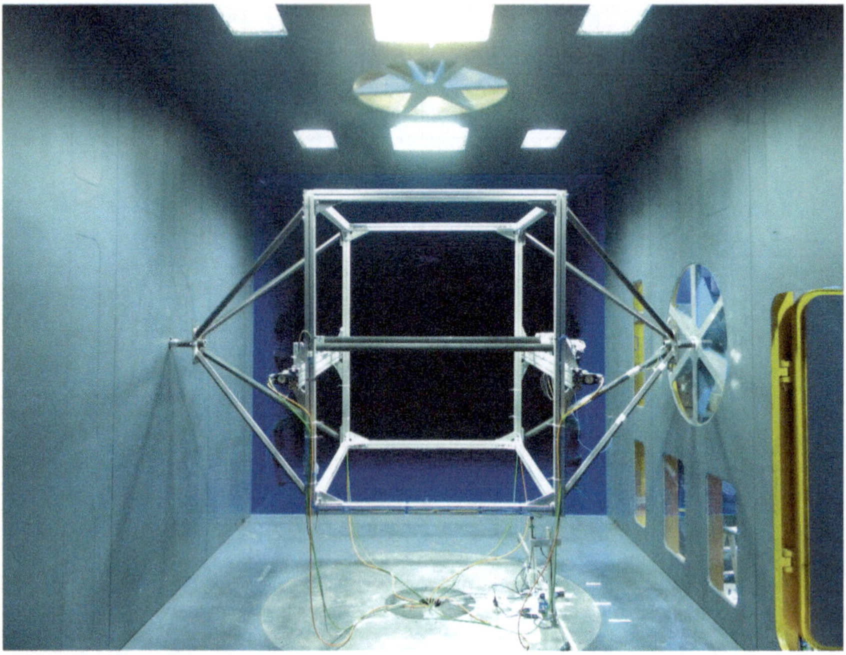

Fig. 4.62 Model set-up in the wind tunnel test section

in the transmission lines. All the dimensions (cylinder diameter D and cylinders separation I) are in a 2:1 scale: this choice corresponds to the necessity of simplifying the pressure measurements increasing the cylinder diameter. As consequence the frequency must be different from the real one to maintain the same reduced velocity V_r

$$V_r = \frac{V}{f\ I}$$

with a wind speed V typical of a real situation. During the tests the oscillation frequency has been set between 0.5 and 1 Hz and the flow velocity between 10 and 20 m/s.

A series of circumferential holes placed in a central section of each one of the two cylinders (Fig. 4.63) allows for the measurement of the pressure along the cylinder circumference and, from this, the aerodynamic forces acting on the cylinders due to the flow and the imposed motion can be derived Zasso et al. (2008).

The sensors controlling the motion imposed by the motors give the cylinders horizontal displacements while the vertical displacements associated to the frame rotation are measured through laser displacement transducers. The energy introduced in one oscillation cycle by the aerodynamic forces can be evaluated from the measure of forces and displacements. Main purpose of the described test set up is to compare

Fig. 4.63 Pressure taps located on the cylinder surface (rough cylinder)

the measured energy to that evaluated through the mathematical model described in Sect. 4.5.1, as a function of the oscillation amplitude and the reduced velocity.

As it will be clearly evidenced by the results of this comparison, another important parameter, besides the reduced velocity, is the Reynolds number (Re). The Reynolds number significantly affects the value of the drag coefficient, as a function of the cylinders surface roughness. Due to this fact, the tests have been conducted both with smooth cylinders—to reproduce the sub-critical Re zone—and with rough cylinders—to reproduce the critical and super critical Re zone (Zdravkovich 2003; Diana and Gasparetto 1972; Buresti 1981).

4.5.3 *Measurement Results and Comparison with the Model Results*

In this section the results of numerical and experimental comparison are reported in order to prove the accuracy of the numerical model described in Sect. 4.5.1. Firstly a description and evaluation of static and dynamic tests is provided to highlight the peculiar behavior of the cylinders, hence the improvements made to the numerical model, based on experimental results, are described.

4.5.3.1 Static Tests

The static tests were performed both on smooth and rough surface cylinders. The aim of the tests was, as mentioned in Sect. 4.5.1, to evaluate the static forces acting on the cylinders while changing their relative position and as a function of the wind angle of attack. The tests were carried out placing the cylinders in the desired position and switching on the wind tunnel until the testing speed was reached, then the pressures in the two measuring sections of the cylinders (see Sect. 4.5.2) were acquired over a time period of 30 s and properly post-processed in order to calculate the forces acting on the cylinders. Finally from the force time histories it was possible to identify both the drag and lift coefficients for each cylinder configuration and cylinder surface type.

Static tests on smooth cylinders

The smooth cylinders were placed at a given longitudinal and transversal separation so that it was possible to identify the forces acting on the windward cylinder and on the leeward one, as a function of their relative position, i.e. as a function of the position of the leeward cylinder in the wake generated by the windward one. The position is reported to the cylinder diameter and is then identified by x/D (longitudinal) and y/D (transversal).

The static tests demonstrated that there is no variation of the drag coefficient on the upstream cylinder, which has, whatever the speed, a drag coefficient of 1.2, whereas a dependency over speed and position is found for the leeward cylinder. Figure 4.64 reports the drag coefficient for the leeward cylinder, as a function of y/D (for x/D= 15 and for different wind speeds): it is possible to note that C_d is decreasing as the wind speed increases (i.e. as Re increases). This behavior is clarified by the consideration that the wake is turbulent and shifts towards lower Re numbers the critical point: the experimental tests considered in Fig. 4.64 are carried out in the critical region where the C_d is decreasing while increasing Reynolds. The red line in Fig. 4.64 is taken from the results reported by Diana and Giordana (1972) which are related to $\text{Re} = 2 \cdot 10^4$, while Re of the actual experimental tests, when the speed is 10 m/s, is $\text{Re} = 4 \cdot 10^4$.

Static tests on rough cylinders

In order to obtain a full understanding of the subspan oscillation phenomenon, static tests were also carried out considering rough cylinders, see Fig. 4.63. During experiments, the rough cylinders were placed at the same separation as the smooth ones, and the same wind speeds were investigated. Figure 4.65a shows the drag coefficient, measured during static tests, on the upstream cylinder. It can be observed that the trend is not constant: increasing the speed, and consequently the Reynolds number, the drag coefficient tends to decrease from the nominal value of 1.2 down to a minimum value of 0.85, which occurs when the wind speed is 14 m/s.

The shown dependency over Reynolds of the aerodynamic mean forces on the first cylinder, is well known and in accordance with the results obtained by Wardlaw and Cooper (1973)—Fig. 4.65b—on the stranded conductor "Chuckar", 40 mm diameter. In this particular case, the chosen roughness allows to properly replicate, during experiments, the typical behavior of a stranded cable, even with a shift in the Re values, that must be taken into account.

Analyzing Fig. 4.65 it is possible to identify the critical region where the slope of the drag coefficient is negative and to observe that during tests performed at 10 m/s the cylinder works in the critical region, where the boundary layer of the windward conductor changes from laminar to turbulent. The negative slope of the drag coefficient curve corresponds to a decrease of the energy dissipated by the windward cylinder, with respect to the case of the smooth cylinder. The destabilizing effect of the drag coefficient negative slope is well known and had been already put in evidence by Richards (1963) studying the instability observed on the Severn crossing stranded conductor, where the combined effect of the drag coefficient negative slope and of the lift coefficient due to the yaw angle produced large amplitudes of vibration.

The drag coefficient of the cylinder in the wake is reported in Fig. 4.66 for x/D = 21, y/D = 0 and for different wind speeds. It is interesting to observe that the turbulence in the wake shifts the critical region towards lower speed: the critical point is close to 5 m/s instead of 10 m/s as shown in Fig. 4.65. Figure 4.67 reports the value of Cd on the leeward cylinder as function of y/D, for x/D = 15 and for variable wind speed: increasing the wind speed from 6 to 10 m/s, the Cd increases, while

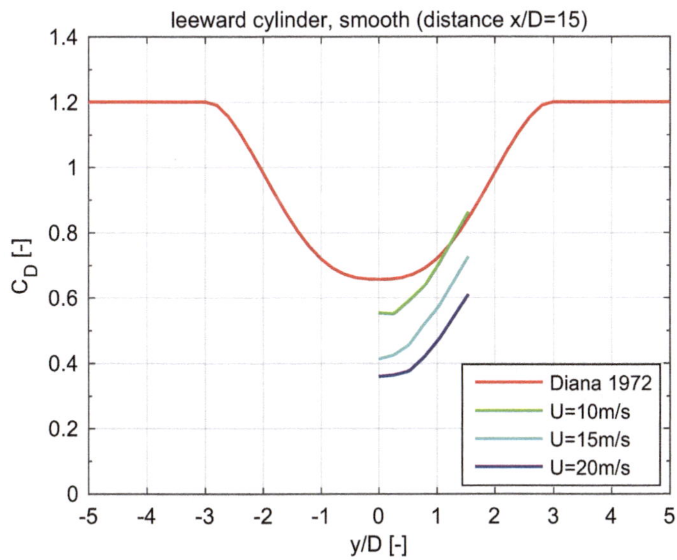

Fig. 4.64 Drag coefficient on leeward smooth cylinder as function of y/D and for x/D =15: red line—experimental results of Diana and Giordana (1972) green line Cd for 10 m/s wind speed, cyan line Cd for 20 m/s wind speed, violet line for 15 m/s wind speed

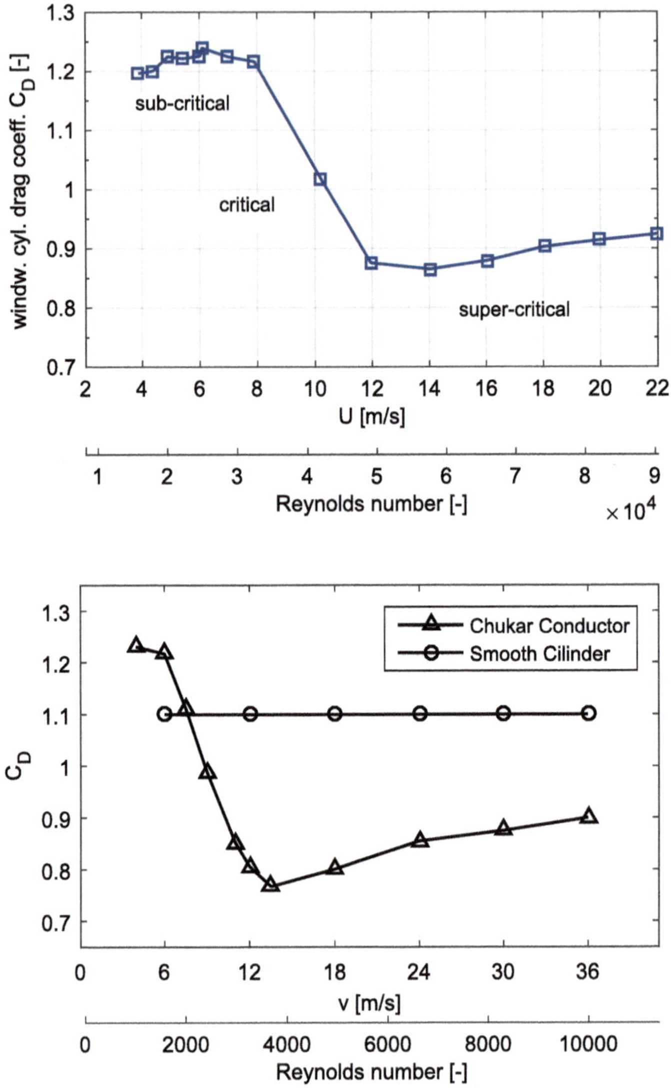

Fig. 4.65 Drag coefficient on windward rough cylinder as function of speed/Reynolds number: **a** wind tunnel experimental tests **b** Wardlaw and Cooper studies (adapted from Wardlaw and Cooper (1973))

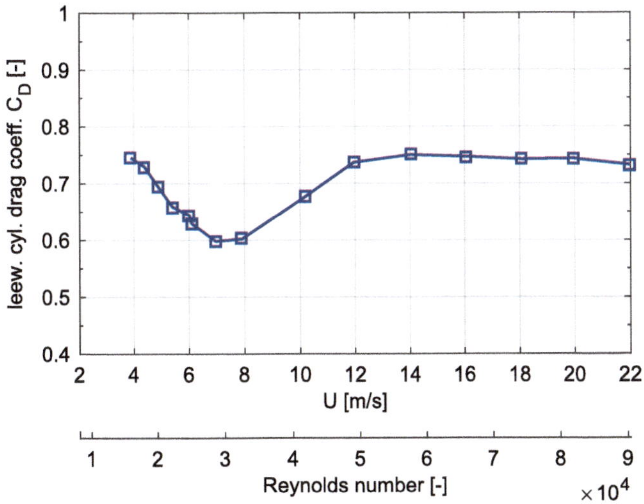

Fig. 4.66 Drag coefficient on leeward rough cylinder as function of speed/Reynolds number for x/D = 21 and y/D = 0

for further increases of wind speed the drag coefficient appears unchanged. Such a behavior is completely different with respect to the one observed for the smooth cylinder in Fig. 4.64 and it is associated to the shifting of the critical region towards lower speeds as explained in Fig. 4.66.

In Fig. 4.68a, b, c and d, as an example, a comparison is proposed for Cd and Cl coefficients on rough leeward cylinder and on smooth leeward cylinder for x/D = 15, as function of y/D and for different wind speeds. The red curve on the diagrams represents the static characterization of the cylinder in the wake described in Diana and Giordana (1972), while the other curves represent the dependency over speed of the coefficients, in the wake zone explored during the tests.

The results described for the static tests highlight the strong influence of the Reynolds number on the behavior of a cylinder: not only the position but also the wind speed significantly affect the leeward conductor aerodynamic coefficients; it is hence important to include this effect while studying subspan oscillations.

4.5.3.2 Dynamic Tests

In this section a description of dynamic tests performed during the experimental campaign described in Sect. 4.5.2 is reported and the major results discussed. The dynamic tests have the purpose to better understand the mechanisms causing subconductor oscillations, identifying the aerodynamic forces as a function of cylinder motion and performing an evaluation of the accuracy of the QST generally adopted when dealing with this phenomenon. Additionally, the effect of conductor roughness

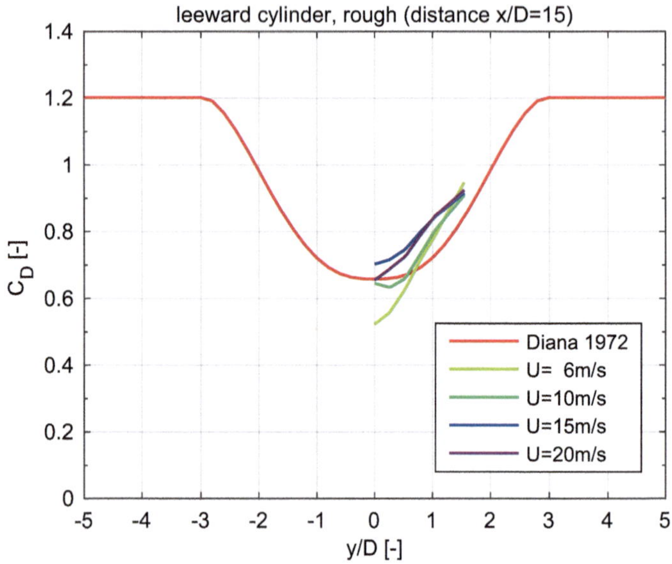

Fig. 4.67 Drag coefficient on leeward rough cylinder as function of y/D and for x/D = 15: light green line C_d for 6 m/s wind speed, green line C_d for 10 m/s wind speed, cyan line C_d for 15 m/s wind speed, violet line for 20 m/s wind speed

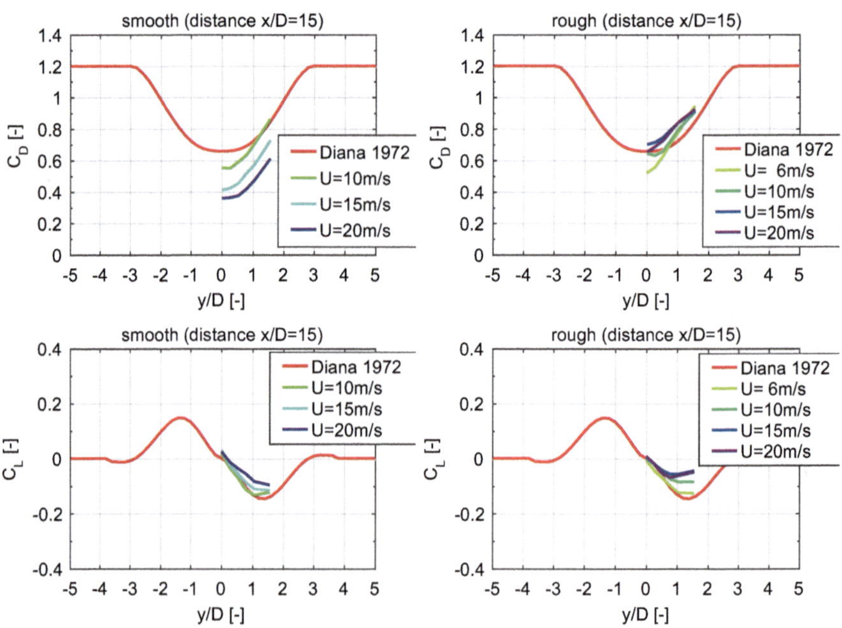

Fig. 4.68 Drag and lift coefficients on leeward cylinder as function of y/D and for x/D = 15: **a** smooth cylinder drag coefficient **b** rough cylinder drag coefficient **c** smooth cylinder Lift coefficient **d** rough cylinder Lift coefficient

Table 4.2 Dynamic tests condition for smooth/rough cylinder

	Smooth cylinders	Rough cylinders
Wind speed range [m/s]	10–20	10–20
Translation motion amplitude x/D	0.5–2.5	0.5–2.5
Rotation amplitude [deg]	3–4	3–4
Frequency [Hz]	0.5–0.7–1	0.5–0.7–1
Static angle of attack [deg]	3–4	3–4
Cylinder static separation in diameters	15	15

is evaluated testing rough cylinders. To reproduce the subspan motion, the tests have been carried out imposing an elliptical motion to the cylinders by means of a combination of:

- Harmonic translational movements of cylinders with different amplitude and frequency;
- Harmonic rotational movements of the testing structure with different angles. The frequency of the imposed harmonic rotation is synchronized to the cylinder 1 translational movements with a phase lag of 90°.

Different wind speeds were considered as well. Many of the tests were performed with a cylinders separation equal to 15 diameters, to replicate the typical configuration of OHTL conductors. Table 4.2 summarizes the main experimental test data both for smooth and rough cylinders.

The outputs of the test are represented by:

- The displacement of the cylinders during tests, which is a combination of the harmonic translational and rotational movements imposed by the controllers. In reality, they are not perfectly harmonic, also due to the effects of the actuator control on the motion of the system.
- The energy associated to the motion of the cylinders, both considering only the first harmonic of the forces/displacements, and all the time history, in order to appreciate the contribution of higher harmonics.
- The aerodynamic forces (modulus and phase) acting on the leeward cylinder (first harmonic component).

Dynamic tests on smooth cylinders

In the following the main results of the dynamic tests are reported. In Fig. 4.69, as an example, the energy associated to the smooth cylinders is reported as a function of the x/D amplitude (i.e. normalized to the cylinder diameter) and for different frequencies (different lines colors), for a wind speed of 10 m/s, a static inclination angle of the structure of about 3° (static angle of attack of the wind), and a dynamic angle amplitude of 3°: Fig. 4.69a and b show the energy of the drag and lift forces acting on the windward cylinder. The continuous line refers to the energy computed accounting for the first harmonic component of the force, whereas the dashed one refers to the energy computed considering all the frequency content of the forces. Firstly, it can be seen that the drag energy of the windward cylinder is systematically a dissipative one Zdravkovich (1997) as expected, while in the case of the leeward cylinder (Fig. 4.69c and d), depending on the frequency of motion and the amplitude of the motion, the energy contribution could be either positive or negative (positive energy introduced by the wind leads to instability, while negative to stability). The energy at a frequency of 0.5–0.7 Hz is positive and becomes negative at 1 Hz, for x/D amplitude greater than 2. We need to remember that 0.5 Hz in the WT experiments corresponds to a frequency of 1 Hz in the real case. It can also be observed that, while on the first cylinder the dissipated energy increases almost linearly with the amplitude of motion, on the leeward cylinder the trend is almost parabolic.

Finally, in Fig. 4.69b and d, it can be observed that the lift energy represents only a small contribution to the overall energy of the cylinders, being one order of magnitude smaller than the drag energy for the same cylinder. The results also demonstrate that harmonic components different from the first one seem to have small influence on the value of the overall energy. As expected, the increase of frequency corresponds, at the same amplitude, to an increase of dissipated energy due to the increase of the velocity of motion. Combining the drag and lift energies and the energy of the windward and leeward cylinder, it is possible to quantify the overall energy of the cylinders and verify the stability of motion. Figure 4.70 shows the cylinders total energy for the smooth cylinders when the wind speed is 10 m/s and the static and dynamic angles are 3deg. It can be observed that, for the tested smooth cylinders, the 1 Hz motion is always stable whatever is the amplitude of the cylinders motion, whereas at 0.7 and 0.5 Hz the cylinders motion is unstable for amplitudes lower than 1.2 and 1.5 respectively.

Figure 4.71a and b show the overall energy of the cylinders for increasing speed, at the same testing conditions as before: the cylinders energy tends to increase, for all the tested frequencies, when the speed is increased.

The experimental tests allowed also to highlight the mechanism typical of the investigated phenomenon: the relative phase with respect to the cylinders motion of the drag and lift forces on the windward and leeward cylinders were analysed for the different testing conditions. For the leeward cylinder, Fig. 4.72 shows the phase of the drag and lift force first harmonic with respect to the x/D and y/D motion, for the base test case of 10 m/s speed and static and dynamic angle of 3°. It is evident the role played by the lift force which has, also in all the other tests, a phase with respect to the y/D motion close to 180deg, such indicating that the lift force is like a positive stiffness, justifying the increase of the vertical frequency observed in subspan oscillations.

Dynamic tests on rough cylinders

Static tests demonstrated the importance of the cylinder roughness on the aerodynamic forces, hence dynamic tests were performed with rough cylinders as well.

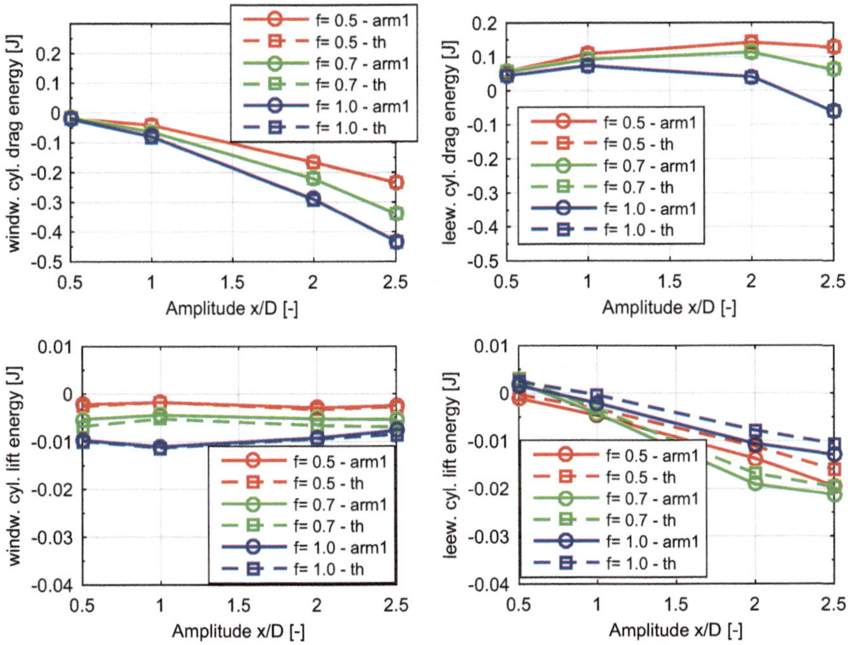

Fig. 4.69 Smooth cylinder energy: continuous line first harmonic calculations and dashed line time history calculations: **a** windward drag energy; **b** windward lift energy; **c** leeward drag energy; **d** leeward lift energy (U = 10 m/s, separation 15 diameters, static and dynamic angle = 3°)

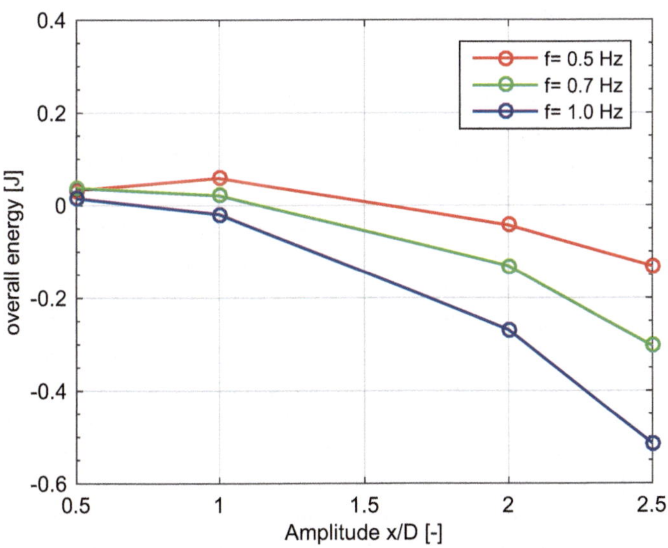

Fig. 4.70 Smooth cylinder global energy. Continuous line first harmonic calculations and dashed line time history calculations (U = 10 m/s, separation 15 diameters, static and dynamic angle = 3°)

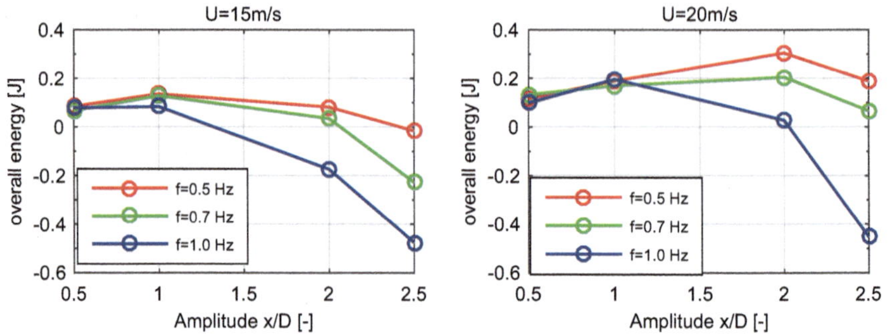

Fig. 4.71 Cylinders energy versus x/D amplitude: continuous line first harmonic calculations and dashed line time history calculations (separation 15 diameters, static and dynamic angle = 3°)

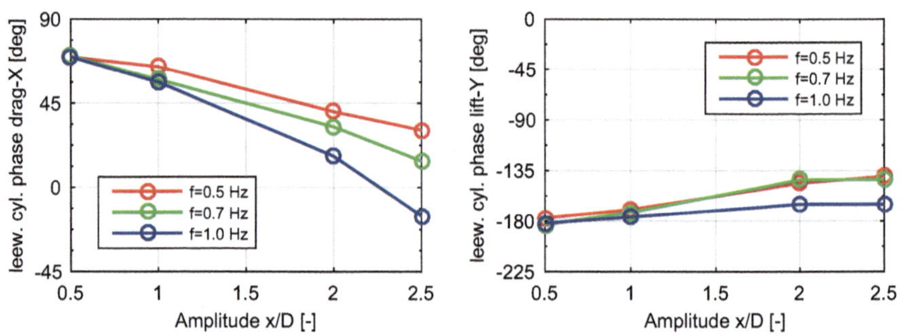

Fig. 4.72 Leeward cylinder force phase versus x/D amplitude (U = 10 m/s, separation 15 diameters, static and dynamic angle = 3°)

The Reynolds effect is demonstrated to play an important role on the mechanism of subspan oscillations: as an example, if the energy related to the windward cylinder is analysed for a wind speed of 10 m/s, which corresponds, for the rough cylinder, to the critical region of the Reynolds number (Fig. 4.65), it is observed that the roughness (dashed lines) leads to a decrease of the dissipated energy up to 3 times the value obtained with the smooth cylinder (see Fig. 4.73); moreover, this occurs whatever is the frequency of motion. This is due to the negative slope of the drag coefficient (Fig. 4.65), which causes a reduction of the drag force damping effect. It can be concluded that the stability of the cylinders is affected by the Re number region where the cylinders themselves are working, i.e. sub-critical, critical or supercritical.

Figure 4.74a, b and c show the variation with speed (i.e. with Reynolds) of the overall energy of the cylinders: the smooth (continuous line) and rough (dashed line) case are compared in the same diagram.

The energies reported in Fig. 4.74 as a function of the cylinder oscillation amplitude and frequency of motion, show substantial differences depending on the surface roughness. More in detail at 10 m/s, which corresponds to a Reynolds number of about $4 \cdot 10^4$ for stranded conductors, the energy of the rough cylinder is greater than the one of the smooth cylinder: this is due to the fact that the upstream con-

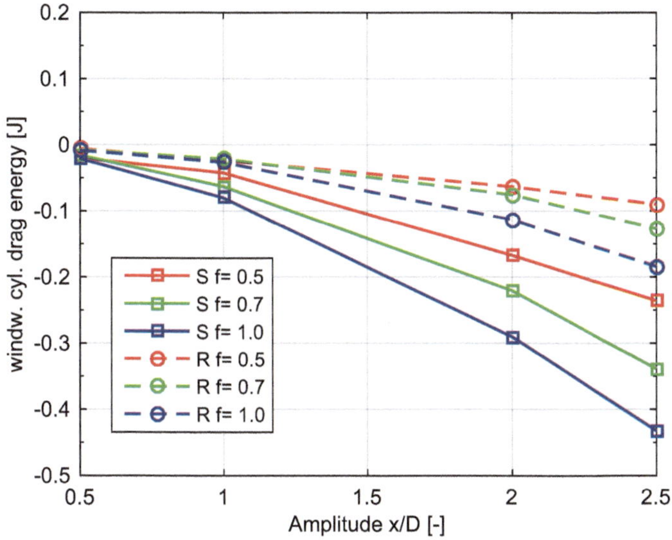

Fig. 4.73 Windward cylinder energy with respect to x/D amplitude for the three tested frequencies: dashed line rough cylinder (R) and continuous line smooth cylinder (S) (U = 10 m/s, separation 15 diameters, static and dynamic angle = 3°)

ductor is in the critical range of Reynolds, while at higher speed the rough cylinders are working in the supercritical zone and the overall energy is lower than that of the smooth cylinders.

The higher the frequency of motion is, the lower is the introduced energy and then the more stable the cylinders appear, due to the effect of the cylinders velocity of motion increase. It can be concluded from the results that the cylinder roughness tends to shift the unstable behavior of the cylinders towards lower wind speeds.

4.5.3.3 Experimental and Numerical Comparison

In Sect. 4.5.1 the numerical model developed within the Dipartimento di Meccanica of Politecnico di Milano, has been described. Hereafter the improvement to the model is described and a validation of the model is proposed.

Experimental tests, both static and dynamic ones, demonstrated the importance of the Reynolds effect on the subspan instability, in fact Reynolds strongly impacts on the aerodynamic field acting on the cylinders and hence on the energy balance. Moreover, dynamic tests are in agreement with the static tests allowing to confirm the validity of the quasi-static theory. The analytical model was modified introducing the Reynolds effect as follows:

- First, from the experimental static tests on the rough cylinder, the drag and lift coefficients are defined for different positions of the leeward cylinder in the wake

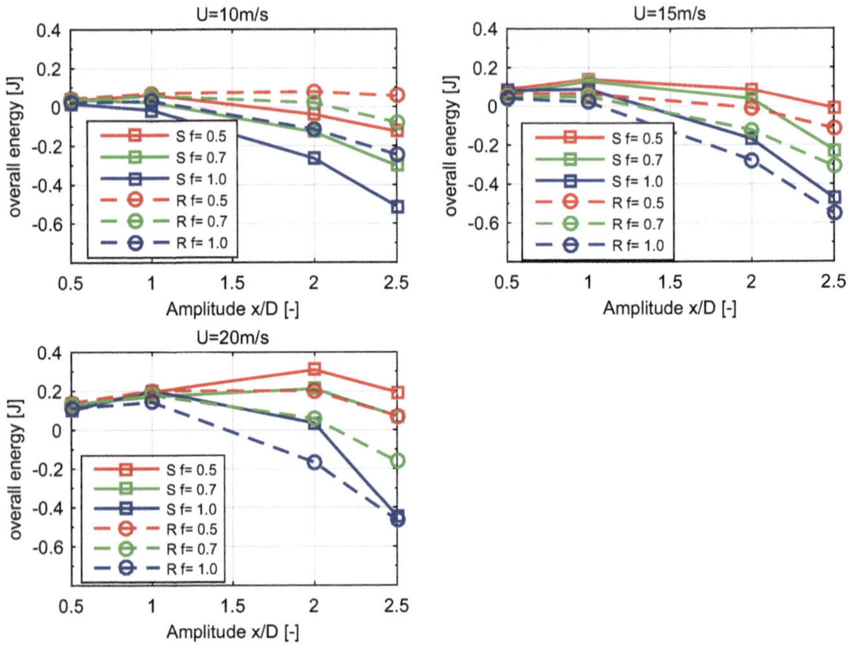

Fig. 4.74 Cylinders total energy with respect to oscillation amplitude for the three tested frequencies and at different wind speeds: dashed line rough cylinder (R) and continuous line smooth cylinder (S) (separation 15 diameters, static and dynamic angle = 3°)

and different Reynolds numbers. An interpolation function is used in order to define the Cd and Cl values for any position in the wake and for any value of the Re number.

- Second, for the windward cylinder, the experimental curve of Fig. 4.65 as a function of Reynolds is introduced for the computation of the Drag force, accounting for the realistic roughness of a stranded conductor as shown in Fig. 4.65b. The Re number is computed considering the relative velocity of the cylinder with respect to the flow.

In the following, a comparison between experimental and numerical results is reported for the test case of 3deg angle and wind speed of 10 and 15 m/s. Considering the scale of the model in the wind tunnel and remembering that the parameter defining the aeroelastic behavior is the reduced velocity Vr, these test cases are representative of the conductor behavior during subspan oscillations.

The trend of the total energy on the two cylinders is described in Fig. 4.75, where the numerical results are reported in continuous line while experimental outputs in dashed line: results show a good agreement, at a speed of 10 m/s (Reynolds $4 \cdot 10^4$). In any case the results support the validity of the quasi-steady theory. In order to improve the model, a more detailed investigation is needed, to better identify the aerodynamic static forces on the leeward cylinder and the dependence on Re. Nevertheless, it can be

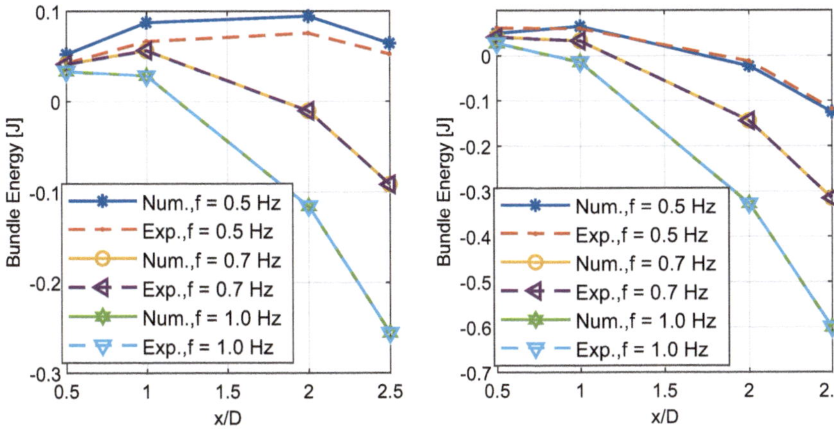

Fig. 4.75 Cylinders wind energy input as a function of the elliptical motion x/D amplitude for the three considered frequencies: experimental—numerical comparison. **a** 10 m/s wind speed (Re = $4 \cdot 10^4$), **b** 15 m/s wind speed (Re = $6 \cdot 10^4$)

concluded that the numerical model is sufficiently accurate to reproduce the subspan oscillation phenomenon.

4.6 Ice Galloping

4.6.1 Introduction

In gross terms, galloping is a vertical oscillation of the conductor span in one or a few loops. However, it usually incorporates other less visible twisting, lateral and longitudinal motions. The most common design of overhead transmission lines consists of many spans of similar length supported on steel lattice tower with suspension insulators. Galloping of these lines is usually in one- or two-loop motions with generally similar amplitudes in each span see (Fig. 4.76).

Aside from the clearly visible vertical motions, there are usually noticeable longitudinal motions of the suspension insulator strings. The upward half of the cycle of motion is larger than the downward motion due to the decrease of tensile load, and the conductor can move above the horizontal. The galloping motions can be large enough to cause flashovers between adjacent phases, especially when the phases are oriented vertically. The vertical motion can also be a combined mode including one- two-, three-, and four-loop motions. The single loop mode can occur even on relatively long spans.

When the galloping span is one of a series of suspension spans in a section of the line between dead ends, there are movements at the suspension points longitudinal to the line. These movements are caused by the variations in tension that can accompany

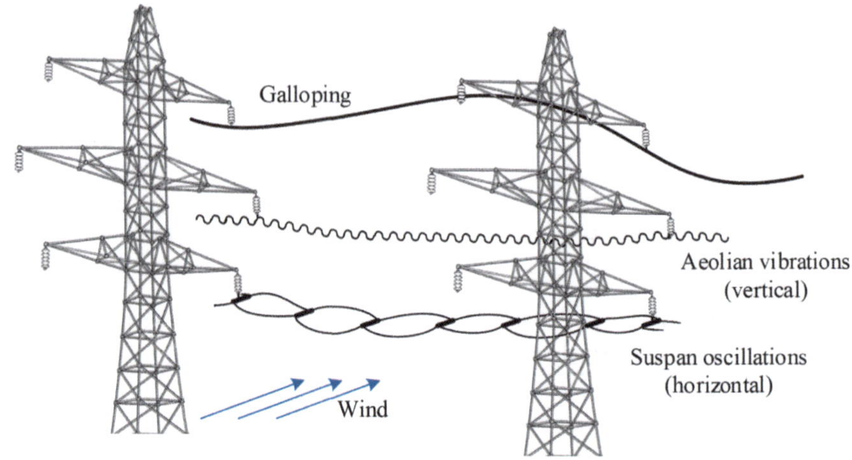

Fig. 4.76 The three main types of power line vibration (Zanelli et al. 2022)

galloping, and they act to couple the galloping motions in adjacent spans. The spans are then not autonomous. Instead, the section as a whole has normal modes, each of which involves participation of every span in the section. There is generally a multitude of such modes, each at a different frequency, and each displaying a different combination of relative amplitudes in the various spans.

The conductor may twist during galloping. This is seldom visible for single conductors, but is readily apparent in bundled conductors through rocking motion of the spacers. The twist is synchronized with the vertical motion. The vertical motion of galloping may be accompanied by horizontal motion, but this is usually, and perhaps generally, small compared to the vertical component. Short line sections between dead ends are the next most common design of overhead lines.

Figure 4.77 illustrates the four lowest-frequency modes for a line section having four suspension spans of 160, 180, 190 and 195 m between dead ends. Calculations, using a linear analysis of the motions, show that for the modes at 0.386 and 0.403 Hz, there is only small variation in tension during galloping. This is because, when one span is at its upward extreme of motion, there is another span at the downward extreme. The variations in arc length of the two spans compensate each other by means of swinging of the suspension support between them.

In the mode at 0.516 Hz in Fig. 4.77, all spans move in phase so there is less ability for arc length compensation between spans to occur. As a result, this mode displays significant tension variations during galloping, rather like the pseudo-fundamental in a dead ended span, which we discuss below. The least common design of overhead transmission line span is that with dead ending at both ends. When the galloping takes place in such a span, the motions are independent of what is taking place in adjacent spans. The galloping may display modes with 1, 2, 3, etc., loops. The modes with even numbers of loops conform in frequency and mode shape to simple taut-string

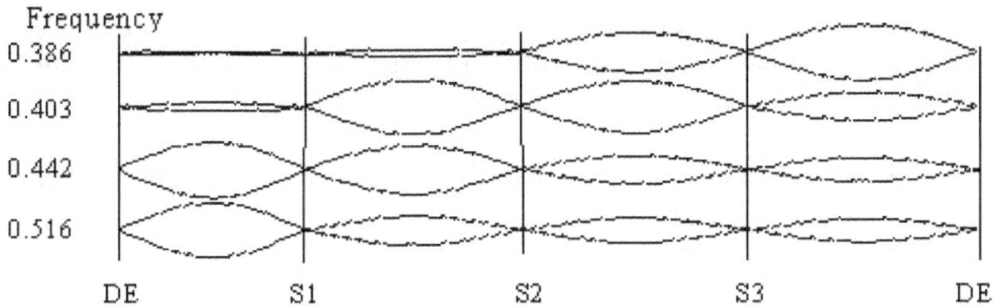

Fig. 4.77 Typical one loop mode shapes (4 span section). The three first ones are "up-and-down" modes (some span are moving up when others are moving down with a global compensation effect). The fourth one is the "up-up" mode when all spans are synchronously going up or down. This last is related to large tension variation and its shape may be non purely sinusoidal

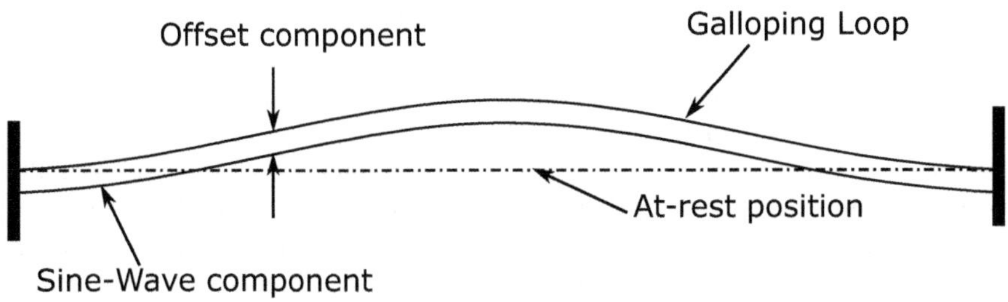

Fig. 4.78 There is a one loop mode shape ("up-up") for which the shape is more complex than a sine wave. The difference with the sine wave is related to "Irvine parameter" which is dependent of many structural parameters, mainly the sag/span ratio and extensional stiffness (taking into account the support flexibility) of the whole section length. It is the so-called "first in plane mode" or "pseudo-one loop" mode

theory. The odd-numbered modes, however, have higher frequencies than predicted by string theory, and their mode shapes take the form of sine waves with an offset, as illustrated in Fig. 4.78. They are called pseudo-modes because of these differences.

Spans are often observed to gallop in a combination of two or more of the modes that are available to them. For example, a suspension span may move simultaneously in a mode of the section and in its own autonomous 2-loop mode.

The vertical component of galloping, and the longitudinal motions at suspension supports, are important relative to violation of electrical clearances, both in spans and at supports. They are also closely associated with the conductor tension variations, which can be large, and the dynamic forces transmitted to insulators and supporting structures. The above discussion neglects the torsional component of galloping motion, as well as motions lateral to the span. Both of these also have normal modes by span and by line section. Those components can have important effects in relation to aerodynamic mechanisms that cause galloping.

Advances during the last 30 years fall in several categories. Of particular importance is the development of several countermeasures that appear to be reliable enough to find common usage on operating lines. These include the conductor construction known as "T2" in which two conventional conductors are twisted together; airflow "spoilers"; offset masses; and detuning pendulums. Other devices, showing promise in development programs, include torsional dampers for use on bundled conductors. Although they are not a recent development, interphase spacers have found increasing use and acceptance in recent decades.

4.6.2 Causes of Galloping

Galloping is a large amplitude (several meters), low frequency (fraction of Hz), wind-induced oscillation. In the vast majority of cases, an ice accretion is present on the conductor: this has the effect of modifying the conductor's cross-sectional shape such that it becomes aerodynamically and/or aeroelastically unstable. An aerodynamic instability is just like a negative damping whose origin is coming from aerodynamic lift, drag and moment coefficient.

An aeroelastic instability of flutter type is much more complex. The instability is only possible due to a strong interaction between aerodynamic lift and moment coefficients, wind speed and structural data (frequencies, stiffness, masses, inertia). There exist some coupling (due to fluid action) between different movement (vertical, horizontal, torsion) and some inappropriate design coupled with specific mixture of ice shape and wind speed, may lead to very spectacular amplitude. The overhead lines are particularly sensible of such phenomenon as bundle conductor lines cannot avoid easily to be designed with very close frequency in both three movement (vertical, horizontal, torsion) and this is valid for any number of loops in the modes (even if less true for single span section, considering the first mode)

Aerodynamic properties of conductor+ice may change completely torsional stiffness and frequencies of typical overhead line conductor in the presence of wind and such changes may induce the "inappropriate" structural data which favour galloping. Galloping is NOT a forced oscillation, it is a self-excited phenomenon. It may (and do) occur during constant wind, but not only. There are very tenuous link between galloping, Aeolian vibration (vortex shedding) and wake-induced vibration (in bundle). All these phenomenon have their own physics and their own range of frequencies, galloping covering the smallest one (0.1 to 1 Hz), thus having the "opportunities" to get the larger amplitudes (up and over the sag, some have been recorded at 15 m peak-to-peak).

Last but not least, the effects of non perpendicular wind may not be easily split in two components (perpendicular and longitudinal to the conductor) as fluid behaviour is not so simple.

We may then summarize factors influencing galloping as the following:

- Ice accretion type and shape (eccentricity, weight, aerodynamic properties)
- Wind speed (with limited effects of turbulence and orientation as detailed)

- Conductor self damping (vertical, torsion) in the low frequency range (including end span effects)
- Span lengths (including all spans of a section) and section length
- Longitudinal stiffness (anchoring tower) at fixation point
- Yoke plate assembly (anchoring and suspension tower) (torsional stiffness effect)
- Number of sub-conductors and their arrangement
- Sub-conductor spacing
- Sagging conditions (vertical frequencies effect)
- Spacers (kind of spacer, location, eccentric weight effect, conductor fixation effect)
- Presence of retrofit devices (all kind including interphase methods)
- Position of ice in the presence of wind
- Ratio vertical/torsional frequency for each mode, in the presence of wind.

4.6.3 Simulation of the Cable Response to Ice Galloping

An analytical approach—based on the QST—to simulate ice galloping is described in this paragraph. The structural model is simplified to the case of a single conductor in order to put in evidence the aerodynamic effect. In case of bundles the approach is similar but we need to introduce the lift, moment and drag coefficients of the bundle. The relative subconductors motion is negligible with respect to the bundle motion in vertical, horizontal and torsional directions. The bundle modes of vibration are those at low frequency, lower than that of the subspan first frequency, which is around 1 Hz. At these very low frequencies the subconductors have very small relative motions.

- The cable is modelled by a finite element scheme (Diana et al. 1974, 1990, 1989);
- The aeroelastic forces are introduced with a quasi steady approach Diana et al. (1974), Diana et al. (1990), and Diana (1991) using the aerodynamic coefficient of the cable with ice;
- Aerodynamic coefficients are derived from wind tunnel tests and are measured for different positions of the ice in respect to the wind speed direction.

4.6.3.1 Definition of the Ice Shapes on the Cable and Aerodynamic Coefficients

The assumed ice shape, used for this example, is reported in Fig. 4.79 and the ratio between ice thickness t and cable diameter has been assumed equal to: $t/D = 0.5$, which can be considered the most unfavourable design value.

The aerodynamic coefficients measured on a rough model (0.5–1 mm roughness), as a function of the angular position of the ice, following the experimental sign conventions reported in Fig. 4.80, have been considered for the simulations.

Fig. 4.79 Particular of the cable + rough ice sectional model

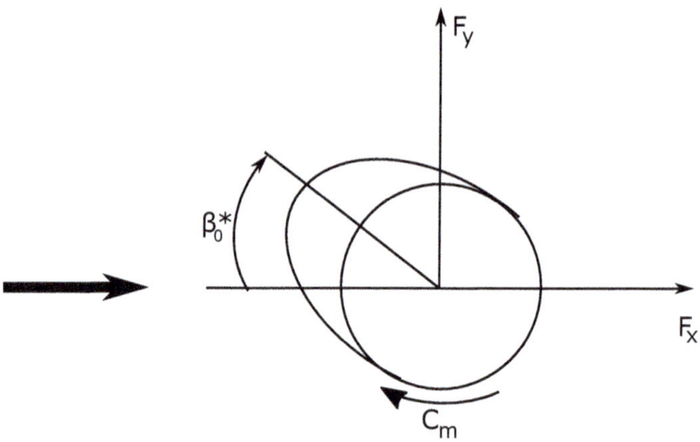

Fig. 4.80 Sign conventions for the ice angular position

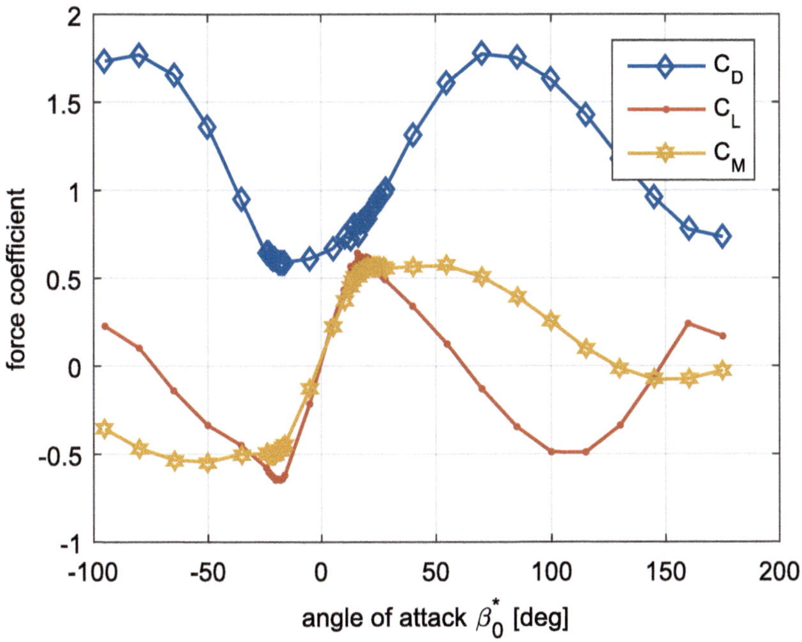

Fig. 4.81 Rough cylinder—aerodynamic coefficients (β_0^* angle, experimental conventions, Fig. 4.80)

Figure 4.81 shows the aerodynamic coefficients measured on the rough cylinder for the ice profile shown in Fig. 4.79, varying the ice angular position. As it can be seen, aerodynamic coefficients variations with negative lift coefficient derivatives for angles in the range (+20°, +100°) and (−20°, −100°) are observed. Their values are an index of unstable dynamic behaviour of the cable with ice, particularly if, as happens in this case, they are coupled with decreasing drag stabilising terms (low values of drag are found in the range (−30°, + 30°).

In general terms, when the slope of the lift coefficient is negative a one degree of freedom instability may occur, with large oscillation amplitudes in the vertical direction and, more precisely, when the derivative of the lift coefficient with respect to the β_0^* angle ($\frac{\partial C_\ell}{\partial \beta_0}$) is greater than C_D (see Chapter about Bridges): in this case the ice shape is like in Fig. 4.82a. When the slope of the lift coefficient is positive, a typical two degrees of freedom flutter problem may occur and this type of galloping is more frequent in bundles: in this case the ice shape can be of the type in Fig. 4.82b.

Fig. 4.82 One degree of freedom and two degrees of freedom instability ice shapes

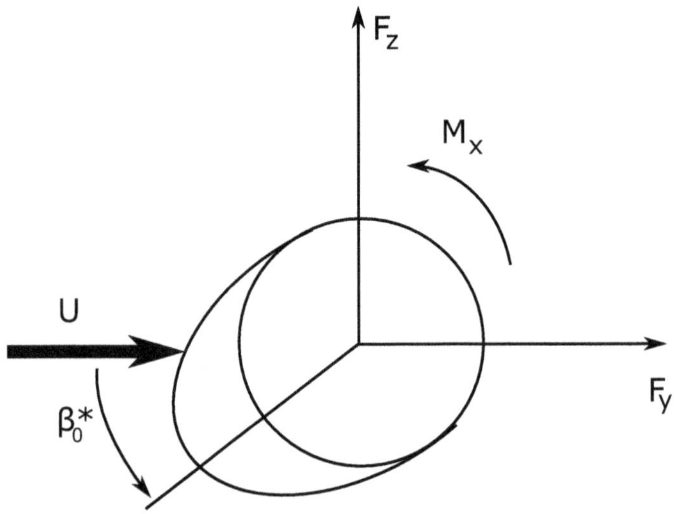

Fig. 4.83 Aerodynamic forces—sign conventions in the numerical simulation model

4.6.3.2 Description of the Numerical Model of the Cable and of the Aeroelastic Forces Model

In the numerical simulation model both structural cable characteristics and wind exciting forces, responsible of the ice galloping phenomena, have to be considered. The cable response to the wind has been obtained integrating the non linear differential equations of motion of the model, introducing, as better displayed in the following, the aeroelastic forces as a function of both the wind speed and the induced displacements and velocities of the cable. The cable was schematised by a finite element model, using finite element "tensioned beam", in which both tension (T) contribution and flexural cable stiffness (EJ and GJ_t) were taken into account: the equation of motion can be defined as following:

$$[M_b]\,\underline{\ddot{X}}_b + [R_b]\,\underline{\dot{X}}_b + [K_b]\,\underline{X}_b = \underline{F}_b$$

The $\underline{F}_b$ aeroelastic forces are introduced using the quasi steady theory, hypothesizing laminar wind, constant in time and space.

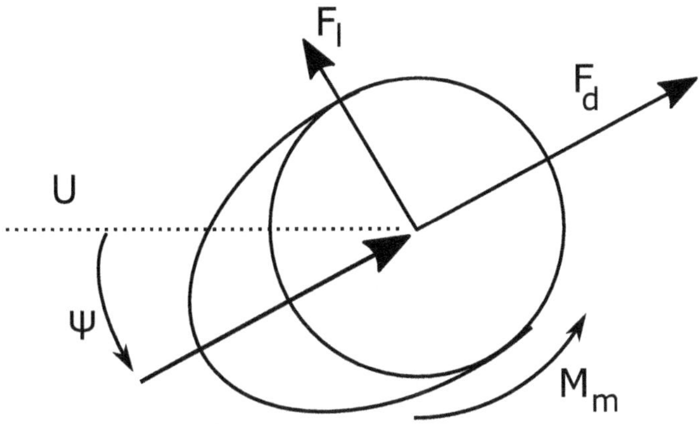

Fig. 4.84 The aeroelastic forces on the cable

The per unit length aerodynamic forces on the cable Fig. 4.83, following the proposed approach, can be defined as:

$$F_y = F_d \cos(\psi) - F_l \sin(\psi) \tag{4.14}$$

$$F_z = F_d \sin(\psi) + F_l \cos(\psi) \tag{4.15}$$

$$M_x = M_m \tag{4.16}$$

where M_m is the aerodynamic moment and F_d ed F_l are the drag and lift components on the cable, whose direction (Fig. 4.84) is, respectively, parallel and perpendicular to the relative wind velocity V_r:

$$F_d = \frac{1}{2}\rho\, V_r^2\, D\, C_d$$

$$F_l = \frac{1}{2}\rho\, V_r^2\, D\, C_l$$

$$M_m = \frac{1}{2}\rho\, V_r^2\, D^2 C_m$$

The relative velocity V_r (Fig. 4.85) depends on the U wind speed and on the cable motion $\dot{y}(t)$ and $\dot{z}(t)$:

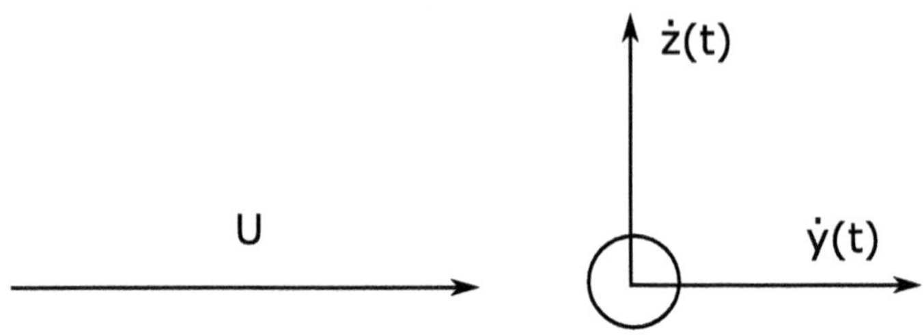

Fig. 4.85 Definition of the relative velocity on the cable

$$V_y = U - \dot{y} \tag{4.17}$$

$$V_z = -\dot{z} + b_1\,\dot{\vartheta} \quad (b_1 = 0) \tag{4.18}$$

$$V_r^{\,2} = V_y^2 + V_z^2 \tag{4.19}$$

$$\psi = a\tan\left(\frac{V_z}{V_y}\right) \tag{4.20}$$

Aerodynamic coefficients:

$$C_d = C_d(\alpha) \tag{4.21}$$

$$C_l = C_l(\alpha) \tag{4.22}$$

$$C_m = C_m(\alpha) \tag{4.23}$$

They depend on the wind angle of attack α of the wind respect to the cable section and α depends on:

- The relative position of the ice profile in respect to the wind direction β_0;
- The cable section rotation θ;
- The angle of the relative wind velocity ψ

through the relationship:

$$a = \vartheta + \beta_o - \psi$$

Taking into account the fluid-elastic forces terms, the equations of motion of the system become:

$$[M_b]\,\underline{\ddot{X}}_b + [R_b]\,\underline{\dot{X}}_b + [K_b]\,\underline{X}_b = \underline{F}_b\left(\underline{\dot{X}}_b, \underline{X}_b, U\right)$$

where the cable forces $\underline{F}_b$ are function both of the wind speed U and of the cable motion.

The dependence of the forces on the cable displacements $\underline{X}_b$ are due to the cable rotation of the cable with ice which cause drag and lift force variations; the dependence of the forces on the cable velocities $\dot{\underline{X}}_b$ modifies the relative wind velocity and the angle of attack and therefore again the aerodynamic forces. These relationships are non linear: for this reason, the equations of motion of the cable with wind are non linear: these equations are integrated using a modified step by step numerical integration method.

References

Belloli M, Diana G, Resta F, Muggiasca S (2006) A numerical model to reproduce vortex induced vibrations of a circular cylinder. In: American Society of Mechanical Engineers, pressure vessels and piping division (publication) PVP, vol 2006

Belloli M, Melzi S, Negrini S, Squicciarini G (2010) Numerical analysis of the dynamic response of a 5-conductor expanded bundle subjected to turbulent wind. IEEE Trans Power Deliv 25(4):3105–3112

Buresti G (1981) The effect of surface roughness on the flow regime around circular cylinders. J Wind Eng Ind Aerodyn 8(1–2):105–114

Carberry J, Sheridan J, Rockwell D (2001) Forces and wake modes of an oscillating cylinder. J Fluids Struct 15(3–4):523–532

Carberry J, Govardhan R, Sheridan J, Rockwell D, Williamson C (2004) Wake states and response branches of forced and freely oscillating cylinders. Eur J Mech B/Fluids 23(1):89–97

CIGRE (1979) Guide on conductor self-damping measurements. In: ELECTRA, vol 62 (identical to IEEE 563–1978 Standard)

CIGRE (1998) CIGRE SC22 WG11 TF1. Modelling of aeolian vibrations of single conductors: assessment of the technology. ELECTRA 198:53–69

CIGRE (2005a) CIGRE WG B2. 11. TF1. Modeling of aeolian vibrations of a single conductor plus damper-assessment of technology. Tech Rep ELECTRA, 223

CIGRE (2005b) CIGRE WG B2.11.04-overhead conductor safe design tension with respect to aeolian vibrations-technical brochure no. 273. Technical report CIGRE Paris

CIGRE (2005c) Modeling of aeolian vibrations of a single conductor plus damper-assessment of technology. ELECTRA, vol 223. Technical report, CIGRE WG B2. 11. TF1

Claren R, Diana G (1969a) Dynamic strain distribution on loaded stranded cables. IEEE Trans Power Appar Syst 88:1678–1690

Claren R, Diana G (1969b) Mathematical analysis of transmission line vibration. IEEE Trans Power Appar Syst PAS-88:1741–1771

Curami A, Diana G, Riva R, Di Giacomo G, Nicolini P (1977) Wake-induced oscillations in bundle systems. Part I-finite element method. analytical and experimental results. IEEE Paper A77:1–218

Diana G (1991) Parameter sensitivity analysis of bundled conductors to ice galloping. In: CIGRE S 33–91, symposium compacting overhead transmission lines, Leningrad URSS, N.200-09

Diana G, Cheli F (2010) Dinamica dei sistemi meccanici. Polipress

Diana G, Falco M (1966) On the forces transmitted to a vibrating cylinder by a blowing fluid-experimental study and analysis of the phenomenon. Meccanica 6(1):9–22

Diana G, Gasparetto M (1972) Energy method for computing the amplitude of vibration of conductor bundles due to wake effect. In: L'Energia Elettrica

Diana G, Giordana F (1972) Sulle vibrazioni di un cilindro in scia di un altro analisi del fenomeno con modello analitico. L'Energia Elettrica 7:448–57

Diana G, Giordana A, Fand M, Nicolini P (1973) Experimental research on subspan galloping. In: IEEE conference paper C74, vol 493, p 3

Diana G, Gasparetto M, Giacomo GD, Nicolini P (1974) Analytical method for computing subspan oscillation analytical and experimental results. In: IEEE summer meeting, Anaheim California (USA)

Diana G, Giacomo GD, Claren R (1979) An approach to vortex shedding under turbulent air-flow on a single vibrating cylinder. In: IEEE SM paper A, vol 79, p 529

Diana G, Falco M, Curami A, Manenti A (1986) A method to define the efficiency of damping devices for single and bundled conductors of EHV and UHV lines. In: IEEE/PES winter meeting paper 86 WM 162–2

Diana G, Cheli F, Collina A, Manenti A, Tavano F (1989) Sensitivity analysis of three bundle system to ice galloping. CIGRE SC22-WG11 TF on Galloping. Opatija, Yugoslavia, pp 89–19

Diana G, Cheli F, Manenti A, Nicolini P, Tavano F (1990) Oscillation of bundle conductors in overhead lines due to turbulent wind. IEEE Trans Power Deliv 5(4):1910–1922

Diana G, Cheli F, Manenti A, Resta F, Bocciolone M (1999) The aeroelatic behaviour of the OHTL expanded bundles. In: Proceedings of the 3rd international symposium on cable dynamics, Trondheim (Norway)

Diana G, Falco M, Cigada A, Manenti A (2000) On the measurement of over head transmission lines conductor self-damping. IEEE Trans Power Delivery 15(1):285–292

Diana G, Resta F, Belloli M, Muggiasca S (2005) Experimental analysis on vortex induced vibration of a long flexible cylinder. In: Proceedings of the sixth international symposium on cable dynamics. Charleston, SC

Diana G, Zasso A, Belloli M, Giappino S, Muggiasca S (2006a) Vibrazioni indotte dal fenomeno di distacco di vortici su un cilindro a sezione circolare. In: IX Convegno Nazionale di Ingegneria del Vento IN-VENTO

Diana G, Zasso A, Vigevano L, Auteri F, Formaggia L, Nobile F (2006b) Flow-structure interaction of the oscillating circular cylinder in the lock-in region: computational versus experimental approach comparison. In: Proceedings of CWE2006 4th international symposium on computational wind engineering. Yokohama, Japan

Diana G, Belloli M, Giappino S, Muggiasca S, Zasso A (2007) The effect of Reynolds number on aeolian vibrations. In: Seventh international symposium on cable dynamics, proceedings of the seventh international symposium on cable dynamics, iNVITED LECTURE

Diana G, Belloli M, Giappino S, Muggiasca S (2008) Vortex induced vibrations at high Reynolds numbers. In: VI international colloquium on bluff bodies aerodynamics & applications BBAA VI

EPRI (2009) Transmission line reference book: wind induced conductor motion, II Edn. EPRI N. 1018554, Palo Alto, USA

Govardhan R, Williamson C (2000) Modes of vortex formation and frequency response of a freely vibrating cylinder. J Fluid Mech 420:85–130

Govardhan R, Williamson C (2004) Critical mass in vortex-induced vibration of a cylinder. Eur J Mech B/Fluids 23(1):17–27

Hard AR, Holben RD (1967) Application of the vibration decay test to transmission line conductors. IEEE Trans Power Appar Syst PAS-86(2):189–199

Hardy C, Van Dyke P (1995) Field observations on wind-induced conductor motions. J Fluid Struct 9(1):43–60

Hardy C, Leblond A, Goudreau S, Cloutier L (1999) Application of contact mechanics to cable self-damping. In: Third international symposium on cable dynamics. Trondheim (Norway), pp 1–6

IEEE (1978) Std. 563. Guide on conductor self-damping measurements. Technical report. IEEE Power Engineering Society

IEEE (1993) IEEE Std. 664. Guide for the laboratory measurements of the power dissipation characteristics of aeolian vibration damper for single conductor (revision of IEEE Std. 664, 1980). Technical report. IEEE Power Engineering Society

Ikegami R (1971) Structural analysis high voltage power transmission systems. In: Boeing Co., Seattle BPA Ctr 14–03-1362N

Kalombo R, Miranda T, Rocha P, Ferreira J, da Silva C, Araújo J (2018) Fatigue performance of overhead conductors tested under the same value of h/w parameter. Procedia Eng 213:346–358

Ko RG (1973) Theoretical investigation for hydro-quebec into the aerodynamic stability of bundled power line conductors, part I: two dimensional stability analysis of a conductor in the wake of a fixed conductor. Technical report. NRC (Canada) Lan Tech Rep LA-122

Morse PM (1948) Vibration and sound, vol IV. McGraw-Hill, New York

Papailiou KO (2000) Discussions and closure by authors. IEEE Trans Power Delivery 15(4):1329–1331

Rawlins CB (1958) Recent developments in conductor vibration research. In: Alcoa Technical Paper No. 13

Rawlins CB (1988) An efficient method for measuring dissipation by dampers in laboratory spans. IEEE Trans Power Delivery 3(3):1146–1155

Rawlins CB (1998) Model of power imparted to a vibrating conductor by turbulent wind. In: ALCOA Technical Note No. 31

Richards DJW (1963) Aerodynamic properties of the severn crossing conductor. In: Proceedings of the conference held at the National Physical Laboratory, Teddington, UK, pp 26–28

Simpson A (1970) Stability of subconductors of smooth circular cross-section. Proc IEEE 117(4):741–750

Simpson A (1971) On the flutter of a smooth circular cylinder in a wake. Aeronaut Q 22(1):25–41

Slethei TO, Huse J (1965) Conductor vibration theoretical and experimental investigations on a laboratory test span. Proc IEEE 112:1173–1179

Snegovskiy D (2010) Wake-induced oscillations in cable structures: finite element approach. PhD thesis, ULG, Liege, Belgium

Sturm RG (1936) Vibrations of cables and dampers-I and II. Electr Eng 55:455–466

Tompkins JS, Merrill LL, Jones BL (1956) Quantitative relationships in conductor vibration using rigid models. In: IEEE transactions, pp 879–894

Tsui YT (1975) Two dimensional stability analysis of a circular conductor in the wake of another. IEEE Paper A75:576–579

Wardlaw RL, Cooper KR (1973) A wind tunnel investigation of the steady aerodynamic forces on smooth and stranded twin bundled power conductors. Technical report, Aluminum Company of America NRC (Canada) Lab Tech Rep LA-117

Wardlaw RL, Cooper KR (1974) Wind tunnel and analytical investigations into the aeroelastic behaviour of bundled conductors. IEEE Trans Paper T74:367–368

Zanelli F, Mauri M, Castelli-Dezza F, Tarsitano D, Manenti A, Diana G (2022) Analysis of wind-induced vibrations on HVTL conductors using wireless sensors. Sensors 22(21):8165–8193

Zasso A, Manenti A, Belloli M, Giappino S, Muggiasca S (2005) Energy input by the flow on a vibrating smooth circular cylinder in cross flow at re=5x104. In: Sixth international symposium on cable dynamics. A.I.M., Liege (Belgium)

Zasso A, Belloli M, Giappino S, Muggiasca S (2006) On the pressure and force field on a circular cylinder oscillating in the lock- in region at sub-critical Reynolds number. In: 2006 ASME pressure vessels and piping conference, vol 9. ASME Pressure Vessels and Piping Division; Conference code: 70069, pp 919–927

Zasso A, Belloli M, Giappino S, Muggiasca S (2008) Pressure field analysis on oscillating circular cylinder. J Fluids Struct 24(5):628–650

Zdravkovich M (1997) Flow around circular cylinders, vol. 1: fundamentals. Oxford University Press

Zdravkovich M (2003) Flow around circular cylinders, vol 2: applications. Oxford University Press

Chapter 5
Wind Effects on Bridges

5.1 Introduction

Wind blowing over a bridge induces both static and dynamic actions, as is the case with all structures.

For short bridges (approximately 100 m or less), designed primarily to withstand loads from road and railway traffic, wind actions generally do not pose significant challenges. However, specific phenomena like vortex shedding may cause localized issues.

As the span length increases and depending on the bridge's typology, the structure becomes progressively more flexible, with natural frequencies decreasing accordingly. In such cases, wind-induced bridge motion becomes increasingly significant. The motion of bridge components excited by wind can, in turn, alter aerodynamic forces, potentially leading to instability. This phenomenon will be explored further in subsequent sections. For this reason, the focus of the following chapters will be on long-span bridges, where wind action plays a critical role.

Long bridges typically fall into two main categories: cable-stayed bridges and suspension bridges. By contrast, arch bridges, being inherently more massive, are generally less susceptible to wind-related problems. The aforementioned considerations, however, do not apply to pedestrian bridges, which can be highly flexible even with relatively short spans.

To fully understand the challenges posed by wind actions on bridges, it is essential to define the aerodynamic forces acting on the various bridge components, including towers, decks, and cables. This chapter will primarily address the aerodynamic forces induced on the bridge deck by its motion and by incoming wind turbulence. Notably, the deck is the most wind-sensitive component of a bridge.

To analyze these forces, wind tunnel testing is often employed, complemented by numerical simulations that model the bridge's response to turbulent wind conditions.

D. Rocchi et al., *Wind Engineering*, Research Topics in Wind Energy 9,
https://doi.org/10.1007/978-3-032-17423-9_5

5.2 Overview of Long Span Bridges

Long bridges are defined as those with a main span on the order of 1000 m. When the span exceeds 1500 m, the bridge is classified as a very long bridge.

The longest bridge in the world is the Akashi Kaikyō Bridge, with a main span of 2000 m. The longest bridge currently at the executive design stage is the Messina Bridge, with a planned main span of 3300 m.

For spans up to approximately 1100 m, cable-stayed bridges are feasible. Beyond this length, suspension bridges become the preferred solution. The longest cable-stayed bridge is the Russky Bridge, completed in 2012 across the Eastern Bosphorus Strait near Vladivostok, with a main span of 1104 m. An example of a hybrid solution is the Third Bosphorus Bridge, a combination of cable-stayed and suspension design, spanning 1410 m over the Bosphorus Strait.

The key characteristics of these two bridge types (suspension and cable-stayed) will be briefly analyzed in the following sections.

5.2.1 Cable Stayed Bridges

From a structural perspective, the dead load of the deck and the traffic load are transferred from the deck to the towers via the stay cables, as illustrated in Fig. 5.1.

The deck must withstand significant longitudinal forces resulting from the horizontal components transmitted by the cables. The load on the deck increases progressively from mid-span towards the towers. As the span length increases, the load near the towers grows substantially, necessitating a larger deck cross-section, which in turn increases the overall weight of the bridge. This also leads to a need for taller towers.

The length of the stay cables cannot increase indefinitely, as excessive lengths result in pronounced catenary effects, which adversely impact structural performance. For these reasons, the maximum main span of cable-stayed bridges is typically limited to around 1100 m.

In such bridges, deck stiffness is a critical factor influencing both the static and dynamic performance of the structure. For example, the first natural frequencies of a cable-stayed bridge are shown in Fig. 5.2.

The ratio between the torsional and vertical natural frequencies is primarily influenced by the torsional stiffness of the deck. Generally, this stiffness is sufficiently high to ensure that the ratio exceeds 2. Maintaining this ratio is crucial for controlling the bridge's flutter stability, a topic that will be analyzed in detail in subsequent sections.

Fig. 5.1 Force flow of a cable stayed bridge

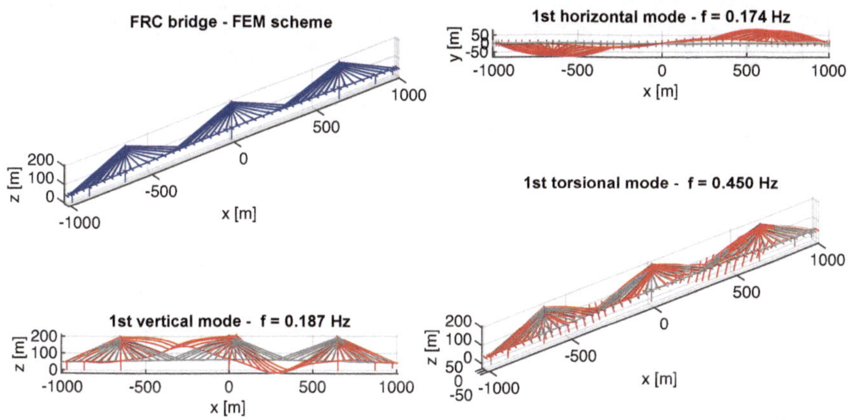

Fig. 5.2 First natural frequencies of a cable stayed bridge (Forth Replacement Crossing bridge)

5.2.2 Suspension Bridges

A suspension bridge is a fundamentally simple structure, characterized by a *serial type* static behavior in load distribution. In simple terms, the loads are applied to the deck, transferred to the main cables through the hangers, and ultimately carried to the ground. Horizontal components are absorbed by the anchor blocks, while vertical components are transmitted mainly through the towers, as illustrated in Fig. 5.3.

The main parameters defining a suspension bridge are the span length l, the tensile load on the cables T, the sag of the main cables s, and the tower height h.

Fig. 5.3 Suspension bridge scheme with force flow. Author: Rich Niewiroski Jr. Adapted from: https://commons.wikimedia.org/wiki/File:GoldenGateBridge-001.jpg

When subjected to live loads, such as vehicular traffic, a suspension bridge exhibits interaction between its structural elements, characteristic of redundant systems. The forces are distributed based on the relative stiffness of the deck and the main cables. The deck stiffness primarily depends on flexure, whereas the main cable stiffness is geometric, determined by the tensile force, which itself is dependent on the bridge's weight. This geometric cable stiffness is often referred to as *gravity stiffness*.

The balance between the deck's bending stiffness and the main cable's tensile stiffness is a critical factor governing the behavior of suspension bridges. With a constant tensile load, the cable stiffness decreases as the span length increases. Simultaneously, the deck stiffness, assuming a constant cross-section, decreases at a much faster rate—approximately proportional to the cube of the span.

For shorter spans, typical of nineteenth century suspension bridges, the stiffness of the deck and cables is comparable, enabling both to share significant portions of the applied loads. This concept led to the role of the deck as a *stiffening girder*, collaborating with the main cable to carry live loads.

In contrast, as spans increase, the decrease in cable stiffness due to length is partially compensated by the increased tensile force, which necessitates larger cable dimensions. However, the deck stiffness decreases rapidly with span, and its stiffening role for global loads becomes negligible. In long-span suspension bridges, the deck primarily acts as a collector of live loads, distributing them to the hangers, while the main cables carry nearly all global loads.

For large suspension spans, lateral wind loads on the deck are predominantly transferred to the supporting towers via the *pendulum effect* of inclined hangers, rather than through the deck's lateral bending stiffness. These wind loads are carried to the tower tops, significantly influencing the design of tower cross-bracing and lateral supports. Similarly, eccentric live loads inducing torsional moments around the bridge axis are absorbed mainly by the cables. The distance between the main cable suspension planes becomes a decisive parameter for the bridge's torsional (or rotational) stiffness, as it provides the lever arm needed to translate deck rotation into vertical cable forces and stiffness.

The dominant role of cable stiffness for static behavior extends to the dynamic response of suspension bridges, interacting with their inertial properties, including the distribution and magnitude of mass. Figures 5.4, 5.5, and 5.6 illustrate the typical first mode frequencies for a suspension bridge, based on the analysis of the Messina Bridge. These modes, which may not necessarily occur in the order listed, include:

- the first two **Lateral modes**:
  - (a) Symmetric mode: Associated with a lateral "pendulum" motion of the deck and cables, dominated by the tensile load of the cables. This mode often corresponds to the first absolute structural mode of a long suspension bridge (i.e., the mode with the lowest natural frequency).
  - (b) Antisymmetric mode: Also characterized by lateral pendulum motion, dominated by cable tension.
- the first two **Vertical modes**:
  - (c) Symmetric mode: Involves both cable tensile and axial stiffness.
  - (d) Antisymmetric mode: Dominated by the tensile load of the cables. Depending on the cable configuration, either the symmetric or antisymmetric mode may have the lower frequency, usually corresponding to the second absolute bridge mode. These are referred to as "vertical modes" to avoid the misleading term "bending modes," as cable stiffness dominates over deck stiffness.
- the first two **Rotational modes**:
  - (e) Symmetric mode: Exhibits greater participation of cable axial stiffness.
  - (f) Antisymmetric mode: Dominated by the tensile load of the cables. These modes, typically higher in frequency than lateral or vertical modes, are sometimes referred to as "torsional modes" due to the form of deck deformation. However, the term "rotational modes" is preferred here to emphasize the dominance of cable stiffness over deck stiffness.

The stiffness of the main cables plays a dominant role in determining the dynamic properties of any suspension bridge, as is also true for its static behavior. However, for

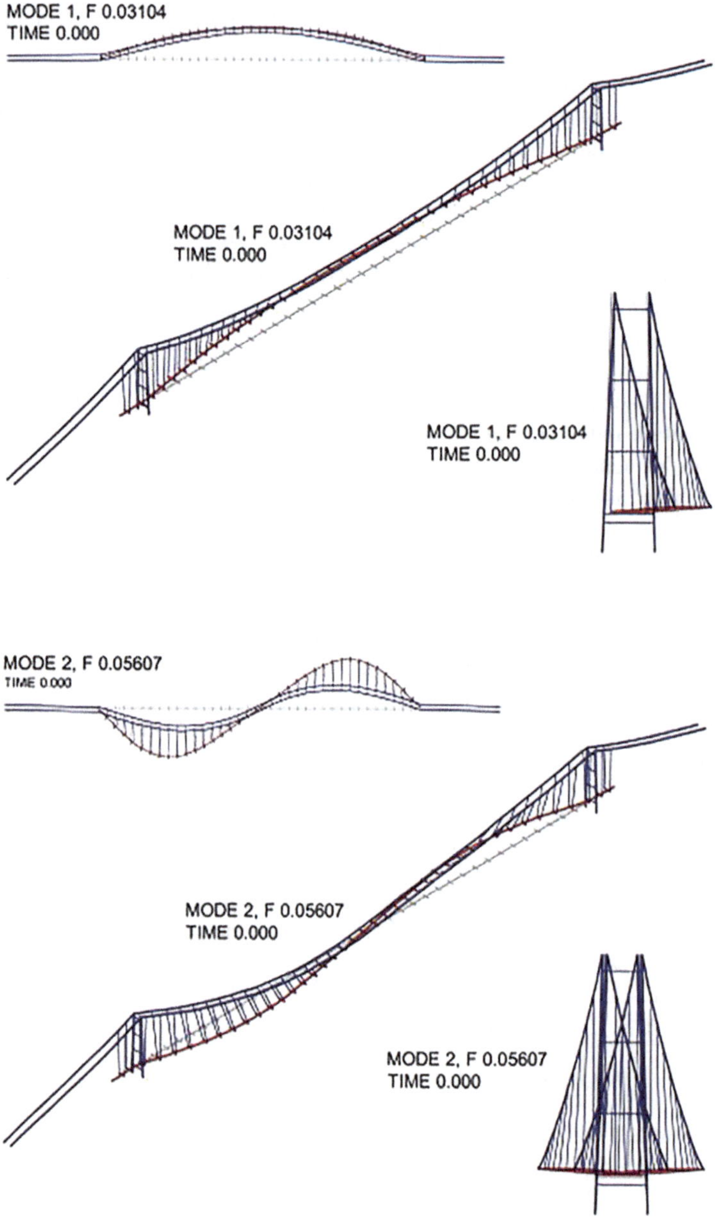

Fig. 5.4 Suspension bridge first lateral modes

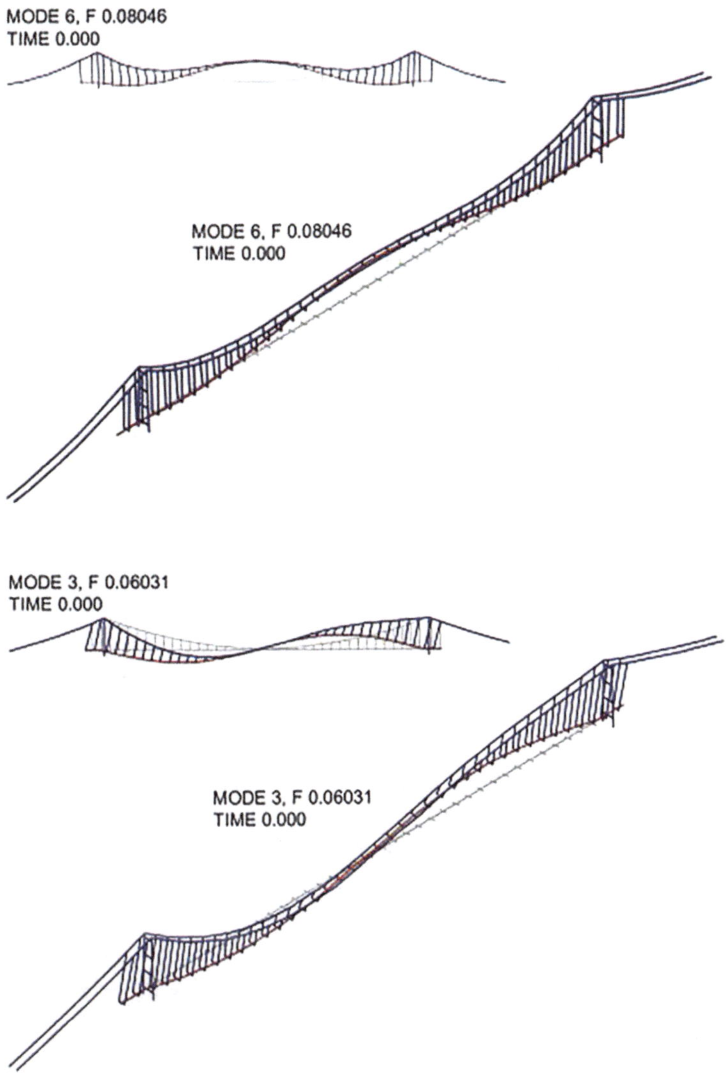

Fig. 5.5 Suspension bridge first vertical modes

very long spans, the situation changes: the main cables become the largest elements in terms of weight and mass, and as a result, they dominate the distribution of inertia. In other words, the mode shapes and frequencies of very large span suspension bridges increasingly resemble those of stand-alone cables, with the influence of the other structural elements becoming progressively smaller.

This is particularly significant for the vertical and rotational modes, whose frequencies tend to converge as the span length increases. In the theoretical extreme, if

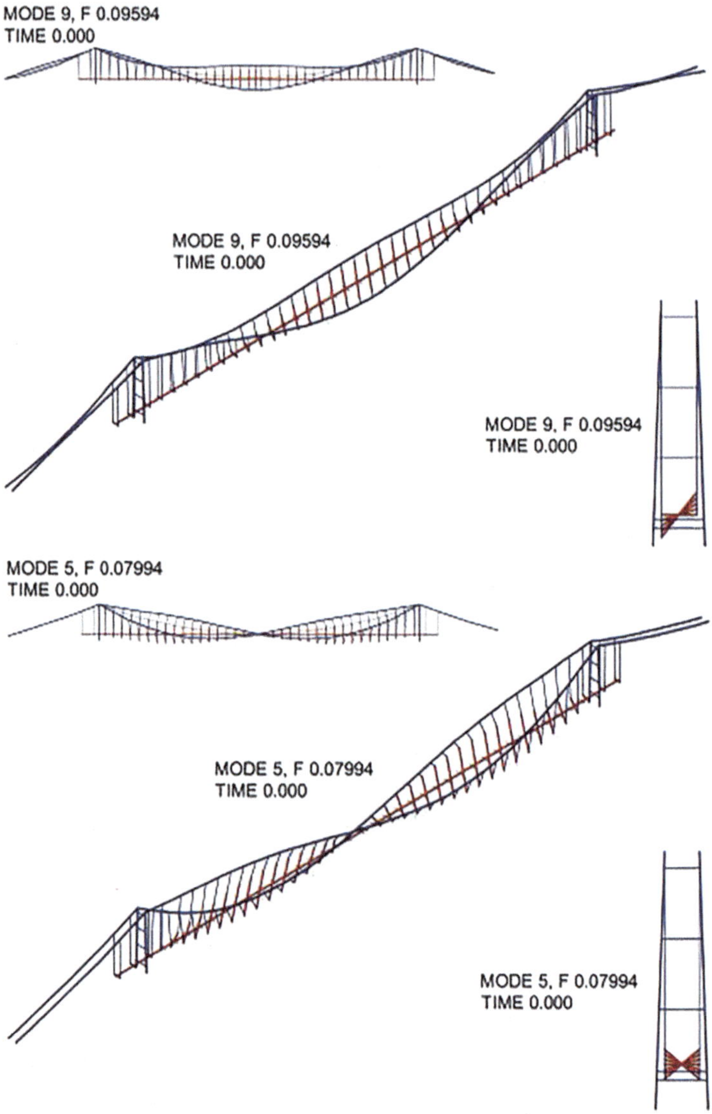

Fig. 5.6 Suspension bridge first rotational modes

the two main cables were to exist as stand-alone, perfectly restrained elements, the two modes would have identical frequencies, corresponding to in-phase and out-of-phase oscillations of the two identical, perfectly restrained cables.

For a given configuration and sag-to-span ratio, several factors contribute to maintaining a frequency separation between the vertical and rotational modes, although their relative importance varies. Among these factors, the deck's torsional stiffness and the ratio of the deck's moment of inertia to its mass play the most critical roles in ensuring that the torsional frequency remains higher than the corresponding vertical frequency. Figure 5.7 shows the ratio between torsional and vertical frequencies for different suspension bridges.

5.2.2.1 Historical Overview of Suspension Bridges

The development of suspension bridges was greatly accelerated in the early nineteenth century by advancements in material properties and technological skills. A significant milestone was the completion of the George Washington Bridge in 1931, which had a span length of 1067 m, making it one of the longest bridges of its time. Just six years later, the Golden Gate Bridge was completed in San Francisco, with a span length of 1280 m, surpassing the George Washington Bridge.

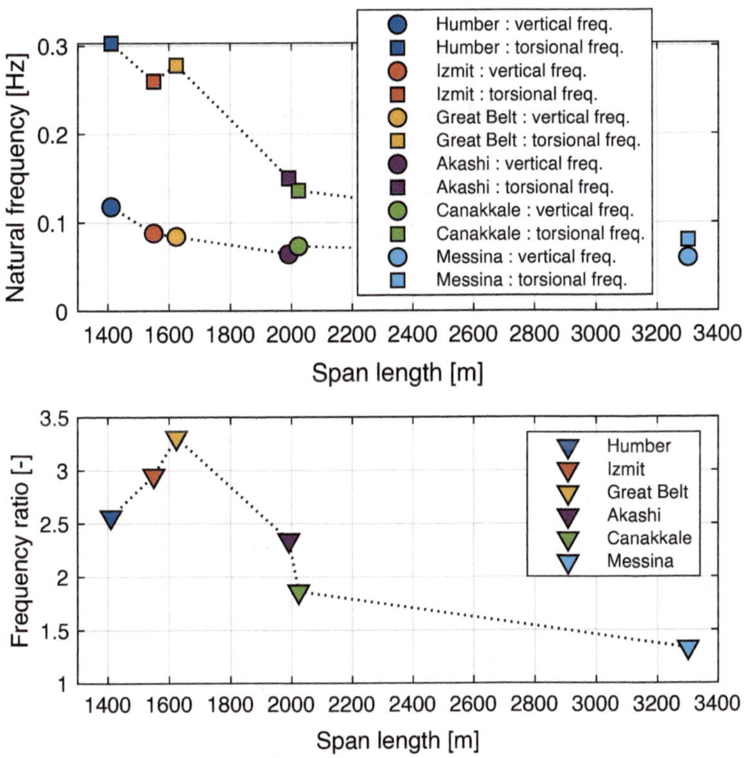

Fig. 5.7 Suspension bridge: reduction of bending and torsional natural frequencies with span length, and frequency ratio between torsional and vertical frequencies

It wasn't until the end of the nineteenth century that another major increase in span length occurred. In the interim, several suspension bridges were built worldwide, each with spans ranging from 1200 to 1400 m.

The Akashi Kaikyō Bridge marked another milestone, taking the span length to 1991 m. Following the completion of the Akashi Bridge, several other suspension bridges were inaugurated, with span lengths in the range of 1400–1650 m, and knowledge regarding this type of bridge construction became increasingly sophisticated.

A third major leap in span length is represented by the Messina Strait Bridge, which has been fully designed but not yet constructed. Once completed, it will become the longest suspension bridge in the world, with a span length 1.5 times greater than that of the Akashi Bridge.

Recently, the Çanakkale 1915 Bridge in Turkey (opened in 2022), with a main span of 2030 m, surpassed the Akashi Kaikyō Bridge as the world's longest suspension bridge. The Çanakkale 1915 Bridge feature a deck width of 36 m, supporting three traffic lanes in each direction.

Figure 5.8 summarizes the key milestones in suspension bridge construction.

5.3 Wind Induced Problems on Bridges

In general, we classify long bridges as those with a main span on the order of 1000 m and very long bridges as those with a main span exceeding 1500 m, as previously mentioned. As the span length increases, the challenges associated with wind excitation are generally exacerbated.

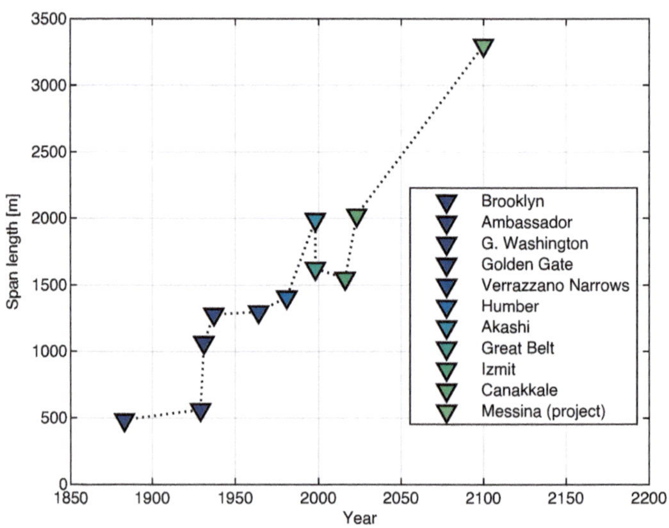

Fig. 5.8 Span length over the last decades

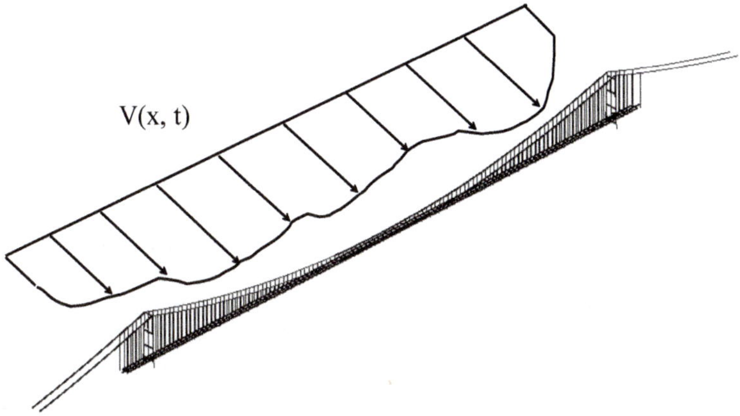

Fig. 5.9 Incoming wind profile on a suspension bridge

To better understand the issues related to wind excitation, we recall the structure of wind, as described in detail in Chap. 2. Referring to Fig. 5.9, the wind acting on the bridge is a function of both time and space. The bridge deck is the most sensitive component to wind action.

Both the static and dynamic responses of the bridge to wind loads are primarily determined by the deck's aerodynamic properties. Referring to a generic bridge deck section, the aerodynamic forces acting on the section as a function of the wind are illustrated in Fig. 5.10.

The aerodynamic forces can be expressed as follows:

$$F_L = \frac{1}{2}\rho V_{rel}^2 S C_L(\alpha) \tag{5.1}$$

$$F_D = \frac{1}{2}\rho V_{rel}^2 S C_D(\alpha) \tag{5.2}$$

$$M = \frac{1}{2}\rho V_{rel}^2 S B C_M(\alpha) \tag{5.3}$$

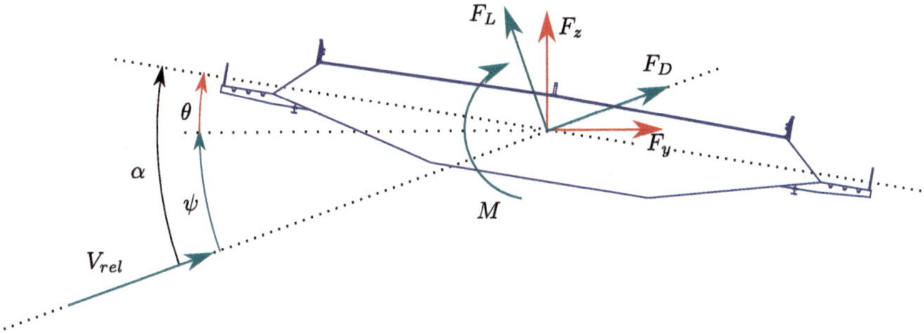

Fig. 5.10 Aerodynamic forces on a deck section

where ρ is the air density, V_{rel} is the relative air velocity, S is a reference body area, typically the deck's upper surface, and B is a reference body dimension, usually the deck's width. The coefficients $C_D(\alpha)$, $C_L(\alpha)$, and $C_M(\alpha)$ represent the Drag, Lift, and Moment coefficients, respectively, which are functions of the angle of attack α of the relative wind direction with respect to a reference axis.

For reference, the C_L, C_D, and C_M coefficients for various deck sections are shown in Fig. 5.11.

Returning to the issues caused by wind action on the bridge, these can be classified into two categories: static and dynamic.

Static problems are associated with the static loads applied to the bridge due to the average wind speed. These loads on the deck depend on the aerodynamic coefficients and the angle of attack. To fully define the overall loads on the bridge, the aerodynamic coefficients of the towers and cables must also be considered.

Dynamic problems, on the other hand, are associated with the turbulence of the incoming wind and the aerodynamic forces induced by the bridge's motion.

5.3.1 Static Problem

As previously mentioned, the average wind generates a static load applied to all parts of the bridge: the deck, the towers, and the cables.

For very long bridges, the load on the deck is the most significant. This load is transferred through the hangers to the main cables and then to the tops of the towers, resulting in a very high bending moment that heavily influences both the tower design and the overall bridge design.

From this perspective, it is essential to minimize the deck drag to reduce the load on the tops of the towers. This tower load is proportional to the cable tension and the angle β (in the horizontal plane) between the vertical plane containing the main cable at the top of the tower and the plane connecting the two towers (Fig. 5.13). The magnitude of this angle, for a given span length, is related to the horizontal deflection of the bridge deck and cables at mid-span, as illustrated in Fig. 5.14.

To better understand this issue, we can compare the maximum static horizontal deflection of the Akashi Bridge with that of the Messina Bridge. Figure 5.15 shows the mid-span deflections for a mean wind speed of 60 m/s for both bridges. These values were computed and validated through wind tunnel tests on aeroelastic models of the two bridges. The maximum deflection for the Messina Bridge is approximately 10 m, compared to about 30 m for the Akashi Bridge. These differences are primarily attributed to the larger deck drag of the Akashi Bridge relative to that of the Messina Bridge (see Fig. 5.11).

In summary, this analysis highlights that the deck drag coefficient is a critical parameter that must be minimized for very long suspension bridges.

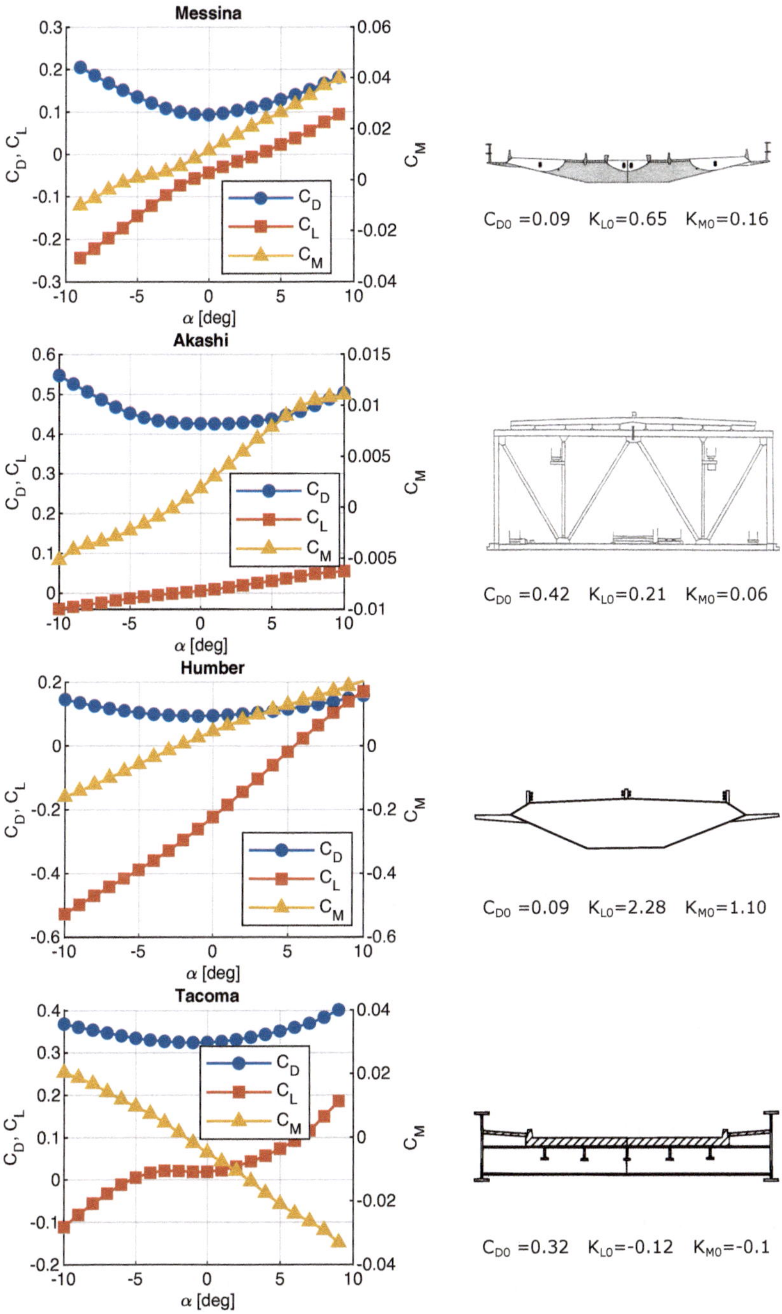

Fig. 5.11 Aerodynamic force coefficients for different deck sections

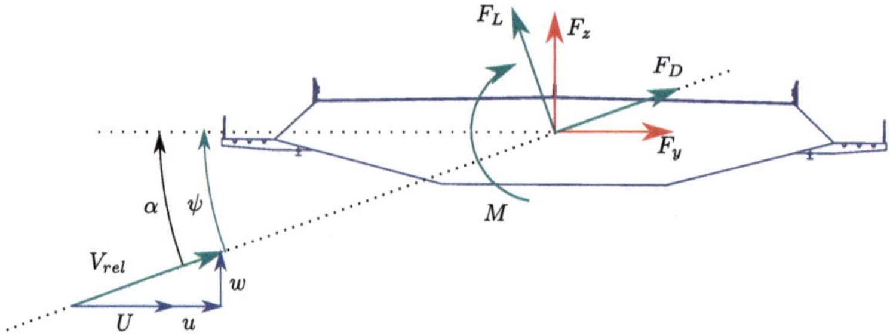

Fig. 5.12 Turbulent wind components effect

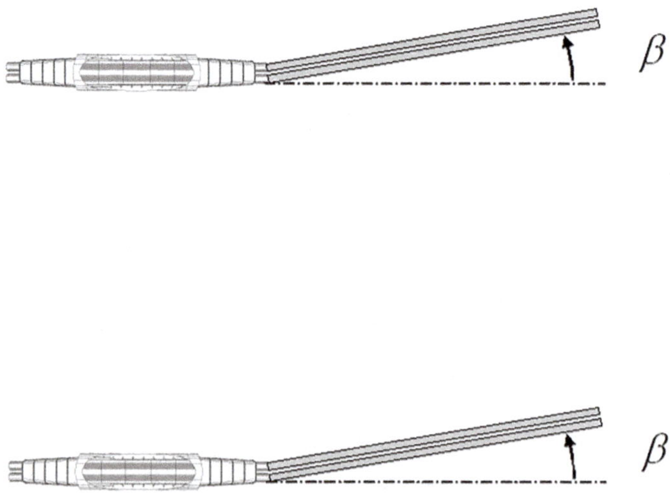

Fig. 5.13 Lateral deflection of the cables. Tower top plan view showing main cable deflection in the horizontal plane and definition of the β angle

5.3.2 *Dynamic Problems*

As already said, the dynamic problems are related to stability, to buffeting, and to vortex induced vibrations.

5.3.2.1 Buffeting

More in detail, if reference is made to a deck section, the wind has a vertical ($w(t)$) and a horizontal ($u(t)$) turbulent component that are added to the average wind, as shown in Fig. 5.12.

The wind angle of attack $\alpha(t)$, is a function of the turbulent wind vertical and horizontal components and changes with time. As a consequence, the aerodynamic

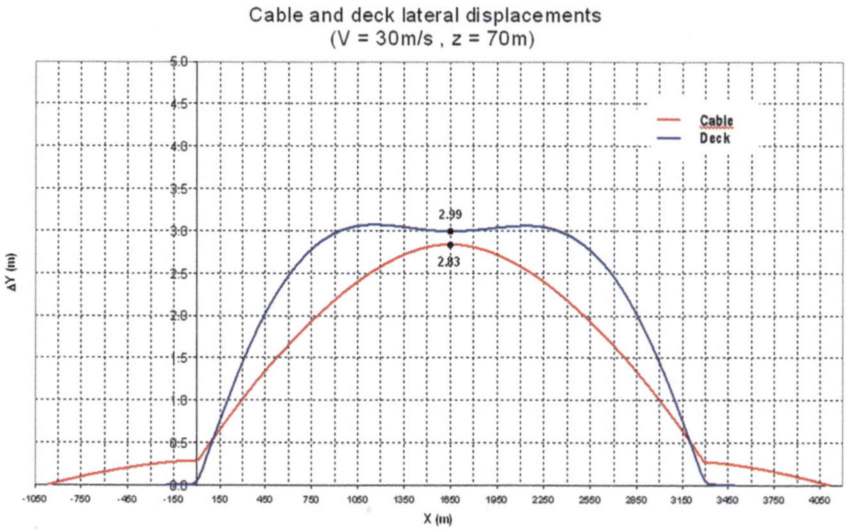

Fig. 5.14 Lateral deflection of the main cables and deck. Plan view showing main cable deflection in the horizontal plane. Note lateral deflection of tower tops

Fig. 5.15 Comparison of Akashi (left) and Messina (right) horizontal deflection due to wind in the wind tunnel. Equivalent wind speed = 60 m/s

lift, drag and moment forces are changing with the angle of attack as it is clearly shown in Fig. 5.12, showing the lift, drag and moment coefficients as a function of the wind angle of attack. The aerodynamic forces variation produces a bridge motion induced by turbulence. This type of motion is called buffeting and is a problem always exhibited by any structure subjected to the wind action.

5.3.2.2 Stability

If the deck or any part of the bridge is moving with a given velocity in the wind flow, the forces applied to the body are functions of the relative velocity of the wind with respect to the body and, in the simplified approach, the aeroelastic forces can be computed introducing V_{rel} instead of U.

If for instance a deck is moving vertically with a given velocity $\dot{z}$ as in Fig. 5.16, being subject to a horizontal wind of velocity U, the relative velocity that is that one seen by an observer moving with the deck is V_{rel} with an angle of attack α, function of $\dot{z}$. If the angle of attack is function of $\dot{z}$, also the Drag, Lift and Moment forces will be functions of $\dot{z}$ or of the motion of the deck.

If the deck is for any reason subject to a rotation, as shown in Fig. 5.17, the rotation θ of the deck will change the angle of attack α, being $\alpha = \theta$. So we can more generally say that a motion of the deck will produce a variation of the angle of attack and making reference to the diagrams representing $C_D(\alpha)$, $C_L(\alpha)$ and $C_M(\alpha)$ will produce as a consequence a variation of the aerodynamic forces function of the motion.

For this reason, we can summarize that the aerodynamic forces are functions not only of the incoming wind, but also of the motion of the deck. In other words the aerodynamic forces are motion dependent. As a function of the shape of the deck, if these aerodynamic forces produced by the motion act in favour of the motion (or introduce energy) the motion is amplified and the bridge becomes unstable.

Different kinds of instability can occur: one degree of freedom instability in the vertical or torsional mode, two degrees of freedom instability, resulting from the coupling of vertical and torsional motions. This last type of instability is also called flutter instability.

5.3.2.3 Vortex Induced Vibrations

Vortex-induced vibrations (VIV) may affect any structural element of bridges, namely deck, towers, and cables. Indeed, having a bluff section, these elements are prone to vortex shedding. When the vortex shedding frequency, which is proportional to the wind speed, becomes equal to one of the natural frequencies involving the motion of one of these parts, vibrations can be excited with vibration amplitudes that depend on the structural damping levels, and on the amplitude of vibration itself. In general VIV are to be avoided for fatigue problems and for traffic safety (in case of deck motion).

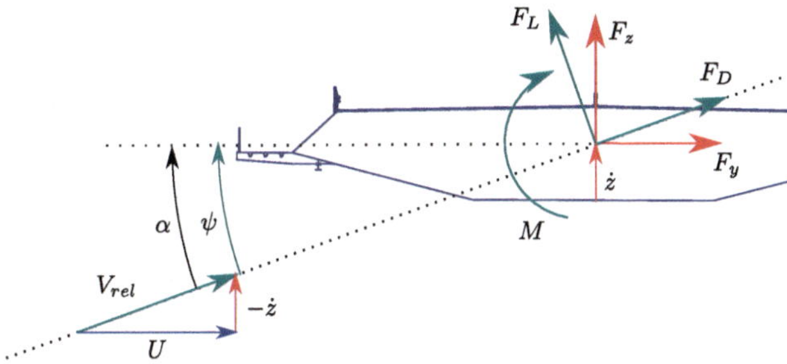

Fig. 5.16 Deck section moving vertically

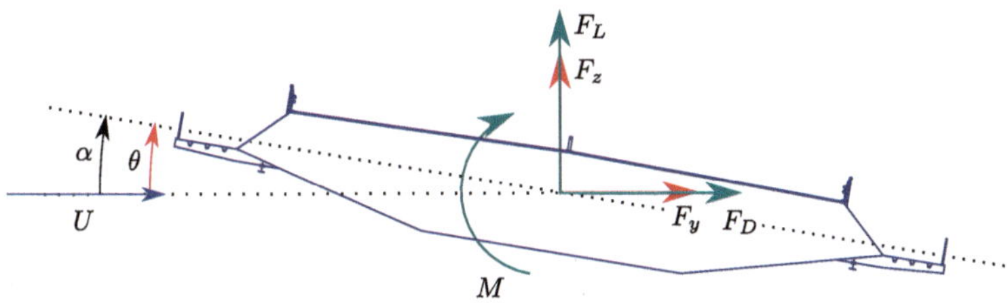

Fig. 5.17 Rotated deck section

5.4 Analytical Methods

In order to better understand the aeroelastic problems, we need to consider the analytical approaches available to simulate these types of phenomena. Different approaches are available and they will be described in the following subsections.

5.4.1 *Quasi-Steady Theory*

The Quasi-Steady Theory (QST) is, among the different approaches, the most suitable to better understand the physics of the problems, related to buffeting and stability. The QST well reproduces the aerodynamic forces on a deck, function of motion and wind turbulence, if the reduced velocity U^* is greater than 10–20. The reduced velocity is defined as:

$$U^* = \frac{U}{fB} \tag{5.4}$$

where U is the fluid velocity, f is the vibration frequency and B is the deck width.

The reduced velocity U^* can be interpreted as the ratio between the period associated to the body vibration ($T = 1/f$) and the time B/U needed by the fluid particle to move through the deck width B. High reduced velocity means that the time needed by a particle to cross the body is very small compared to the period of oscillation of the body: thus the aerodynamic forces are in a small amount influenced by the motion frequency and aerodynamic coefficients defined by static tests can be used.

The frequency f in the reduced velocity expression can also represent—in case that the turbulence effect is considered—the frequency of fluctuation of the turbulence spectrum. In other words, the turbulence frequency must be small compared to the frequency associated to the fluid particle velocity and to the deck dimension B.

As already observed we will analyze the deck, which is the most sensitive part of the bridge to the wind action. More in general the wind blowing on the bridge (Fig. 5.9) produces aerodynamic forces on all the bridge structure. The equation of motion can be written as:

$$[M_s]\ddot{\underline{x}} + [R_s]\dot{\underline{x}} + [K_s]\underline{x} = \underline{F}_a\left(\underline{x}, \dot{\underline{x}}, t\right) \tag{5.5}$$

Equation (5.5) represents a FEM model, where $\underline{x}$ is the displacement of the nodes of the structure, being $[M_s]$, $[R_s]$ and $[K_s]$ the mass, damping a stiffness matrices of the bridge structure obtained by a linearization around the equilibrium position due to dead loads. $\underline{F}_a\left(\underline{x}, \dot{\underline{x}}, t\right)$ are the aerodynamic forces function of the bridge motion $\underline{x}$ and associated velocity $\dot{\underline{x}}$ and of the time t, being the incoming turbulence a function of the time.

As already said the definition of the aerodynamic forces $\underline{F}_a$ on the deck represents the most important part of the problem. Several approaches are available to define $\underline{F}_a\left(\underline{x}, \dot{\underline{x}}, t\right)$ and they will be treated in Sects. 5.4.2, 5.4.3, and 5.4.4.

The *QST* is one of these approaches. To solve Eq. (5.5) a modal approach is generally used and the first vertical, horizontal and torsional modes are introduced in a simplified but often employed numerical procedure (see Figs. 5.4, 5.5, and 5.6), in which the different modes are reported.

To better understand the model, here we will consider a deck section (with length L), subjected to a turbulent wind characterized by a vertical (w) and an horizontal (u) component, with motion defined by vertical (z), horizontal (y) and torsional (θ) components. In the Appendix 5.7.1 more details are given about the relationship between the equations of motion of a full-bridge and those of a single-section model.

The *QST* supposes that the unsteady aerodynamic forces, acting on the bridge deck, can be defined using the same static force coefficients that are measured in the wind tunnel on still sectional model, expressed as:

$$F_D = \frac{1}{2}\rho V_{rel}^2 BLC_D(\alpha) \tag{5.6}$$

$$F_L = \frac{1}{2}\rho V_{rel}^2 BLC_L(\alpha) \tag{5.7}$$

$$M = \frac{1}{2}\rho V_{rel}^2 B^2 LC_M(\alpha) \tag{5.8}$$

where V_{rel}^2 represents the square of the wind velocity, relative to the deck (see Fig. 5.19), that can be computed taking into account the deck velocity component

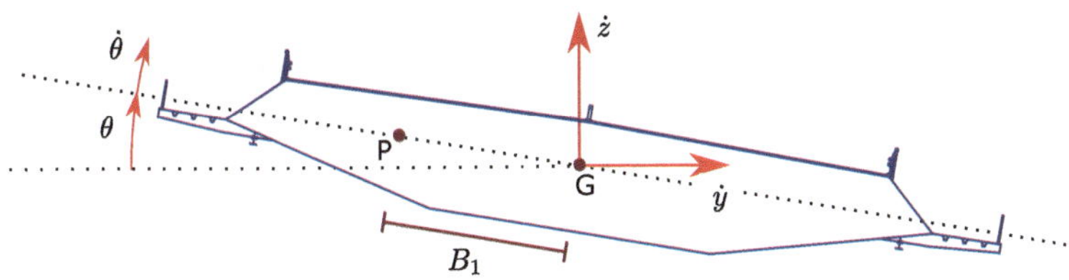

Fig. 5.18 Deck velocity components

while α represents the angle between the wind relative velocity and the deck during its motion.

Under these hypothesis, calling y and z respectively the horizontal and vertical displacements of the bridge deck motion, θ its rotation, and $\dot{y}$, $\dot{z}$ and $\dot{\theta}$ the corresponding velocities, the relative wind speed is easily expressed.

Actually, as reported in the Fig. 5.18, the velocity of the deck point P has an horizontal component equal to:

$$U_{P,y} = \dot{y}$$

and a vertical component equal to:

$$U_{P,z} = \dot{z} + B_1\dot{\theta}$$

As a consequence, according to the choice of the position of the considered point P a different relative velocities may be defined We decide to assume, as reference point, the position of a point at a distance B_1 upwind the deck centre G. This distance will be later defined while treating the Corrected Quasi Steady Theory (*QSTC*).

The relative velocity, defined as the velocity seen by an observer that is moving together with a point at a distance B_1 from G, may be, therefore, defined by its components along the y and z direction:

$$U_{Rel,y} = U_y - \dot{y} \quad (5.9)$$

$$U_{Rel,z} = U_z - \dot{z} - B_1\dot{\theta} \quad (5.10)$$

being U_y and U_z the horizontal and vertical components of the incoming turbulent wind.

If the wind is turbulent, it is possible to define a mean value of the wind speed U, an horizontal turbulent component u and a vertical turbulent component w.

In Fig. 5.19 the vector of the relative velocity is reported, taking into account also the turbulent wind components.

The square of the modulus of the relative velocity can be expressed as:

$$V_{rel}^2 = (U + u - \dot{y})^2 + \left(w - \dot{z} - B_1\dot{\theta}\right)^2 \quad (5.11)$$

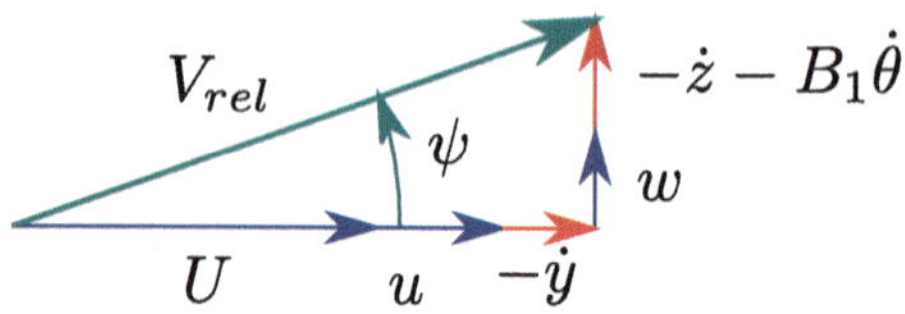

Fig. 5.19 Relative wind vector definition

while ψ represents the angle of attack of the relative wind speed[1]:

$$\psi = \arctan \frac{w - \dot{z} - B_1\theta}{U + u - \dot{y}} \quad \text{where } B_1 = B_{1y} \text{ or } B_{1z} \text{ or } B_{1\theta} \tag{5.12}$$

The aerodynamic forces acting on the bridge deck according to the QST are:

$$F_D = \frac{1}{2}\rho V_{rel}^2 BLC_D(\alpha) \tag{5.13}$$

$$F_L = \frac{1}{2}\rho V_{rel}^2 BLC_L(\alpha) \tag{5.14}$$

$$M = \frac{1}{2}\rho V_{rel}^2 B^2 LC_M(\alpha) \tag{5.15}$$

where α is the angle of attack between the relative velocity and the deck:

$$\alpha = \psi + \theta \tag{5.16}$$

and the sign convention is shown in Fig. 5.21. The bridge has its natural vibration modes defined as vertical, horizontal and torsional motions. To understand the bridge behaviour under turbulent wind conditions and the flutter instability, it is sufficient to consider the first vertical, torsional and horizontal modes.

The problem can be therefore modelled as in Fig. 5.20 where y, z and θ stand for the horizontal, vertical and torsional displacement components. The elastic links grant to have the same natural frequency of the real structure for the first modes. This model also represent the set up that is used in a Wind Tunnel when a an elastically suspended deck section is tested for stability analysis and turbulent wind response. In this case y, z and θ are physical coordinates.

The aerodynamic forces acting on the bridge deck are F_y, F_z and F_θ for a given incoming wind. The equations of motion of the system are:

[1] The choice of B_1 and the definition of B_{1y}, B_{1z}, $B_{1\theta}$ will be better explained later on.

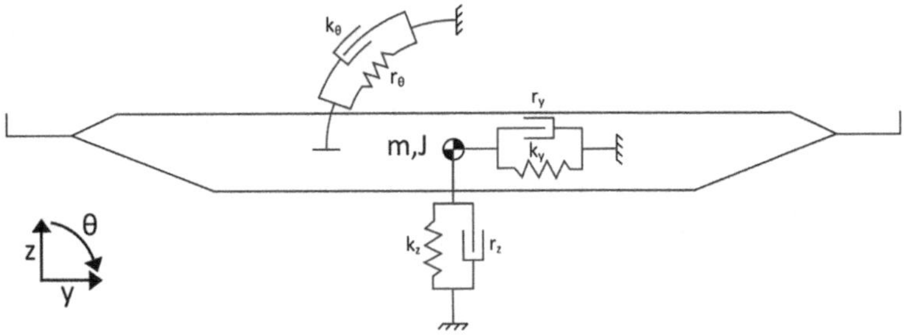

Fig. 5.20 Elastically suspended deck section

$$m_y \ddot{y} + r_y \dot{y} + k_y y = F_D \cos \psi - F_L \sin \psi = F_y \tag{5.17}$$

$$m_z \ddot{z} + r \dot{z} + kz = F_D \sin \psi + F_L \cos \psi = F_z \tag{5.18}$$

$$J_G \ddot{\theta} + r_\theta \dot{\theta} + k_\theta \theta = M = F_\theta \tag{5.19}$$

where m_y, m_z and J_G are the masses, r_y, r_z and r_θ the damping, and k_y, k_z and k_θ the stiffness of the springs.

By substituting the previously presented non linear relations in the expressions of F_y, F_z, F_θ the equations become:

$$m_y \ddot{y} + r_y \dot{y} + k_y y = \frac{1}{2} \rho S V_{rel}^2 \left(C_D(\alpha) \cos \psi - C_L(\alpha) \sin \psi\right) \tag{5.20}$$

$$m_z \ddot{z} + r_z \dot{z} + k_z z = \frac{1}{2} \rho S V_{rel}^2 \left(C_D(\alpha) \sin \psi + C_L(\alpha) \cos \psi\right) \tag{5.21}$$

$$J_G \ddot{\theta} + r_\theta \dot{\theta} + k_\theta \theta = \frac{1}{2} \rho B S V_{rel}^2 C_M(\alpha) \tag{5.22}$$

with:

$$V_{rel}^2 = (U + u - \dot{y})^2 + \left(w - \dot{z} - B_1 \dot{\theta}\right)^2 \tag{5.23}$$

$$\alpha = \theta + \psi = \theta + \arctan \frac{w - B_1 \dot{\theta} - \dot{z}}{U + v - \dot{y}} \tag{5.24}$$

where B_1 assumes different values as a function of the considered component of the aerodynamic forces (i.e. B_{1y} for F_y, B_{1z} for F_z, $B_{1\theta}$ for F_θ. This formulation is referred to as $QSTC$ (Corrected QST).

Using the displacement vector:

$$\underline{x} = \begin{Bmatrix} y \\ z \\ \theta \end{Bmatrix}$$

the equation of motion of the system may be written using a matrix formulation as:

$$[M_s]\ddot{\underline{x}} + [R_s]\dot{\underline{x}} + [K_s]\underline{x} = \underline{F}_A\left(\underline{x}, \dot{\underline{x}}, U, u(t), w(t)\right) \tag{5.25}$$

This equation reproduces by a non linear approach the bridge response to the incoming turbulent wind. As already said, the QST is valid for high reduced velocity $U^* = \frac{U}{fB} > 15$.

5.4.2 QST Linearised Theory

A simplified approach, for the study of the aeroelastic problem, may be obtained by considering the linear formulation of the aerodynamic forces, acting on a generic section of a bridge deck. The aerodynamic forces, as we have seen in the previous paragraph, are generally a non linear function of the bridge motion, of the wind speed and of the turbulent wind velocity components:

$$\underline{F}_A = \underline{F}_A\left(\underline{x}, \dot{\underline{x}}, U, v(t), w(t)\right) = \underline{F}_A\left(\underline{x}, \dot{\underline{x}}, U, \underline{b}\right)$$

using a matrix formulation where:

$$\underline{F_A} = \begin{Bmatrix} F_y \\ F_z \\ F_\theta \end{Bmatrix}; \quad \underline{x} = \begin{Bmatrix} y \\ z \\ \theta \end{Bmatrix}; \quad \dot{\underline{x}} = \begin{Bmatrix} \dot{y} \\ \dot{z} \\ \dot{\theta} \end{Bmatrix}; \quad \underline{b} = \begin{Bmatrix} u \\ w \end{Bmatrix}$$

A linear formulation of the aerodynamic forces may be considered, if the hypothesis of small variations of the parameters $\underline{x}$, $\dot{\underline{x}}$ and $\underline{b}$ is assumed.

Starting from the static deformation, reached under mean wind speed conditions (equilibrium position), the value of x_0 can be obtained solving the static problem:

$$[K_s]\underline{x}_0 = \underline{F}_A\left(\underline{x}_0, U\right) \tag{5.26}$$

$$x_0 = \begin{Bmatrix} y_0 \\ z_0 \\ \theta_0 \end{Bmatrix}; \quad \dot{\underline{x}}_0 = \begin{Bmatrix} 0 \\ 0 \\ 0 \end{Bmatrix}; \quad \underline{b}_0 = \begin{Bmatrix} 0 \\ 0 \end{Bmatrix}$$

under the hypothesis of small variation both of deck motion and turbulent wind fluctuations around the equilibrium values, the linear expression of the aerodynamic forces can be obtained by:

$$\underline{F}_A\left(\underline{x};\, \dot{\underline{x}};\, \underline{b}\right) = \underline{F}_A\left(\underline{x}_0;\, \underline{0};\, \underline{0}\right) + \left.\frac{\partial \underline{F}_A}{\partial \underline{x}}\right|_0 \left(\underline{x} - \underline{x}_0\right) + \left.\frac{\partial \underline{F}_A}{\partial \dot{\underline{x}}}\right|_0 \dot{\underline{x}} + \left.\frac{\partial \underline{F}_A}{\partial \underline{b}}\right|_0 \underline{b}$$

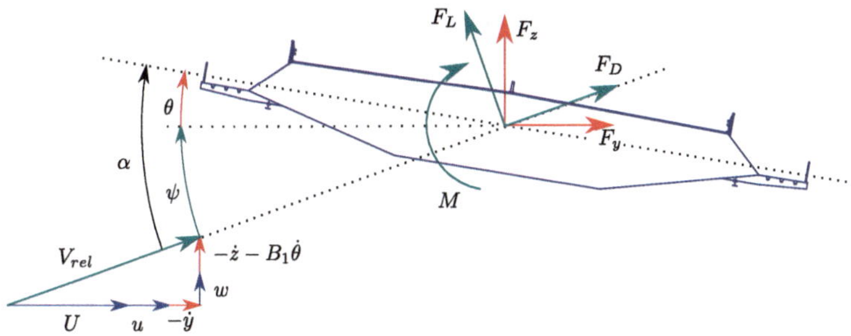

Fig. 5.21 Absolute and apparent reference frame

where $\underline{x} - \underline{x}_0$, $\underline{\dot{x}}$, and $\underline{b}$ represent the variations of the corresponding quantities, starting from the values assumed at the equilibrium position:

$$\underline{\tilde{x}} = \begin{Bmatrix} y(t) - y_0 \\ z(t) - z_0 \\ \theta(t) - \theta_0 \end{Bmatrix}; \quad \underline{\dot{x}} = \begin{Bmatrix} \dot{y}(t) \\ \dot{z}(t) \\ \dot{\theta}(t) \end{Bmatrix}; \quad \underline{b} = \begin{Bmatrix} v(t) \\ w(t) \end{Bmatrix}$$

Considering the QSTC formulation of the aerodynamic forces:

$$F_y = \frac{1}{2}\rho V_{rel}^2 BL \left(C_D(\alpha)\cos\psi - C_L(\alpha)\sin\psi\right) \tag{5.27}$$

$$F_z = \frac{1}{2}\rho V_{rel}^2 BL \left(C_D(\alpha)\sin\psi + C_L(\alpha)\cos\psi\right) \tag{5.28}$$

$$F_\theta = \frac{1}{2}\rho V_{rel}^2 B^2 L C_M(\alpha) \tag{5.29}$$

according to the sign convention in Fig. 5.21, with:

$$V_{rel}^2 = (U + u - \dot{y})^2 + \left(w - \dot{z} - B_1\dot{\theta}\right)^2$$

$$\psi = \arctan\frac{w - \dot{z} - B_1\theta}{U + v - \dot{y}} \quad \text{where } B_1 = B_{1y} \text{ or } B_{1z} \text{ or } B_{1\theta}$$

$$\alpha(t) = \theta(t) + \psi(t)$$

We need to linearize the different terms of the previous equation around y_0, z_0, and θ_0, defining:

$$\begin{cases} \tilde{y} = y - y_0 \\ \tilde{z} = z - z_0 \\ \tilde{\theta} = \theta - \theta_0 \end{cases}$$

We linearize the different terms, obtaining:

$$\tilde{\psi} = \frac{w - \dot{\tilde{z}} - B_1\dot{\tilde{\theta}}}{U} \tag{5.30}$$

$$\tilde{\alpha} = \alpha - \alpha_0 \tag{5.31}$$

$$\sin\psi = \tilde{\psi} \tag{5.32}$$

$$\cos\psi = 1 \tag{5.33}$$

$$V_{rel}^2 = U^2 + 2Uu - 2U\dot{y} \tag{5.34}$$

$$\alpha_0 = \theta_0 \tag{5.35}$$

$$C_D(\alpha) = C_{D_0} + \left(\frac{\partial C_D}{\partial \alpha}\right)_{\alpha_0} \tilde{\alpha} = C_{D0} + K_{D0}\tilde{\alpha} \tag{5.36}$$

$$C_L(\alpha) = C_{L_0} + \left(\frac{\partial C_L}{\partial \alpha}\right)_{\alpha_0} \tilde{\alpha} = C_{L0} + K_{L0}\tilde{\alpha} \tag{5.37}$$

$$C_M(\alpha) = C_{M_0} + \left(\frac{\partial C_M}{\partial \alpha}\right)_{\alpha_0} \tilde{\alpha} = C_{M0} + K_{M0}\tilde{\alpha} \tag{5.38}$$

If we substitute the linearized expressions in the F_y, F_z and F_θ equations, we obtain:

$$F_y = \frac{1}{2}\rho\left(U^2 - 2U\dot{y} + 2Uu\right)BL\left((C_{D0} + K_{D0}\tilde{\alpha}) - (C_{L0} + K_{L0}\tilde{\alpha})\tilde{\psi}\right) \tag{5.39}$$

$$F_z = \frac{1}{2}\rho\left(U^2 - 2U\dot{y} + 2Uu\right)BL\left((C_{D0} + K_{D0}\tilde{\alpha})\tilde{\psi} - (C_{L0} + K_{L0}\tilde{\alpha})\right) \tag{5.40}$$

$$F_\theta = \frac{1}{2}\rho\left(U^2 - 2U\dot{y} + 2Uu\right)B^2L\left(C_{M0} + K_{M0}\tilde{\alpha}\right) \tag{5.41}$$

Developing and neglecting the second order terms:

$$F_y = \frac{1}{2}\rho BLU^2C_{D0} + \frac{1}{2}\rho BLU^2\left[K_{D0}\left(\tilde{\theta} + \frac{w - \dot{\tilde{z}} - B_{1y}\dot{\tilde{\theta}}}{U}\right) - C_{L0}\left(\frac{w - \dot{\tilde{z}} - B_{1y}\dot{\tilde{\theta}}}{U}\right)\right] + \frac{1}{2}\rho BL2UC_{D0}(u - \dot{\tilde{y}}) \tag{5.42}$$

$$F_z = \frac{1}{2}\rho BLU^2 C_{L0} + \frac{1}{2}\rho BLU^2 \left[K_{L0}\left(\tilde{\theta} + \frac{w - \dot{\tilde{z}} - B_{1z}\dot{\tilde{\theta}}}{U}\right) + C_{D0}\left(\frac{w - \dot{\tilde{z}} - B_{1z}\dot{\tilde{\theta}}}{U}\right)\right]$$
$$+ \frac{1}{2}\rho BL2VC_{L0}(u - \dot{\tilde{y}}) \tag{5.43}$$

$$F_\theta = \frac{1}{2}\rho B^2 LU^2 C_{M0} + \frac{1}{2}\rho B^2 LU^2 \left[K_{M0}\left(\tilde{\theta} + \frac{w - \dot{\tilde{z}} - B_{1\theta}\dot{\tilde{\theta}}}{U}\right)\right]$$
$$+ \frac{1}{2}\rho B^2 L2VC_{M0}(u - \dot{\tilde{y}}) \tag{5.44}$$

If we consider the vector $\underline{F}_A$, with components F_y, F_z, and F_θ :

$$F_A = \begin{Bmatrix} F_y \\ F_z \\ F_\theta \end{Bmatrix}$$

and the vector $\underline{\tilde{x}}$ with components $\tilde{y}$, $\tilde{z}$ and $\tilde{\theta}$:

$$\underline{\tilde{x}} = \begin{Bmatrix} \tilde{y} \\ \tilde{z} \\ \tilde{\theta} \end{Bmatrix}; \quad \underline{\dot{\tilde{x}}} = \begin{Bmatrix} \dot{\tilde{y}} \\ \dot{\tilde{z}} \\ \dot{\tilde{\theta}} \end{Bmatrix}; \quad \underline{b} = \begin{Bmatrix} v \\ w \end{Bmatrix}$$

Defining with $F_{A,st}$ the static aerodynamic forces, and $F_{A,dyn}$ the forces function of $\underline{\tilde{x}}$ and $\underline{\dot{\tilde{x}}}$, the static and aerodynamic forces can be written in matrix form:

$$\underline{F}_{A,st} = \frac{1}{2}\rho BLU^2 \begin{bmatrix} C_{D0} \\ C_{L0} \\ BC_{M0} \end{bmatrix} \tag{5.45}$$

$$\underline{F}_{A,dyn} = \frac{1}{2}\rho BLU^2 \begin{bmatrix} 0\;0 & K_{D0} \\ 0\;0 & K_{L0} \\ 0\;0 & BK_{M0} \end{bmatrix} \underline{\tilde{x}}$$
$$- \frac{1}{2}\rho BLU \begin{bmatrix} 2C_{D0} & K_{D0} - C_{L0} & B_{1y}(K_{D0} - C_{L0}) \\ 2C_{L0} & K_{L0} + C_{D0} & B_{1z}(K_{L0} + C_{D0}) \\ 2C_{M0}B & BK_{M0} & B_{1\theta}BK_{M0} \end{bmatrix} \underline{\dot{\tilde{x}}}$$
$$+ \frac{1}{2}\rho BLU \begin{bmatrix} 2C_{D0} & K_{D0} - C_{L0} \\ 2C_{L0} & K_{L0} + C_{D0} \\ 2C_{M0} & BK_{M0} \end{bmatrix} \underline{b} \tag{5.46}$$

which, in compact form, can be written as:

$$\underline{F}_{A,dyn} = -[K_A]\underline{\tilde{x}} - [R_A]\underline{\dot{\tilde{x}}} + [A_m]\underline{b}$$

To identify the static equilibrium position defined by y_0, z_0, θ_0, the non linear Eq. (5.26) must be solved.

In this way, it is possible to identify y_0, z_0, θ_0, and $\alpha_0 = \theta_0$. C_{D_0}, C_{L_0}, and C_{M_0} are the $C_D(\alpha)$, $C_L(\alpha)$, and $C_M(\alpha)$ values for $\alpha = \alpha_0$ and K_{D0}, K_{L0}, K_{M0} are their derivatives for $\alpha = \alpha_0$.

The linearized equations of motion may be written as:

$$[M_s]\underline{\ddot{\tilde{x}}} + ([R_s] + [R_A])\underline{\dot{\tilde{x}}} + ([K_s] + [K_A])\underline{\tilde{x}} = [A_m]\underline{b} \tag{5.47}$$

where

$$[M_s] = \begin{bmatrix} m_y & & \\ & m_z & \\ & & J \end{bmatrix}; \quad [R_s] = \begin{bmatrix} r_y & & \\ & r_z & \\ & & r_\theta \end{bmatrix}; \quad [K_s] = \begin{bmatrix} k_y & & \\ & k_z & \\ & & k_\theta \end{bmatrix}$$

are the structural modal diagonal matrices and $[R_A]$ and $[K_A]$ are the equivalent damping and stiffness matrices due to the linearized aerodynamic forces. $[A_m]$ is the so called aerodynamic admittance matrix.

The $[R_A]$ and $[K_A]$ matrices multiplied by $\underline{\dot{\tilde{x}}}$ and $\underline{\tilde{x}}$ define the aerodynamic forces function of the motion, while $[A_m]$, when multiplied by $\underline{b}$, defines the aerodynamic forces function of the incoming turbulence. $[K_A]$ may change the natural frequencies of the bridge with respect to those computed in absence of wind, related to $[K_s]$ and $[M_s]$. $[R_A]$ may change the value of the system overall damping.

If we analyze the different terms of the $[R_A]$ and $[K_A]$ matrices, we can easily identify which are the parameters giving rise to 1 d.o.f. and 2 d.o.f. flutter instability. Of course, in order to define if the system is stable or not, we need to solve the homogeneous equation.

5.4.2.1 Stability

In order to identify the instability conditions, the homogeneous equation of (5.47) can be analysed, i.e.:

$$[M_s]\underline{\ddot{\tilde{x}}} + ([R_s] + [R_A])\underline{\dot{\tilde{x}}} + ([K_s] + [K_A])\underline{\tilde{x}} = \underline{0} \tag{5.48}$$

Imposing $\underline{\tilde{x}} = \underline{\Psi}e^{\lambda t}$, and considering $[R_t] = [R_s + R_A]$, $[K_t] = [K_s + K_A]$, we obtain:

$$(\lambda^2 M_s + \lambda R_t + K_t)\underline{\Psi} = [H(\lambda)]\underline{\Psi} = \underline{0} \tag{5.49}$$

The λ values for which the $[H(\lambda)]$ determinant is equal to zero identify the eigenvalues, while the correspondent $\underline{\Psi}$ identify the eigenvectors. Instability occurs when one of the λ values ($i = 1 \dots 6$) have a real positive part, i.e.:

$$\lambda_i = \alpha_i + i\omega_i \quad \text{with} \quad \alpha_i > 0$$

The associated eigenvector value defines the type of unstable motion: one degree of freedom or two degrees of freedom instability. However, in order to better understand, it is possible to identify the instability conditions, making reference to the type of matrices present in Eq. (5.49) and defined in (5.46): symmetry and positive or negative definition of these matrices can be used to identify the type and condition of instability, as better described in Diana and Cheli (2010).

One Degree of Freedom Instability

One Degree of Freedom Instability can be excited if one of the diagonal terms of the $[R_A]$ matrix is negative and greater than the corresponding term in the $[R_s]$ matrix, as an absolute value.

More in detail, if $C_{D0} + K_{L0}$ is negative (i.e. K_{L0} is negative and as an absolute value greater than C_{D0}), we have 1 d.o.f. instability in the vertical mode of vibration if:

$$\frac{1}{2}\rho BLU \left|(C_{D0} + K_{L0})\right| \geq r_z$$

The wind velocity U above which this occurs is called critical velocity U_c.

We have 1 d.o.f. torsional instability when $BK_M B_{1_\theta}$ is negative and the critical velocity is defined by:

$$\frac{1}{2}\rho B^2 LU \left|K_{M0} B_{1\theta}\right| = r_\theta$$

Of course C_{D0} is always positive and then the horizontal motion is always stable.

The 1 d.o.f. instability in the vertical and torsional mode—with the sign conventions used by us—occurs when the slopes of the lift and moment coefficients are negative—see Fig. 5.11, for different deck types. As it can be observed the new bridges decks, which are airfoil type, all have positive slopes. The Tacoma bridge has a moment coefficient negative slope. Generally, the negative slope of the lift and moment coefficients are also associated with vortex shedding, which is what happened for the Tacoma bridge.

The two mechanisms of vortex shedding and 1 d.o.f. instability are interconnected. Vortex shedding occurs at the Strouhal velocity, producing vibration amplitudes generally only inside the synchronization range. When the two phenomena are interconnected, the vibrations, instead of disappearing when the wind speed is above the synchronization range, increase more and more, as much as higher is the wind speed—see Fig. 5.22.

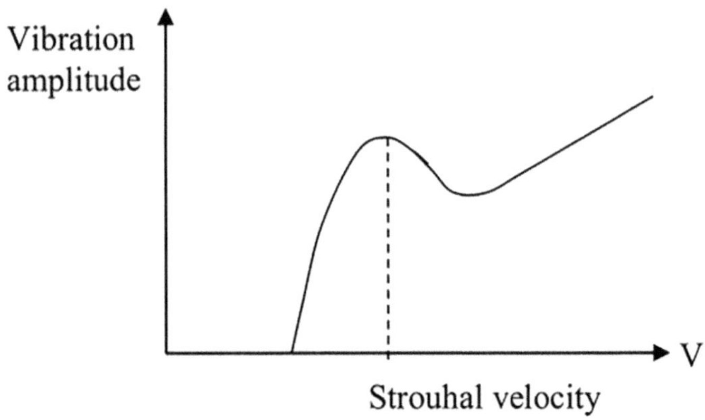

Fig. 5.22 Vortex shedding and 1 d.o.f. instability

To follow the *QST*, it seems that vortex shedding produces an increment of the negative slope of the lift and moment coefficients.

Two Degree of Freedom Instability (Flutter)

Airfoil like decks do not suffer for 1 d.o.f. instability, while they are subjected to 2 d.o.f. instability, also called flutter instability. As explained in detail in Diana and Cheli (2010), flutter instability is produced when the equivalent stiffness matrix of the aerodynamic forces is not symmetric, like in our case, and is in condition to make close the two frequencies that, in absence of wind, are different. The two motions become synchronized and the aerodynamic forces can introduce energy into the system, such producing the flutter type, 2 d.o.f. instability.

Due to the fact that the bridge torsional frequencies are physiologically higher than the corresponding vertical ones, in order to have flutter the slope of the moment coefficient (K_M) must be positive.

Looking at the $[K_A]$ matrix, with $K_M > 0$, the diagonal term: $-1/2\rho U^2 B^2 L K_{M0}$ is negative and, as the wind speed increases, lowers the overall torsional stiffness, decreasing—as a consequence—the torsional frequency until it become closer the corresponding vertical one, as in Fig. 5.23. When this occurs, we may have flutter instability.

To better understand, if we compute the λ_i value in Eq. (5.49) as a function of the wind speed, for an airfoil type deck—which has a positive moment coefficient slope—we obtain the values of α_i/ω_i indicated in Fig. 5.24. In Fig. 5.23 the torsional frequency is decreasing, becoming close to the vertical one: at this point the real part of the eigenvalue λ_i becomes positive, giving rise to instability.

Flutter is the most important problem of very long bridges because, increasing the span length, the ratio between torsional and corresponding vertical frequency decreases and the wind aerodynamic forces are in condition to make these two frequencies closer at wind speed values as much lower as the span length increases—see Fig. 5.7.

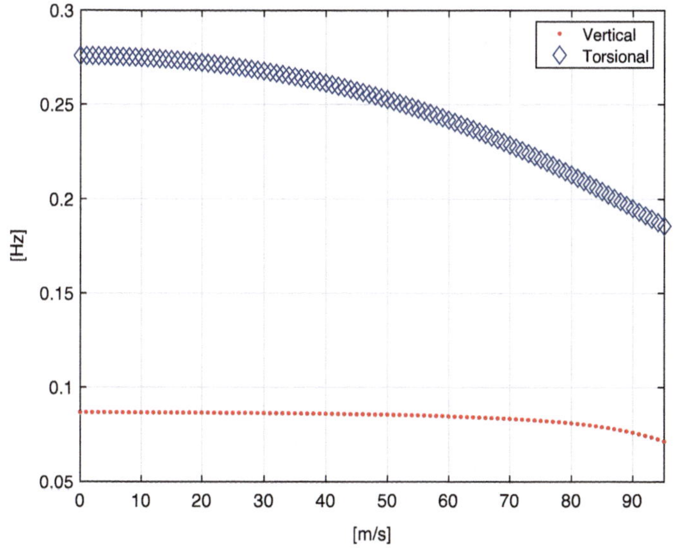

Fig. 5.23 Frequency as a function of the wind speed an airfoil type deck

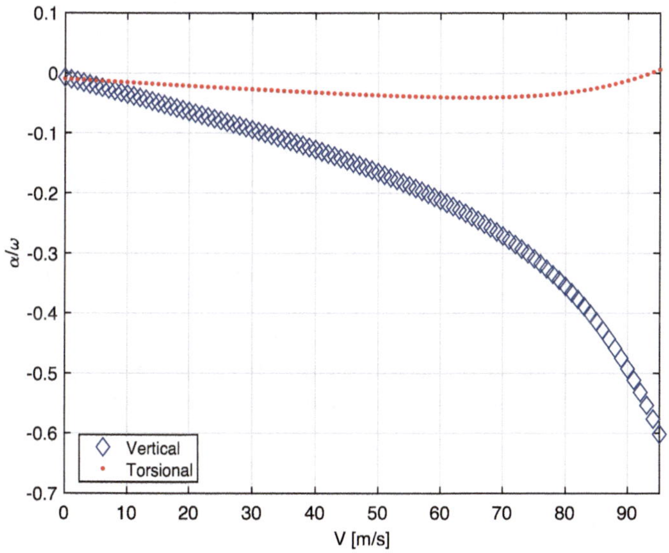

Fig. 5.24 α_i/ω_i for an airfoil type deck

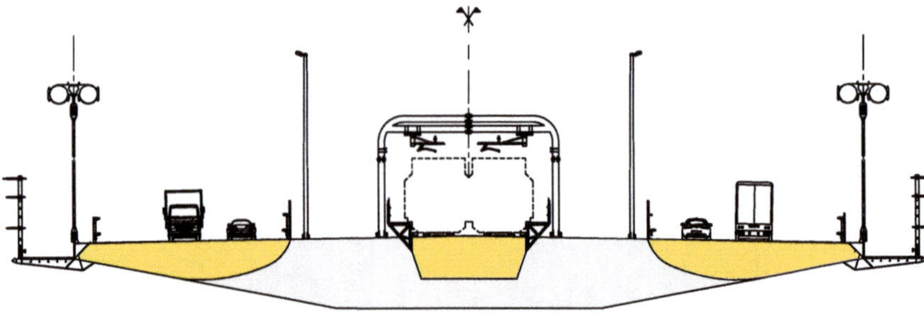

Fig. 5.25 Messina deck section

For this reason the very long bridges, like the Messina, must have a multi-box deck-like that shown in Fig. 5.25—to have a high flutter critical speed.

This deck has been optimized in order to have a positive slope both for the lift and moment coefficients, but with a very small value: of the order of 10 times lower than that of an airfoil, also reported as "flat plate" in Fig. 5.11. The K_L and K_M values for a flat plate are $K_L = 2\pi$ and $K_M = \pi/2$.

The low value of the moment coefficient slope produces a low negative torsional stiffness, which changes in a small amount the torsional frequency of the bridge that becomes closer to the vertical one only at 85 m/s, well above the design speed of 60 m/s.

Another way to solve the problem would be to increase the ratio between torsional and vertical frequencies through structural modifications.

Many attempts have been made to increase the ratio between torsional and vertical frequencies by structural means, and some of them are here reported:

- To increase the horizontal distance between the main cables and try to concentrate the deck weight along a central strip. This action reduces the contribution of the deck to the overall rotational moment of inertia.
- Use of transversely inclined cross hangers together with or instead of vertical ones (Fig. 5.26). This is another proposed solution, but it presents practical difficulties, including restricted headroom clearance near the centre of the span.

5.4.2.2 Buffeting Response

As already explained, the fluctuations of wind due to turbulence produce a motion of the bridge which is called buffeting, or bridge motion induced by turbulence. This is always present in all bridges, and generally in all structures and cannot be avoided. The vibration amplitudes associated with buffeting can be controlled by increasing the aerodynamic damping or equivalently, by increasing the stability of the bridge, as will be explained in the following.

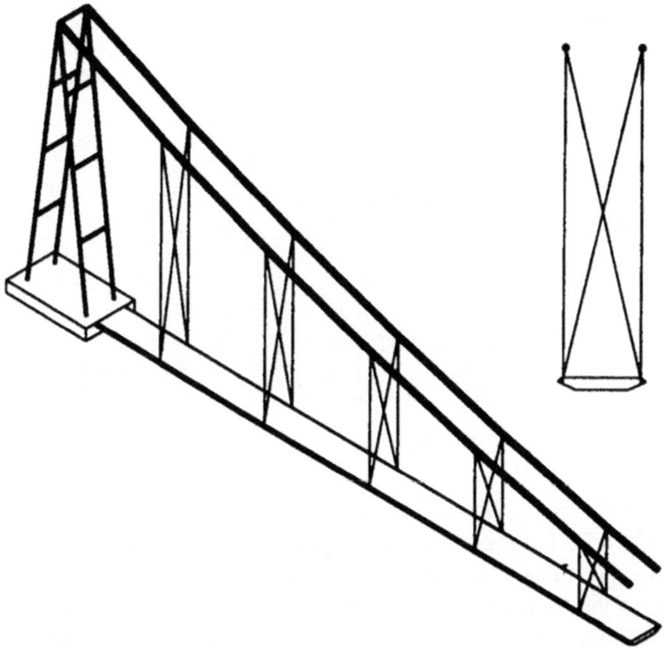

Fig. 5.26 "Cross Hangers" structural scheme

The aerodynamic forces due to the incoming turbulence can be identified through wind tunnel tests or by an analytical approach. In the previous sections, the QST was also used to identify the aerodynamic forces due to turbulence.

The most important effect is associated to the variation of the wind angle of attack produced by the turbulence vertical component. In order to compute the bridge motion due to the incoming turbulence, we need to reproduce the space and time function of the wind blowing on the bridge. This can be achieved through the method described in Chap. 2, on the base of the wind spectrum and correlation functions.

5.4.3 Unsteady Linearised Theory

The unsteady linearized theory introduces the dependence upon the reduced velocity of the aerodynamic matrices $[K_A]$, $[R_A]$ and $[A_m]$.

5.4.3.1 Aeroelatic Matrices

The sectional 3×3 $[K_A]$ and $[R_A]$ matrices contain the flutter derivatives coefficients (a^*, h^*, p^*) with a generally accepted formulation[2]:

$$[R_a] = \frac{1}{2}\rho U B L \begin{bmatrix} p_5^* & p_1^* & p_2^* B \\ h_5^* & h_1^* & h_2^* B \\ a_5^* B & a_1^* B & a_2^* B^2 \end{bmatrix} \tag{5.50}$$

$$[K_a] = -\frac{1}{2}\rho U^2 B L \begin{bmatrix} p_6^* \frac{\pi}{2V_\omega^{*2}} \frac{1}{B} & p_4^* \frac{\pi}{2V_\omega^{*2}} \frac{1}{B} & p_3^* \\ h_6^* \frac{\pi}{2V_\omega^{*2}} \frac{1}{B} & h_4^* \frac{\pi}{2V_\omega^{*2}} \frac{1}{B} & h_3^* \\ a_6^* \frac{\pi}{2V_\omega^{*2}} & a_4^* \frac{\pi}{2V_\omega^{*2}} & a_3^* B \end{bmatrix} \tag{5.51}$$

where the asterisk * indicates the dependence upon the reduced velocity V^* and $V_\omega^* = V^*/(2\pi)$.

It is important to compare the $[K_A]$ and $[R_A]$ matrix coefficients to those evaluated through the QST. We have seen, for all the tested decks, that at high values of V^*, the $[K_A]$ and $[R_A]$ matrix coefficients, measured by dynamic tests, converge to the expression of the $[K_A]$ and $[R_A]$ values of the QST. Thanks to that, it is possible to define the values of B_{1y}, B_{1z}, and $B_{1\theta}$ of the QSTC (see $[K_A]$ and $[R_A]$ matrices of the linearised QSTC) by the ratio of the $[R_A]$ matrix coefficients experimentally identified at high reduced velocity. More precisely:

$$B_{1y} = \frac{p_2^*}{p_1^*} B \tag{5.52}$$

$$B_{1z} = \frac{h_2^*}{h_1^*} B \tag{5.53}$$

$$B_{1\theta} = \frac{a_2^*}{a_1^*} B \tag{5.54}$$

This formulation, that uses the flutter derivatives to correct the QST, is what we call Correct Quasi Steady Theory QSTC—see Diana et al. (1995) for more details.

It is useful to report—as a reference—the values of the flutter derivatives for a flat plate. For this particular section, Theodorsen developed an analytical approach, based on the potential flow theory, to define the static and aerodynamic forces. No indication is of course given for the motion in horizontal direction, so, with Theodorsen, it is possible to give only the 8 flutter derivatives related the vertical, z, and the torsional, θ, motion.

[2] A different formulation is also used, introduced by Simiu and Scanlan (1996).

$$h_1^* = 2\pi F(k) \qquad a_1^* = \frac{\pi}{2} F(k) \tag{5.55}$$

$$h_2^* = -\frac{\pi}{2}\left[1 - F(k) - \frac{2G(k)}{k}\right] \qquad a_2^* = \frac{\pi}{8}\left[1 - F(k) - \frac{2G(k)}{k}\right] \tag{5.56}$$

$$h_3^* = 2\pi\left[F(k) - \frac{kG(k)}{2}\right] \qquad a_3^* = \frac{\pi}{2}\left[F(k) - \frac{kG(k)}{2}\right] \tag{5.57}$$

$$h_4^* = \left[1 + \frac{2G(k)}{k}\right] \qquad a_4^* = \frac{G(k)}{2k} \tag{5.58}$$

being k a reduced frequency ($k = \frac{B}{2}\frac{\omega}{U}$), and $F(k)$ and $G(k)$ the real and imaginary parts of the Theodorsen circulatory function.

The values of the flutter derivatives are reported in Fig. 5.27 as a function of the reduced velocity.

These values can be compared to those of other decks for better understanding their aerodynamic behavior. From these flutter derivatives, making reference to the QSTC, the $B_{1\theta}$ and B_{1z} values can be also defined (see Fig. 5.28):

$$B_{1\theta} = \frac{a_2^*}{a_1^*} B \quad B_{1z} = \frac{h_2^*}{h_1^*} B$$

5.4.3.2 Aerodynamic Admittance Matrix

The buffeting forces $\boldsymbol{F}_{\text{buff}}$ per unit length due to incoming turbulent wind u and w are defined in the frequency domain through the admittance functions:

$$\mathcal{F}\left\{\begin{bmatrix} F_{D,\text{buff}} \\ F_{L,\text{buff}} \\ M_{\text{buff}} \end{bmatrix}\right\} = \frac{1}{2}\rho U B \begin{bmatrix} \chi_{Du}^* & \chi_{Dw}^* \\ \chi_{Lu}^* & \chi_{Lw}^* \\ B\chi_{Mu}^* & B\chi_{Mw}^* \end{bmatrix} \mathcal{F}\left\{\begin{bmatrix} u(t) \\ w(t) \end{bmatrix}\right\} \tag{5.59}$$

where $\mathcal{F}$ is the Fourier transform operator; χ^* are the admittance functions, depending on the reduced velocity V^* and mean angle of attack.

Typically, the admittance functions χ^* are measured as a function of V^* and the mean angle of attack through wind tunnel tests on sectional models, similar to what is done for flutter derivatives.

5.4.3.3 Stability

The stability is studied by means of an iterative eigenvalue-eigenvector problem:

$$\left(\lambda^2\left[\boldsymbol{M}_s + \boldsymbol{M}_{\text{se}}(V^*)\right] + \lambda\left[\boldsymbol{R}_s + \boldsymbol{R}_{\text{se}}(V^*)\right] + \left[\boldsymbol{K}_s + \boldsymbol{K}_{\text{se}}(V^*)\right]\right)\boldsymbol{\Phi}e^{\lambda t} = \boldsymbol{0} \tag{5.60}$$

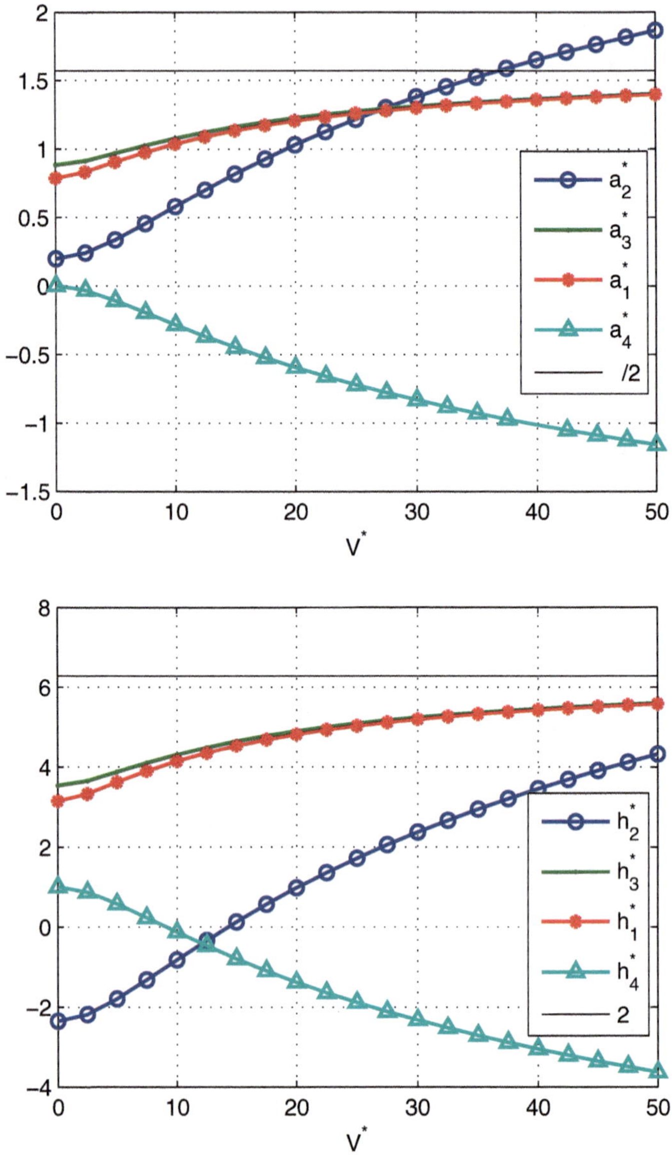

Fig. 5.27 Flutter derivatives for the flat plate (Theodorsen)

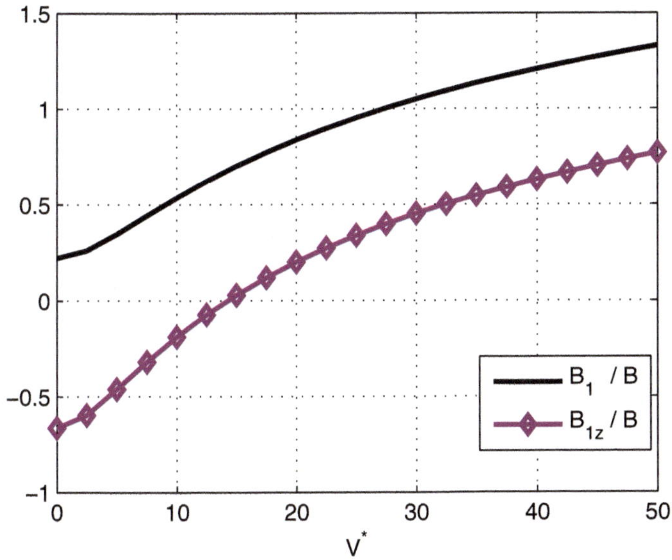

Fig. 5.28 $B_{1\theta}/B$ and B_{1z}/B as a function of the reduced velocity

being the eigenvalues λ in general complex conjugate (or reals), function of the reduced velocity. The algorithm must be iterative since the frequency of motion is the imaginary part of λ (e.g. see Argentini et al. (2014)).

5.4.3.4 Buffeting Response

At a given wind speed U, the equation of motion in frequency domain is (e.g., see Chen and Karen (2001a)):

$$\left(-\omega^2\left[\boldsymbol{M}_s+\boldsymbol{M}_{\text{se}}(V^*)\right]+i\omega\left[\boldsymbol{R}_s+\boldsymbol{R}_{\text{se}}(V^*)\right]+\left[\boldsymbol{K}_s+\boldsymbol{K}_{\text{se}}(V^*)\right]\right)\boldsymbol{X}(f)=\boldsymbol{F}_{\text{buff}}(f) \tag{5.61}$$

where:

- $\boldsymbol{X}(f)=\mathcal{F}\{\boldsymbol{x}(t)\}=\begin{bmatrix}Y(f)\\Z(f)\\\Theta(f)\end{bmatrix}$ is the vector of Fourier transforms of generalized coordinates
- $\boldsymbol{M}_s$, $\boldsymbol{R}_s$, $\boldsymbol{K}_s$ are the 3×3 diagonal structural mass, damping, and stiffness matrices
- $\boldsymbol{M}_{\text{se}}$, $\boldsymbol{R}_{\text{se}}$, $\boldsymbol{K}_{\text{se}}$ are self-excited matrices via flutter derivatives
- $\boldsymbol{F}_{\text{buff}}(f)$ is the buffeting force vector in frequency domain
- $\omega=2\pi f$ is the circular frequency.

The buffeting forces in frequency domain are:

$$\boldsymbol{F}_{\text{buff}}(f) = \frac{1}{2}\rho U B \begin{bmatrix} \chi_{Du}^* & \chi_{Dw}^* \\ \chi_{Lu}^* & \chi_{Lw}^* \\ B\chi_{Mu}^* & B\chi_{Mw}^* \end{bmatrix} \begin{bmatrix} W_u(f) \\ W_w(f) \end{bmatrix} = \boldsymbol{\chi}_A(V^*)\boldsymbol{W}_{uw}(f) \tag{5.62}$$

The system response in frequency domain is:

$$\boldsymbol{X}(f) = \boldsymbol{H}(V^*)^{-1}\boldsymbol{F}_{\text{buff}}(f) = \boldsymbol{H}(V^*)^{-1}\boldsymbol{\chi}_A(V^*)\boldsymbol{W}_{uw}(f) \tag{5.63}$$

where:

$$\boldsymbol{H}(V^*) = -\omega^2[\boldsymbol{M}_s + \boldsymbol{M}_{\text{se}}(V^*)] + i\omega[\boldsymbol{R}_s + \boldsymbol{R}_{\text{se}}(V^*)] + [\boldsymbol{K}_s + \boldsymbol{K}_{\text{se}}(V^*)] \tag{5.64}$$

Taking the complex conjugate transpose of $\boldsymbol{X}(f)$:

$$\boldsymbol{X}(f)^* = \boldsymbol{W}_{uw}(f)^*\boldsymbol{\chi}_A(V^*)^*\boldsymbol{H}(V^*)^{-H} \tag{5.65}$$

The cross-spectrum matrix:

$$\boldsymbol{X}(f)\boldsymbol{X}(f)^* = \boldsymbol{H}(V^*)^{-1}\boldsymbol{\chi}_A(V^*)\boldsymbol{W}_{uw}(f)\boldsymbol{W}_{uw}(f)^*\boldsymbol{\chi}_A(V^*)^*\boldsymbol{H}(V^*)^{-H} \tag{5.66}$$

Then, the cross-power spectral density matrix is:

$$\boldsymbol{S}_X(f) = \boldsymbol{H}(V^*)^{-1}\boldsymbol{\chi}_A(V^*)\boldsymbol{S}_{\text{wind}}(f)\boldsymbol{\chi}_A(V^*)^*\boldsymbol{H}(V^*)^{-H} \tag{5.67}$$

where:

$$\boldsymbol{S}_{\text{wind}}(f) = \begin{bmatrix} S_u & S_{uw} \\ S_{uw} & S_w \end{bmatrix} \tag{5.68}$$

By repeating the computation for each f and U, PSD reference curves can be obtained.

5.4.4 Non Linear Analytical Methods

Even if linear approaches are the widest adopted method to study the bridge aerodynamic stability and response of the bridge to turbulent wind they are not able to deal with the important non linear effect intrinsically present in the aeroelastic problem. As already presented, the dependence of the aerodynamic forces from the deck motion and from the turbulent wind velocity components is fully non linear (see Eq. (5.75)).

$$[M_s]\ddot{\underline{x}} + [R_s]\dot{\underline{x}} + [K_s]\underline{x} = \underline{F}_A\left(\underline{x}, \dot{\underline{x}}, U, \underline{b}\right)$$

The two main causes of non-linearities are the dependence of the aerodynamic forces on the angle of attack and on the reduced velocity. The QST (Minh et al. 2000) is able to take into account the dependence on the mean angle of attack in a non linear way but it is valid only at high reduced velocities. On the contrary the linear approaches are able to take into account the dependence on the reduced velocity through the flutter derivatives and the admittance function coefficients but they represent a solution around a fixed mean angle of attack.

Measurements performed on full scale bridges highlighted that during the common operating condition the relative angle of attack between the bridge deck and the wind direction may reach large values that make the linear hypothesis not applicable. Some non linear approaches have been developed (Diana et al. 2008, 2010; Wu and Kareem 2011) to deal with the fully non linear aeroelastic problem. In this section we can only introduce the basic ideas of the "Band Superposition" approach and of the "rheological model" approach, and refer the reader to the literature for a more exhaustive description.

The "Band Superposition" approach relies on the possibility of separating the low frequency and the high frequency contributions of the wind spectrum. The idea is that the low frequency part of the wind spectrum is responsible for the larger fluctuations of the angle of attack and it acts at high reduced velocity. A QSTC (Diana et al. 1995) is therefore proposed to account for this part taking into account the non linear dependence on the angle of attack. The part of the wind spectrum at high frequency is characterised by smaller amplitudes and lower reduced velocity. To consider the effects induced by this part of the wind spectrum, the aeroelastic forces are linearized around the angle of attack that is instantaneously computed by solving the low frequency part of the problem. The problem becomes linear in the high frequency range, but with coefficients changing with the time.

An alternative approach is to consider the whole wind spectrum effects in a single time domain model as represented by the "rheological model". This approach is based on a numerical model that is able to consider both the angle of attack and the reduced velocity dependence in a single time domain approach. To identify the numerical model parameters, aerodynamic hysteresis loops are measured on deck sectional models in wind tunnel by changing the instantaneous angle of attack. The variation of the angle if attack is produced by moving the sectional model or by using the active turbulence generator with high amplitudes at different frequencies (Diana et al. 2008).

5.5 Wind Tunnel Tests

Wind tunnel tests are made on the different bridge components: deck, towers and cables. Tests are also made on the full bridge model (aeroelastic models).

5.5.1 Sectional Model Tests

The procedure generally adopted for the wind tunnel tests on the deck is herein after described:

1. Static tests on the tower and deck section, in order to measure the lift, drag and moment coefficients as a function of the angle of attack (α).
2. Optimization of the deck shape, in order to fulfill the bridge stability requirements, analyzing the slope of the lift and moment coefficients obtained by the static tests—see the QST explained in the previous paragraphs—(Sect. 5.3.2).
3. Verification of the vortex shedding excitation, analysing the vibration level as a function of the Scruton number. This test is done on a sectional model of the the deck elastically suspended. In the case of the tower, an aeroelastic model should usually be employed. If the vibration level fulfills the specifications at the real bridge Scruton number, the optimization procedure is finished, otherwise the deck shape is modified to control vortex shedding and the procedure must be repeated, starting from step 2).
4. Once the deck optimization has been completed, dynamic tests in order to identify the flutter derivatives—i.e. the K_A and R_A matrices, and the admittance functions should be performed. Verification of the flutter derivatives is also performed defining the aeroelastic damping ratio as a function of the wind speed.

5.5.1.1 Static Tests

Deck

The most important test for the deck, as well as for the other components, is the definition of the static aerodynamic coefficients as a function of the angle of attack. These tests are very important because allow for the definition of the static load on the deck (see Sect. 5.3.1) and, using the QSTC, for the definition of the instability conditions for the bridge.

This test is made in a wind tunnel on a sectional model of the deck (Fig. 5.29) The sectional deck model is placed on a dynamometric system (Fig. 5.29) generally placed outside the wind tunnel test section (Brancaleoni et al. 2009).

The model can rotate around a deck longitudinal axis and the lift, drag and moment aerodynamic forces are measured as a function of the wind angle of attack, by changing the deck angle of rotation. Tests are usually repeated for more than one wind speed to check the independence of aerodynamic coefficients on Reynolds number.

Some laboratory uses a deck sectional model made up of a dynamometric (floating) central part (Fig. 5.30) and two lateral structural parts coupled by axial beams. This type of device is also very useful for measuring the dynamic aerodynamic forces, as it will be better explained in the following.

The static forces, due to the average wind, on all the bridge can be derived from Fig. 5.11 and similar ones are obtained for towers and cables.

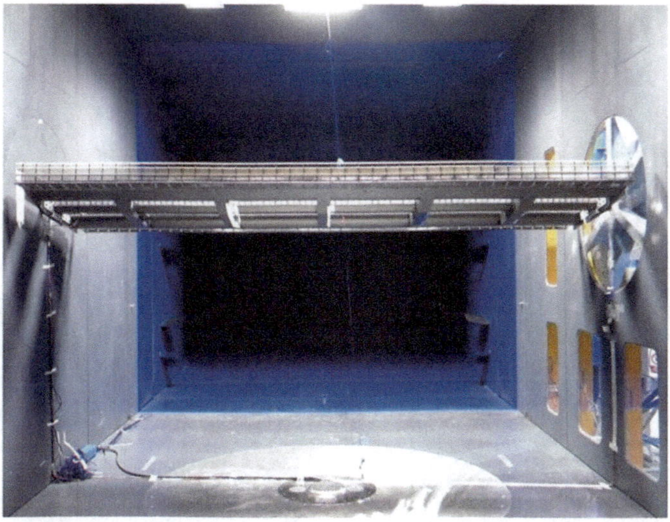

Fig. 5.29 Deck sectional model in wind tunnel (external balance)

Fig. 5.30 Deck sectional model in wind tunnel (internal balance)

Tower

Static coefficients of the tower are measured in a similar way to what is done on the deck, adopting sectional models with internal or external dynamometers and measuring the aerodynamic forces at different angles of attack. Figure 5.31 shows a tower sectional model during wind tunnel tests. In Fig. 5.32, as an example, the drag, lift and moment coefficients of a section of the Messina Bridge tower twin legs are reported, together with the tower drawing. The static forces on all the bridge due to the average wind can be computed in order to define, through a FEM simulation, the deflection of the bridge and the related stresses.

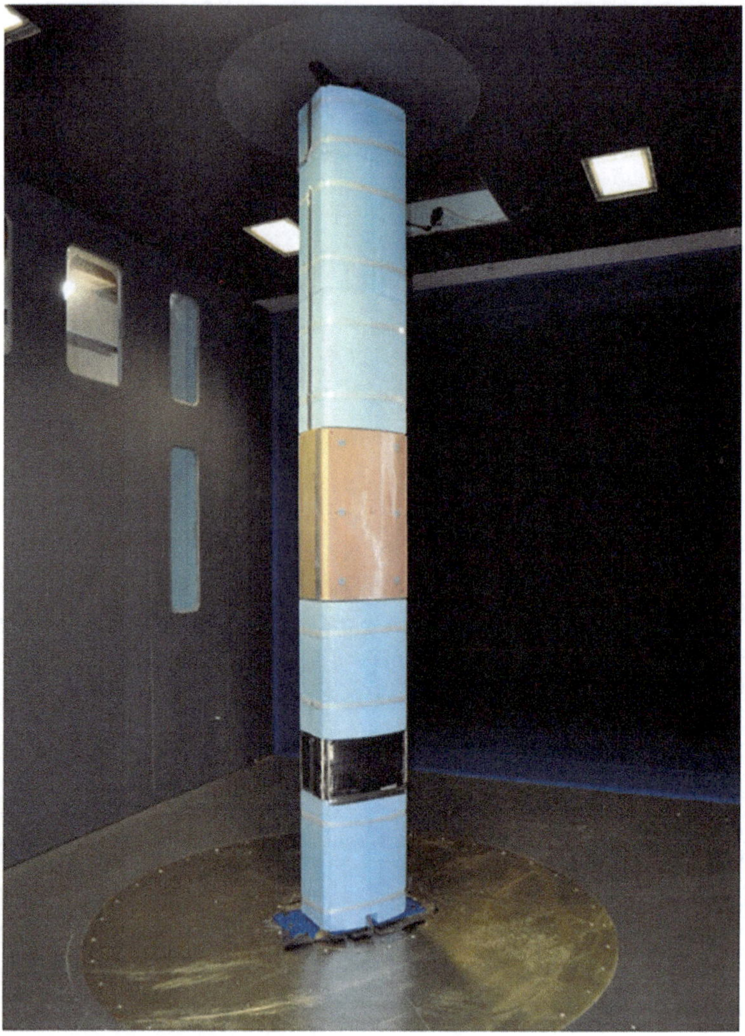

Fig. 5.31 Cable stayed bridge tower sectional model in the wind tunnel test section: measure of static aerodynamic coefficients

For towers with two legs and transverse beams, the aerodynamic coefficients are measured separately on each leg by using a model that represents a tandem arrangement of the two legs where transverse beams are not present.

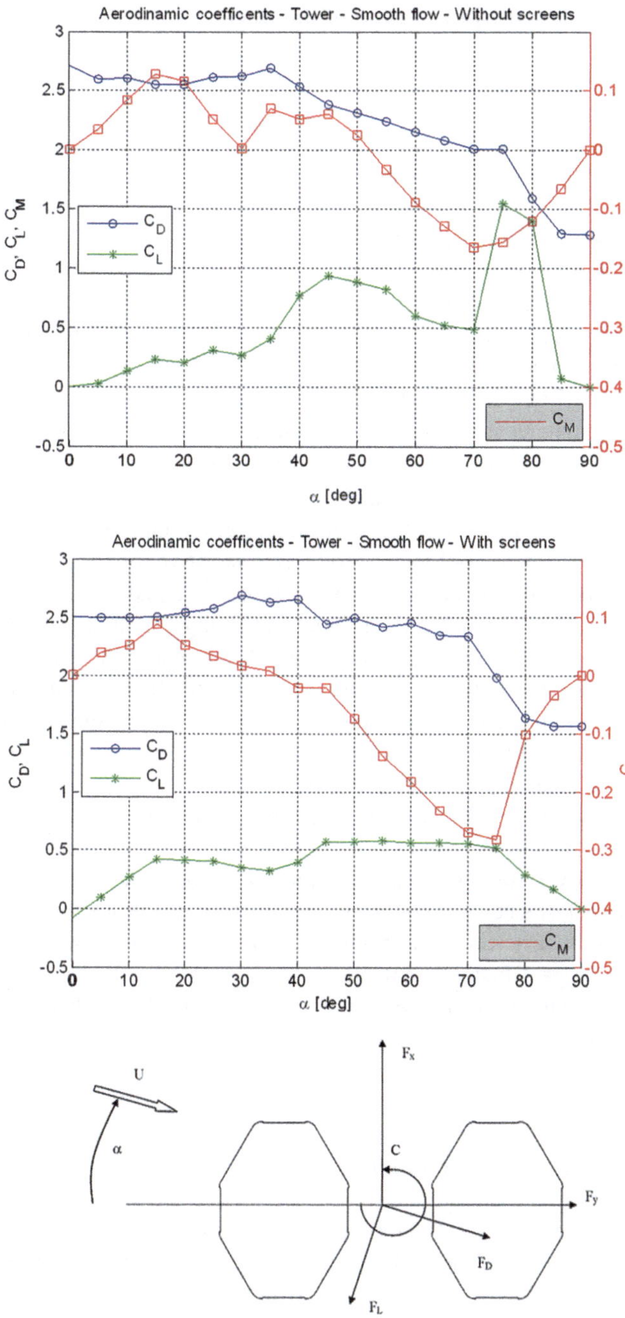

Fig. 5.32 Aerodynamic coefficients of the Messina bridge tower (the distance of the tower legs is not reproduced in the design scale)

5.5.1.2 Aerodynamic Optimization of the Deck

The aerodynamic optimization of the deck represents a trade-off between performance related to flutter instability and vortex shedding. In this context, the lift, drag and moment coefficient figures are very important for the bridge deck optimization. The moment and lift coefficient slopes, in fact, play a huge role in the bridge stability (see Sect. 5.4.2.1).

They must be positive to avoid one-degree-of-freedom instability and, in order to have a high flutter velocity, their slopes must be as much as possible lower than the values of a flat plate: 2π for the lift coefficient and $\pi/2$ for the moment coefficient. The slope value to guarantee stability is a function of the ratio between the torsional and vertical frequencies of the bridge.

As an example, in the Messina bridge, the ratio between the first torsional and the first vertical frequencies is equal to 1.36, leading to a prescription for a value of the moment coefficient derivative comprised between $\frac{1}{5}\frac{\pi}{2}$ and $\frac{1}{10}\frac{\pi}{2}$ and for a value of the lift coefficient derivative comprised between $\frac{1}{5}2\pi$ and $\frac{1}{10}2\pi$.

The final reason for this requirement, as already mentioned, is that the positive slope of the moment coefficient produces a negative torsional stiffness which decreases the torsional frequency, such that it becomes equal to the vertical one and produces the two-degrees-of-freedom flutter instability. Another important issue is to reduce the drag on the deck as much as possible, because the drag forces on the deck are the most important ones and are transferred to the top of the towers by the hangers and the main cables, producing a moment that affects the overall design of the tower. In order to fulfill these requirements, for a very long span bridge—over 2000 m—it is mandatory to use a multi-box section.

A long optimization work is needed to obtain positive lift and moment coefficients with small slopes for all the wind angles of attack, or at least for a $\pm 4^\circ$ portion and avoid VIV. Moreover, the optimization of the aerodynamic details is a compromise between the aeroelastic behavior of the bridge and the traffic protection. Wind barriers must be placed on very long span bridges to protect vehicles from crosswind, and they can have a huge influence on the deck aerodynamic coefficients, depending on the positioning and typology. In the case of the Messina bridge deck, the optimization involved the shape of railway box and the wind barriers, and the porosity of both the road and railway box girders.

5.5.1.3 Vortex Shedding Test on the Deck

Vortex induced vibrations (VIVs) represent one of the most important aspects of the deck shape design. Most of the bridge deck sections can suffer from vortex shedding. The decks with large frontal sections or sharp edges are more prone to this problem than streamlined wing type box sections. Figure 5.33 shows the vortex shedding visualization obtained for the Messina bridge deck model in the wind tunnel.

Fig. 5.33 Messina deck section: vortex shedding visualization

Even though the bridge decks of the last generation have an airfoil-like geometry, flow separation can occur in correspondence of the deck sharp corners, in presence of adverse pressure gradients and of windshields that are usually adopted to protect traffic against the wind action. Vortex shedding excitation can exhibit with the separation of flow from the upstream edge of the deck to produce a vortex that moves downstream. Generally, at the downstream edge, or in another point depending on the deck shape, another vortex can be detached. Depending on the relative position of the two vortices and at a given wind speed (or in non dimensional form at a given reduced velocity $U^* = \frac{U}{fB}$) a motion amplification may occur. Critical values of reduced velocity tend to be around 1 or 0.5.

VIV is a serious problem that contributed in the collapse of the Tacoma Narrow Bridge and must be carefully analyzed. In fact, even if it can be separated from the stability problem for modern bridges, it can lead the deck to reach very large vibration amplitudes in case of low Scruton numbers[3]

[3] This expression is valid for vertical VIV. In the case of torsional VIV, Scruton number is instead defined as

$$Sc = 2\pi \frac{J h_s}{\rho B^4}$$

where J is the torsional inertia per unit length, and h_s the torsional damping coefficient.

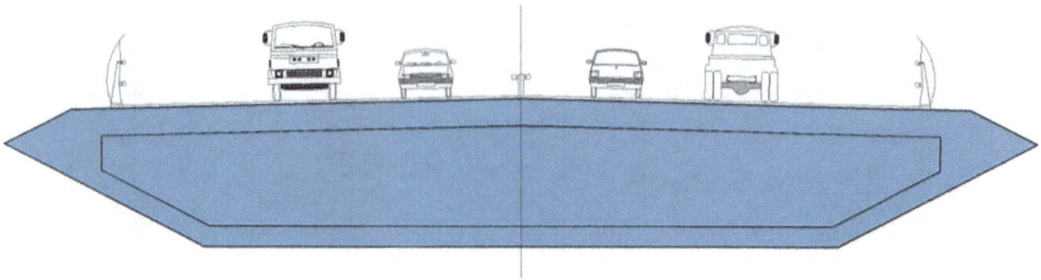

Fig. 5.34 Storebaelt deck section

$$Sc = 2\pi \frac{mh_s}{\rho B^2}$$

m is the mass per unit length, h_s is the structural damping, B is the deck chord and ρ is the air density. Moreover, VIV can be excited at low wind speeds having a high probability of occurrence.

Nevertheless, the problem is widely known and analyzed: past and present experiences highlighted that it represents one of the main concerns in bridge design. In 2010 Russian authorities shut down the Volga River bridge due to the VIV (Ge et al. 2022). As an example, the Storebealt deck section (Fig. 5.34) experiences vortex shedding from the bottom deck corners.

In this case, guide vanes were installed close to the bottom plate/lower side panel joints, to mitigate the vortex-induced oscillations occurred during the final phases of girder erection. The solution was identified after an ad-hoc monitoring activity on the bridge at full scale and a specific wind tunnel activity, and its realization required a specially designed gantry/working platform (Frandsen 2001).

In the 1992 design of the Messina Bridge, a trip wire was placed under the two road boxes to control the derivatives of the moment aerodynamic coefficient. However, wind tunnel testing showed that a large vortex shedding excitation was caused by these trip wires, so they were eliminated from the deck design. For the Messina bridge, boxes with flat plane surfaces and sharp corners as shown in Fig. 5.35 produced poor vortex shedding behaviour and were therefore replaced by boxes with curved surfaces.

If wind barriers are placed along the edges of the deck they can trigger vortex shedding, as can any other similar obstacle or detail close to the deck edges. The solidity of the wind barriers, while needing to be high enough to reduce the wind flows across the bridge deck to reduce the wind forces on the vehicles, must be relatively small in order to limit vortex shedding. Airfoils are also incorporated in order to increase the aerodynamic damping of the deck.

Wind tunnel tests are performed on elastically suspended models (Fig. 5.36). Special attention is paid, in wind tunnel tests, to keep the Scruton number as low as possible by using light models characterized by low structural damping, in order to better highlight vortex shedding features.

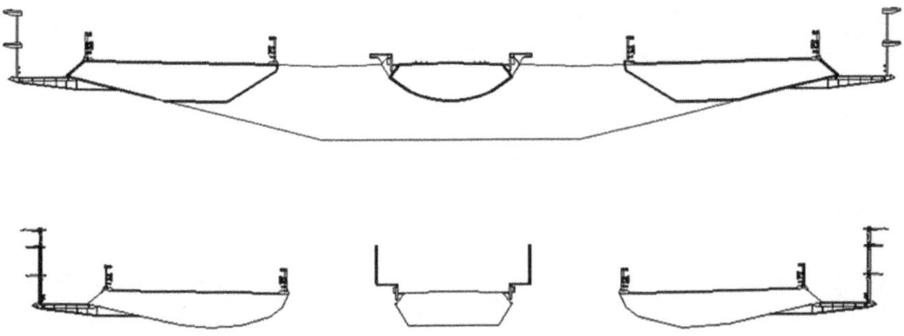

Fig. 5.35 Messina deck section. Boxes with flat surfaces and sharp corners (replaced by boxes with smooth curved surfaces in the final scheme)

The model response in the lock-in range is usually measured in order to define:

1. The maximum oscillation amplitudes reached in the lock-in range.
2. The amount of energy that the flow is able to insert in the system. This may be expressed by a negative value of the aerodynamic damping coefficient measured during build-up tests. Build-ups are performed by releasing the deck model under mean wind conditions and measuring the build-up amplitude due to the vortex shedding.

While single box deck shapes generally show a single mechanism of vortex shedding, basically related to the vortexes detached in the girder wake, multi-box decks may present different vortex shedding phenomena because of the different possibility of flow interactions between the wake of the upwind box and the other boxes. The different vortex shedding mechanism may excite both the heaving and the torsional motion of the bridge.

As an example, Fig. 5.37 presents the two lock-in ranges for the flexural motion of the Messina Bridge multi-box deck section for different Scruton number. From the maximum values of the amplitudes measured in the lock-in range at different Sc, it is possible to define a figure that reports the non-dimensional amplitudes as a function of Sc.

In Fig. 5.38, the vertical amplitudes of the deck divided by the deck chord B are reported as function of Scruton number (configuration MB02). From this type of graph, it is possible to identify what should be the level damping of the real structure suitable to control VIVs.

Making reference to Fig. 5.38, the three vertical dashed lines represent three different values of Scruton number that for the Messina bridge correspond to three different levels of non-dimensional values of structural damping equal to: $2 \cdot 10^{-3}$, $3 \cdot 10^{-3}$ and $5 \cdot 10^{-3}$. To avoid VIV, the structural damping of the Messina bridge should be greater than $3 \cdot 10^{-3}$. To avoid VIVs also for lower values of structural damping, which are not expected, the identified solution is to adopt porous screens between the railway box and the road boxes (solution MB09 reported in Fig. 5.39). In this case no VIVs are produced even for very low damping values.

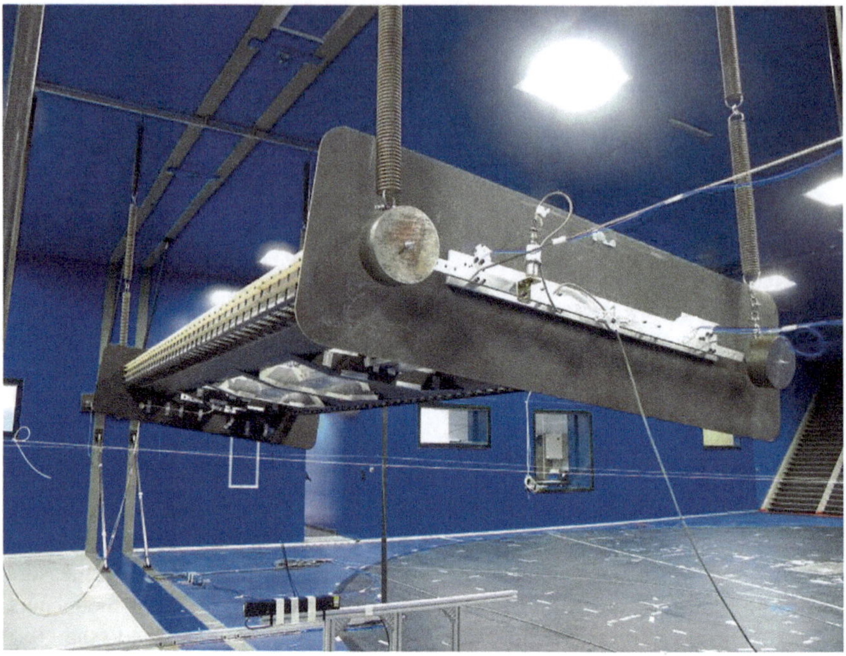

Fig. 5.36 Free motion experimental set-up: Suspension system, tensioned steel stays and spring system

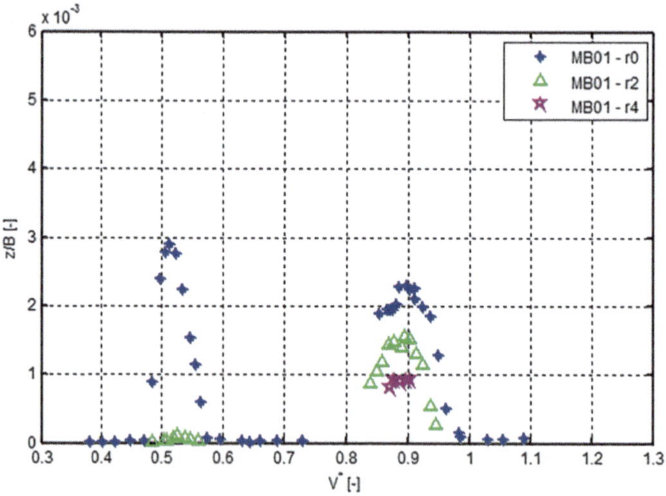

Fig. 5.37 Steady state response: non dimensional oscillation amplitude, flexural motion as function of the reduced velocity varying the Scrouton number (r0=0.007; r2=0.022; r4=0.033)

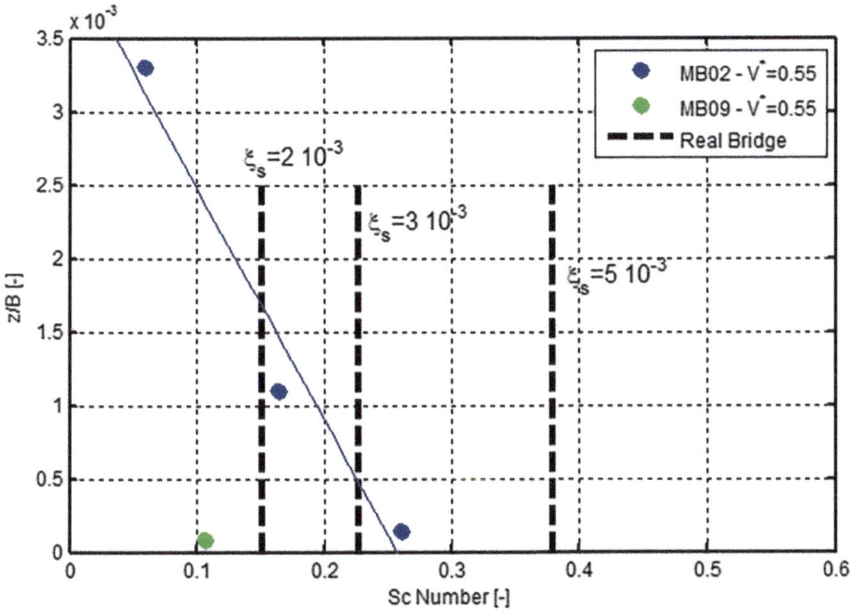

Fig. 5.38 Steady state response: non dimensional oscillation amplitude, MB02 and MB09 configurations, flexural motion as function of Scruton Number

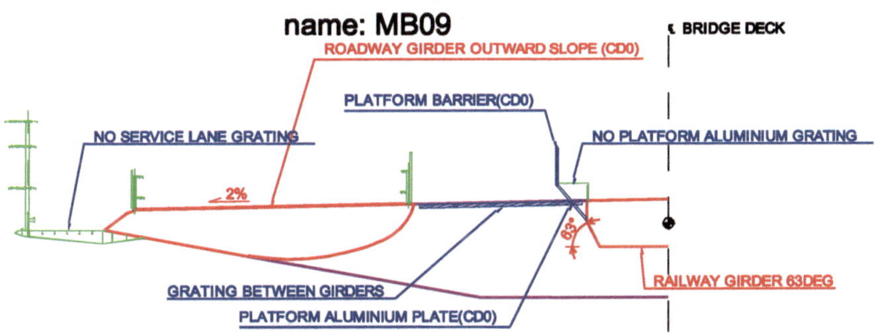

Fig. 5.39 Solution MB09: deck with porous screens between the boxes

Measurements of the surface pressure distribution on each box represent a valuable tool to a deeper understanding of the vortex shedding phenomenon. In Fig. 5.40 the pressure distribution measured on the Messina bridge model when it experiences the largest vibrations (lock-in) is reported in terms of the magnitude of the pressure coefficient C_p fluctuations at the vortex shedding frequency:

$$C_p = \frac{p}{\frac{1}{2}\rho U^2}$$

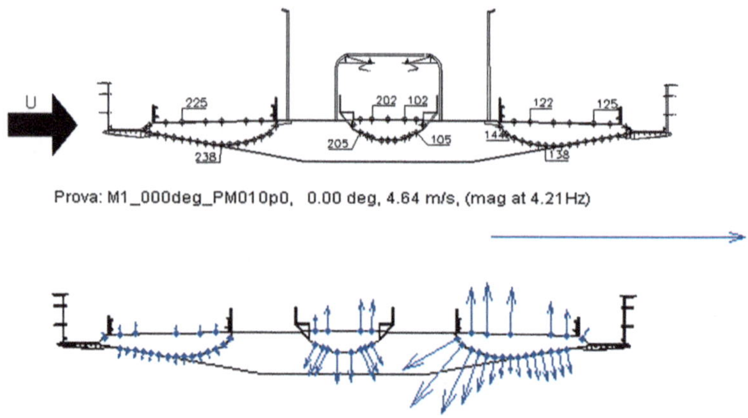

Fig. 5.40 Cp magnitude at f=4.21 Hz

where p is the pressure and U is the wind speed. the arrows pointing outwards to the deck profiles are proportional to the value of the pressure coefficient.

It is possible to appreciate how the vortex shedding is mainly related to the capability of the flow to be attached or separated on the curvilinear profile of the upwind box that is run over by an undisturbed flow.

It is clear where the vortexes are generated and that there is a strong pressure fluctuation in the gap between the downwind roadway girder and the railway girder. The aerodynamics of the downwind boxes is, in fact, strongly influenced by the disturbances introduced in the flow by the wind interaction with the upwind box and by the gap between the boxes.

5.5.1.4 Dynamic Tests on the Deck

To define the aerodynamic forces as a function of the deck motion and of the incoming turbulence the lift, drag, and moment forces have to be identified as a function of the y, z and θ coordinates representing the deck motion and of the w an u velocity components defining the incoming wind turbulence. The linearization of these aerodynamic forces, as a function of motion, gives rise to the K_A and R_A also called flutter derivatives. The linearization with respect to turbulence gives rise to the Admittance matrix.

For what concerns the forces identification as a function of the bridge motion, two methods are available: free motion tests and forced motion tests.

Free Motion Tests

During free motion tests, the deck sectional model is let free to vibrate under the mean wind speed action, by means of an elastic supporting system. The response of the system to the imposed initial condition is measured and flutter derivatives coefficients are defined by computing the variation of the structural response at a given wind velocity compared to the values obtained in still air. To simplify the test, a pure torsional motion is imposed to the deck and the decay at different wind speeds is analyzed. Excitation of both torsional and vertical motion must be considered in order to define the flutter derivatives. The comparison between the measured decay at a given wind and the computed one allows for the identification of the flutter derivatives through a least squares minimization procedure.

This type of model is also used to verify if the deck is stable, both with respect to 1 d.o.f. and 2 d.o.f. (flutter) instabilities. To this purpose, the ratio between the deck model torsional and vertical frequency and the mass properties should be correctly reproduced in the right scale. The test should be performed increasing the reduced velocity $V^* = V/(fB)$ until instability is reached. From this value of reduced velocity, the critical wind speed in the real scale can be defined.

Forced Motion Method

This method is more expensive than the previous, but is much more reliable since it allows a better control of all the parameters. A deck sectional model is forced to vibrate, as in Fig. 5.41, in the three y, z and θ directions and the lift, drag and moment forces are measured for a given constant velocity V_m.

Fig. 5.41 Sectional model on oil dynamic actuators driving its rigid vertical and torsional motion

As already said, the forces are non linear functions of the motion. More in detail, for the deck (see Fig. 5.21):

$$F_y = F_L = \frac{1}{2}\rho U^2 S C_L\left(\underline{x}, \dot{\underline{x}}, U, \underline{b}\right) \tag{5.69}$$

$$F_z = F_D = \frac{1}{2}\rho U^2 S C_D\left(\underline{x}, \dot{\underline{x}}, U, \underline{b}\right) \tag{5.70}$$

$$F_\theta = F_M = \frac{1}{2}\rho U^2 S B C_M\left(\underline{x}, \dot{\underline{x}}, U, \underline{b}\right) \tag{5.71}$$

being:

$$\underline{x} = \begin{Bmatrix} y \\ z \\ \theta \end{Bmatrix}$$

The identification of the aerodynamic coefficients C_L, C_D, C_M, which are non linear functions of $\underline{x}$, $\dot{\underline{x}}$, as already said, is not an easy task and represents a research topic. Methods have been developed by some researchers and are reported in Diana et al. (2008; 2010) for those based on rheological models, and in Wu and Kareem (2011) for those based on neural network algorithms.

The generally adopted procedure is to linearize the forces around a static equilibrium position of the deck identified by $\underline{x}_0$, in laminar flow (Eq. (5.76)):

$$\underline{F}_A = \begin{Bmatrix} F_y \\ F_z \\ F_\theta \end{Bmatrix} = \frac{1}{2}\rho U^2 S\left(-\left[K_a\right]\underline{\tilde{X}} - \left[R_a\right]\dot{\underline{\tilde{X}}}\right)$$

In this formulation, the wind turbulence components are not considered since this effect will be reproduced in the Admittance matrix. The $3 \times 3\,[K_A]$ and $[R_A]$ matrices contain the flutter derivatives coefficients with a generally accepted formulation[4] :

$$[R_a] = \frac{1}{2}\rho U B L \begin{bmatrix} p_5^* & p_1^* & p_2^* B \\ h_5^* & h_1^* & h_2^* B \\ a_5^* B & a_1^* B & a_2^* B^2 \end{bmatrix} \tag{5.72}$$

$$[K_a] = -\frac{1}{2}\rho U^2 B L \begin{bmatrix} p_6^* \frac{\pi}{2V_\omega^{*2}} \frac{1}{B} & p_4^* \frac{\pi}{2V_\omega^{*2}} \frac{1}{B} & p_3^* \\ h_6^* \frac{\pi}{2V_\omega^{*2}} \frac{1}{B} & h_4^* \frac{\pi}{2V_\omega^{*2}} \frac{1}{B} & h_3^* \\ a_6^* \frac{\pi}{2V_\omega^{*2}} & a_4^* \frac{\pi}{2V_\omega^{*2}} & a_3^* B \end{bmatrix} \tag{5.73}$$

where $V_\omega^* = V^*/(2\pi)$

As a matter of fact, the identification of the flutter derivatives is not an easy task. The best results are obtained, imposing a harmonic motion and measuring the aerodynamic forces on a central part of the sectional model.

---

[4] A different formulation is also used, introduced by Simiu and Scanlan (1996).

If we impose, for instance, a vertical harmonic motion $z = Ze^{i\Omega t}$ the horizontal, vertical and moment forces can be measured. The motion amplitude must be small so that the problem can be linearized. The first harmonic of the measured forces—at the motion frequency—is evaluated and the complex transfer functions are identified. These transfer functions define a vector that, multiplied by $z = Ze^{i\Omega t}$ gives the the horizontal, vertical and moment forces:

$$\underline{F}_A = \begin{Bmatrix} F_y \\ F_z \\ F_\theta \end{Bmatrix} = \begin{Bmatrix} a_{yz} + i\Omega b_{yz} \\ a_{zz} + i\Omega b_{zz} \\ a_{\theta z} + i\Omega b_{\theta z} \end{Bmatrix} \cdot Ze^{i\Omega t} \tag{5.74}$$

The real part a of this vector identifies, in this case, the second column of the $[K_A]$ matrix, while the imaginary part identifies the second column of the $[R_A]$ matrix.

Analyzing a motion in $\theta = \Theta e^{i\Omega t}$ and measuring the F_y, F_z and F_θ forces, the third column of the $[K_A]$ and $[R_A]$ matrices can be identified, while first columns are evaluated imposing a motion $y = Ye^{i\Omega t}$, such completing the matrices.

From these tests the flutter derivatives value as a function of the frequency of the given motion, as well as of the given static equilibrium position angle α_0 around which the tests are done, can be derived. The flutter derivatives are expressed as a function of the reduced velocity:

$$U^* = \frac{U}{fB}$$

where B is the deck width, f is the frequency of the imposed motion and V is the velocity at which the test is performed. The coefficients which are very important for stability are the h_1^* and the a_2^*: these must be positive to give 1 d.o.f. vertical and torsional stability. As an example, the value of the h_1^* and the a_2^* for the Messina deck are reported in Figs. 5.42 and 5.43.

The reduced velocity is reported in the abscissa and the different curves presented for the coefficients are related to different angles of attack α_0 around which the harmonic motion is applied. It must be clear that the $[K_A]$ and $[R_A]$ matrix values are function of V^* and of the angle α_0 around which the small motion is imposed.

Identification of the Admittance Matrix

A method used in Politecnico di Milano wind tunnel and also by Chinese researcher (Niu et al. 2011) to identify the aerodynamic forces as a function of the incoming turbulence is to generate a well defined wind turbulence by an active device of the type shown in the Fig. 5.44.

The airfoil can rotate around a fixed axis with a controlled motion, which is generally harmonic, with different amplitudes and frequencies. This generates a given fluctuation of the incoming wind, that can be measured through suitable anemometers, like the hot-wire or multi hole pressure probes anemometers, see Fig. 5.45. The aerodynamic forces are measured on a sectional deck model for a given average angle of attack θ_0 and the coefficients of the admittance matrix are identified, in

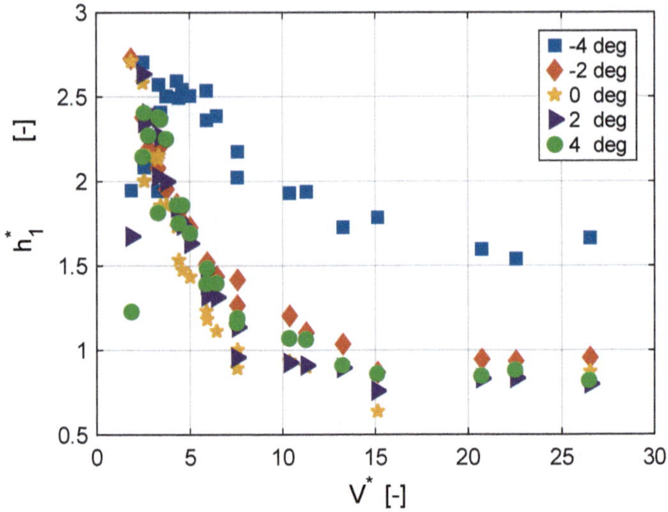

Fig. 5.42 h_1^* flutter derivative of Messina Bridge

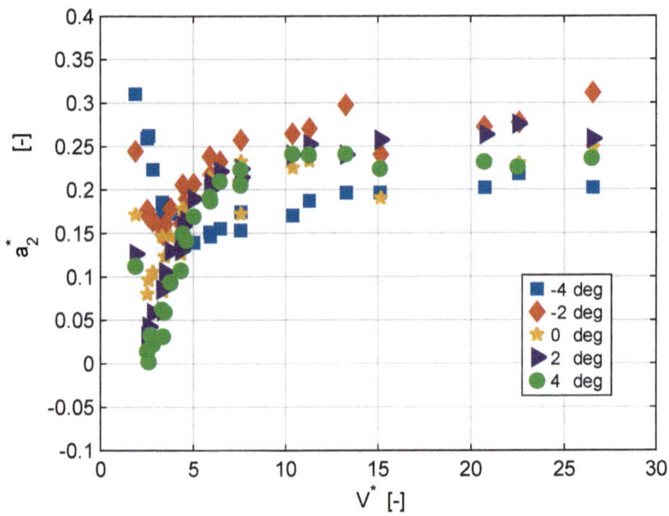

Fig. 5.43 a_2^* flutter derivative of Messina Bridge

Fig. 5.44 Active turbulence generator

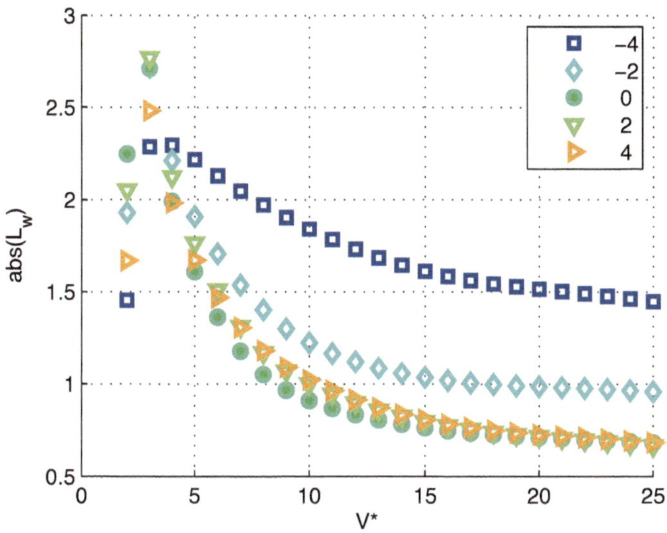

Fig. 5.45 Admittance function (Messina bridge)

a similar way as the flutter derivatives, through the definition of the transfer function between the turbulence components—as an input—and the aerodynamic forces components—as an output.

In these tests the turbulent wind vertical component value is the main produced/considered component. The first harmonic at the imposed frequency of the airfoil motion changes the wind angle of attack α, due to the incoming turbulence w/V. The admittances are functions of θ_0 and of the reduced velocity at which the test is performed. As an example, one of the admittance function coefficients is reported in Fig. 5.45 for the vertical force F_z and the incoming wind turbulent component w.

5.5.1.5 Reynolds Effect

The scaled reproduction of the wind-structure interaction must necessarily consider Reynolds number effects. To keep the correct ratio between the inertial and the viscous forces, the Reynolds number parameter:

$$\mathrm{Re} = \frac{VB}{\nu}$$

where B is a reference length for the deck width and ν is the air kinetic viscosity, should be same of the full scale condition. Due the practical limitations in the possibility to manage the single parameter, this request is commonly not satisfied and in the usual wind tunnel practice a sensitivity analysis on the Reynolds number dependency is performed. Reynolds number dependency is tested by comparing results obtained at different wind speed and/or on models with different scaling factor.

As an example, in Fig. 5.46, the static aerodynamic coefficients of the Messina bridge deck are reported for different Reynolds numbers (i.e. different wind speeds). The deck model is in 1:45 scale, being the model width B equal to 1.33 m. Both the drag coefficient and the aerodynamic moment coefficient show, even in a very small amount, a unique trend of modification increasing the Reynolds number parameter. For the maximum wind speed of 45 m/s, the Re number value is $4 \cdot 10^6$.

In some cases, dealing with curvilinear shapes, it is possible to modify the surface finishing by adopting an increased roughness in order to simulate at lower Reynolds similar fluid dynamic effects of higher Reynolds number conditions (Fig. 5.47). This wind tunnel practice allows to develop on a smaller model similar boundary layer conditions to what is produced at higher Reynolds number. Comparing the results of Fig. 5.46 with those reported in Fig. 5.48, obtained at the same wind speed on the same deck sectional model with increased surface roughness, it is possible to observe how the slope of the lift and of the aerodynamic moment coefficient tends to increase at higher Reynolds number. The values obtained with smooth surface closely match those obtained at low speed (10 m/s) with rough surface. In any case, for the rough surface, at velocities exceeding 36 m/s, corresponding to $\mathrm{Re} = 3.2 \cdot 10^6$ and more, the measured values reach an asymptotic regime, indicating that further increases in Reynolds number do not significantly affect the results. Therefore, these values can be considered representative of the fully developed or realistic flow conditions.

This is mainly due to the different fluid dynamic behavior that is present on the curvilinear part of the upwind box as shown in Fig. 5.49, where the pressure distribution for the nominally smooth and for the rough model are overlapped.

Another parameter that has to be controlled during the wind tunnel tests on scaled models, trying to respect the Reynolds number similitude, is the turbulence characteristics of the flow. Different incoming turbulence conditions may lead on the same model to different results in terms of aerodynamic coefficients obtained at the same wind speed. Nominal wind tunnel smooth flow conditions are actually low turbulence conditions that depends on the quality of the flow of the wind tunnel facility $(0.2 < I_u < 2\%)$.

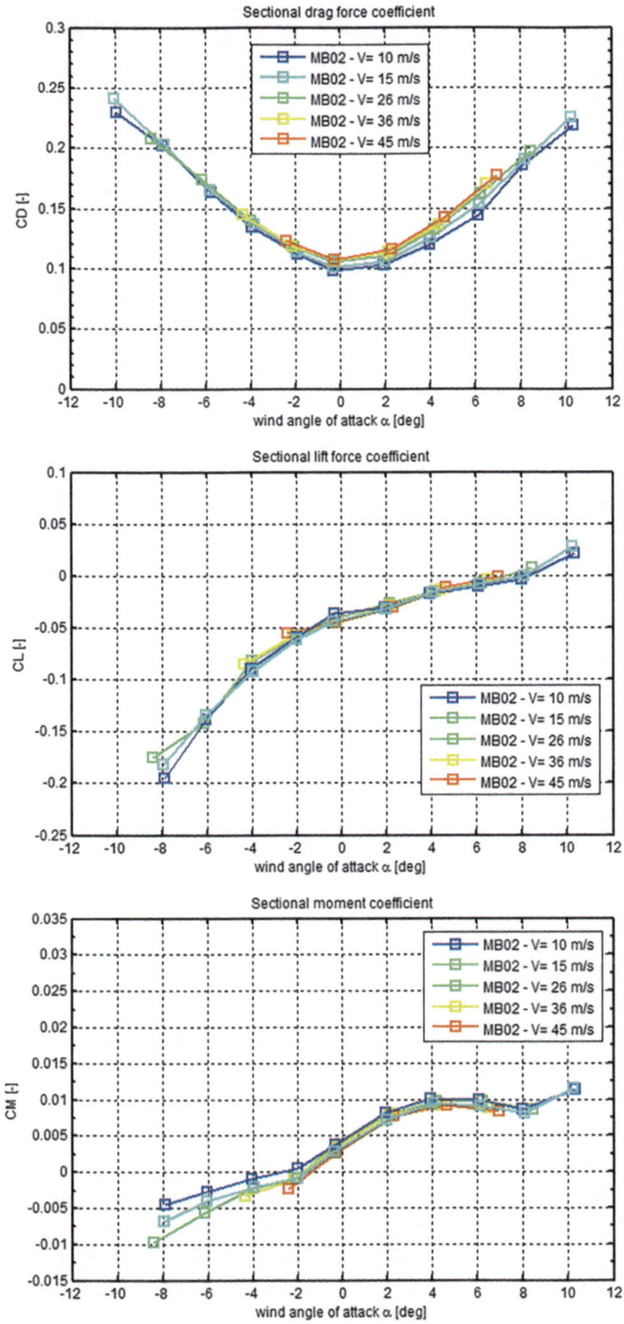

Fig. 5.46 Reynolds number dependency of the deck aerodynamic coefficients varying the wind speed

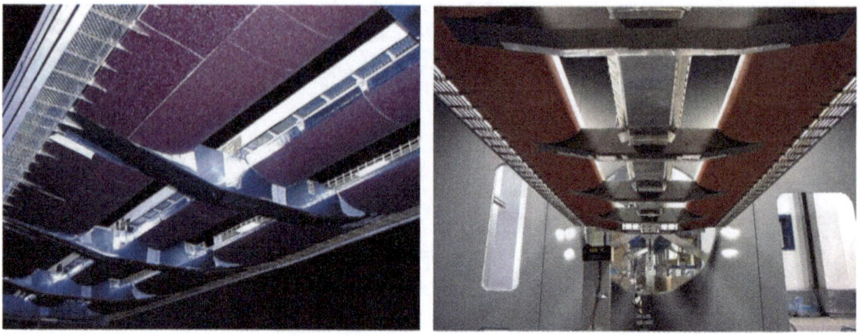

Fig. 5.47 Deck sectional model with increased surface roughness

Although the value of turbulence intensity is very small, in some cases, it may play an important role and it could explain the discrepancies that sometimes arise between results obtained in similar conditions in different wind tunnel plants.

5.5.2 *Tests on Aeroelastic Models*

Aeroelastic models are mainly employed in the final stage of the wind tunnel testing to study the aerodynamic response of the full bridge and the so-called construction stages, i.e. the temporary condition in which the bridge has not yet been completed. In this last case, the most common case study is represented by the aerodynamic characterization of the lone towers which can suffer from VIV issues while they have not yet been connected to the deck by means of stays or hangers.

5.5.2.1 Full Bridge

Full aeroelastic models are used to simulate the complex wind-structure interaction of the completed or semi-completed bridge. In the commonly adopted wind tunnel testing practice, the full aeroelastic model is used as a final check of all the design procedure. The goal is comparing numerical results with the experimental ones, obtained through the full aeroelastic model tests.

Once defined the aerodynamic behavior of each single component of the bridge by means of the previously described wind tunnel tests, the full bridge behavior may be predicted by numerical approaches relying on numerical models or by a scaled simulation of the full aeroelastic structure. Full aeroelastic models are also used to check if undesired dangerous conditions may arise during the construction stage when the structure has not yet reached its final robustness.

The reliability of the results depends on the quality of the model that has to reproduce in a very small scale the dynamic behavior of the real structure. There

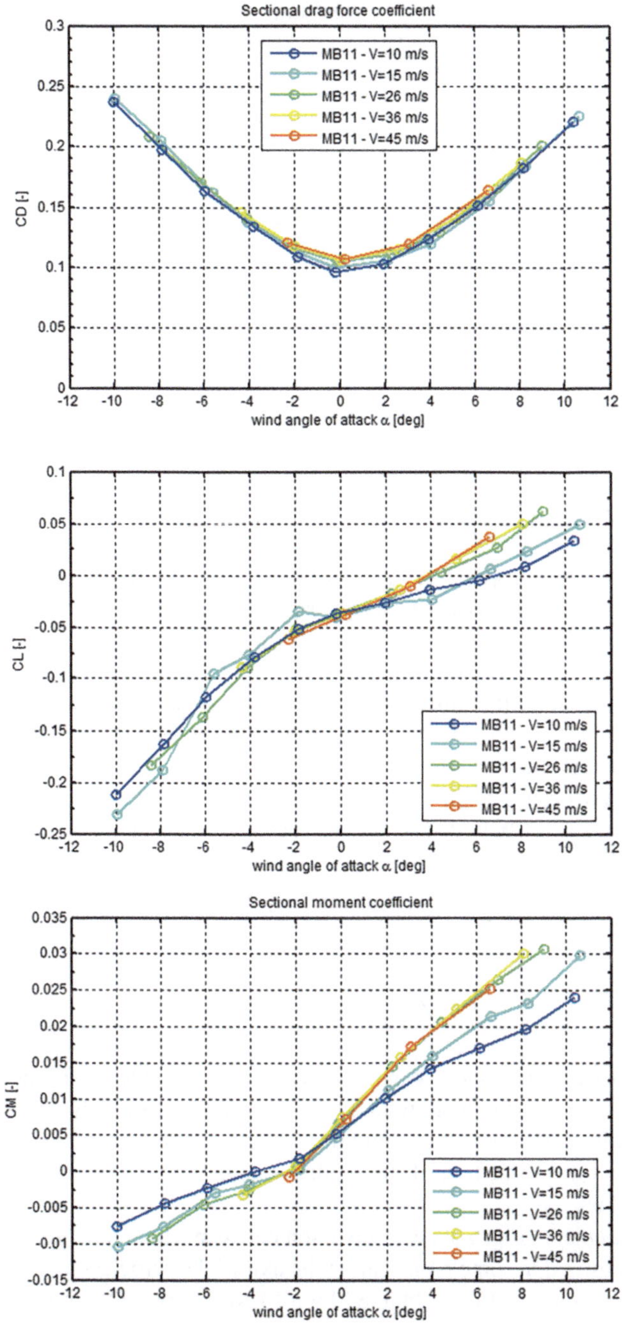

Fig. 5.48 Reynolds number dependency of the deck aerodynamic coefficients varying the wind speed (model with increased surface roughness)

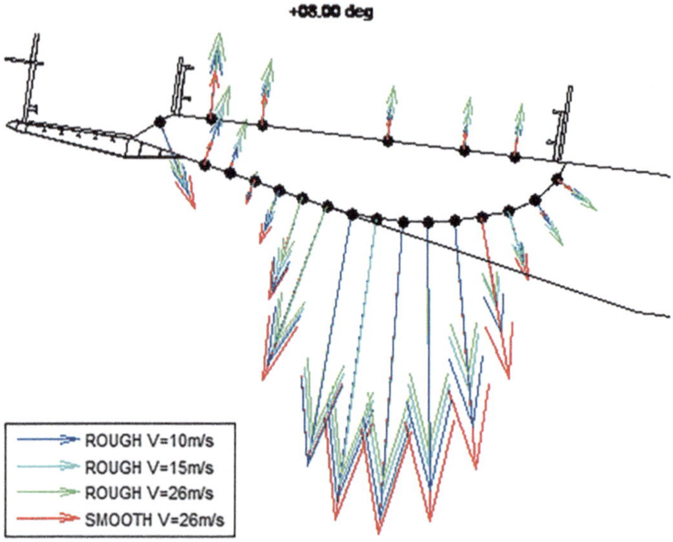

Fig. 5.49 Different boundary layer development (Surface pressure distribution, smooth vs rough model surface, different incoming wing speeds, angle of attack +8 °)

are scaling rules that, if fulfilled, allow reproducing in the correct proportions the aerodynamic and the structural forces in order to be representative of the wind-bridge interaction at full scale (see Sect. 3.1).

Adopting the Froude scaling rule, the velocity scaling factor λ_V and the frequency scaling factor λ_f are bounded to the geometrical scale factor $\lambda_L$according to the following relations:

$$\lambda_U = \frac{U_{FS}}{U_{Model}} = \sqrt{\lambda_L}$$

$$\lambda_f = \frac{f_{FS}}{f_{Model}} = \frac{1}{\sqrt{\lambda_L}}$$

The geometrical scale factor can result very large to be able to reproduce the whole very long structure in the wind tunnel test section. Consequently, the design of the full aeroelastic model can be challenging and a big effort is invested in the model tuning and in the verification of the model structural characteristic with respect to the target values.

As an example, in Fig. 5.50, a comparison between a torsional mode shape experimentally measured in some points along the axis of the aeroelastic full bridge model and the target numerical one obtained by a FEM is reported. Another key feature is the correct reproduction of turbulent characteristics of the atmospheric boundary layer. The goal is reached through the adoption of spires and floor roughness ele-

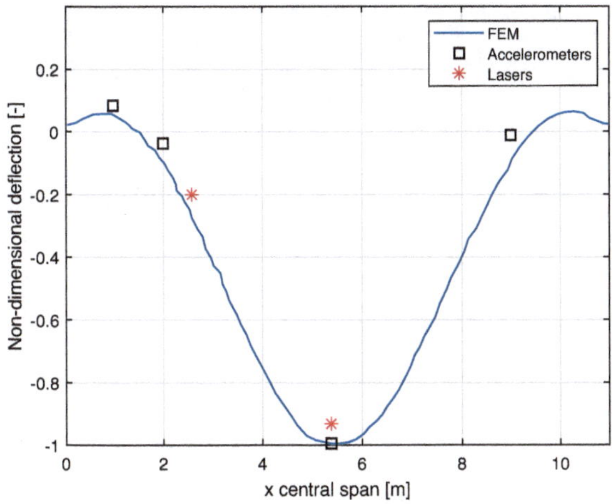

Fig. 5.50 Validation of the aeroelastic model 1:300 of Messina Bridge through the comparison between the 2nd torsional mode shape measured in some points along the bridge axis and the FEM target shape for the second torsional mode shape at 0.095 Hz (1.57 Hz model scale, target frequency 1.64)

ments that, interacting with the wind tunnel flow, produce a turbulent mixing of the incoming wind (see Sect. 3.2.3) (Figs. 5.51 and 5.52).

The turbulent wind characteristics in the wind tunnel must be scaled as described below:

1. Mean wind speed
2. Mean wind profile
3. Turbulence intensities
4. Turbulence profiles
5. Non dimensional Power spectra density
6. Turbulent length scales
7. Spatial correlation.

Wind tunnel tests on aeroelastic full bridge model are aimed at verifying the stability performance of the bridge and the buffeting response.

Figure 5.53 shows the modification of the transfer function between force and displacement that is obtained by exciting the model at the deck with a sinusoidal force at different frequencies and measuring both force and displacement at the same point. The transfer function was obtained in still air and at two different incoming wind speeds in smooth flow conditions. The modification of the transfer function highlights the effect of the aeroelastic coupling. This coupling results in a variation of the natural frequencies of the bridge (with a reduction of the torsional frequency) and in a variation of the slope of the phase diagram in correspondence of the resonance frequency, denoting an aeroelastic damping effect.

Fig. 5.51 Full aeroelastic model of Messina Bridge on the turning table of Polimi Wind Tunnel. Length of the model 14 m

Figure 5.54 reports the trend of the non-dimensional damping measured on the first two horizontal, vertical and torsional modes when the bridge model is run over by smooth flow with different mean wind speeds, and it is released starting from initial conditions corresponding to the mode shape under analysis. It is possible to observe that, for wind speed (reported in model scale) higher than 4 m/s, the second torsional mode go towards negative value of damping denoting the occurrence of the instability.

From the analysis of the bridge response measured under turbulent wind conditions, it is possible to study the buffeting response of the bridge for different mean wind. Figure 5.55 plots the RMS value of the vertical acceleration in the different instrumented sections of the deck versus the mean wind speed. The trend is parabolic, nevertheless vortex shedding phenomena occur, up to conditions where instability problem arise.

An example of the buffeting response to turbulent wind is reported in Fig. 5.56 in terms of the frequency spectrum of the vertical acceleration of the deck section close to mid-span, where the contribution of the different structural modes of the structure is evident.

The comparison of these experimental results with those obtained through the numerical approach, as already said, represents the final check of all the design procedure, for what concerns the bridge response to the wind action.

Fig. 5.52 Full aeroelastic model of a cable stayed bridge

Orography

When the surrounding may influence the wind flow, a scaled reproduction of the most relevant part of the local orography is considered. Figure 5.57 shows an example of a hill positioned close to the region where a cable-stayed bridge has to be built. This presence of this natural element of the surrounding may produce deflections

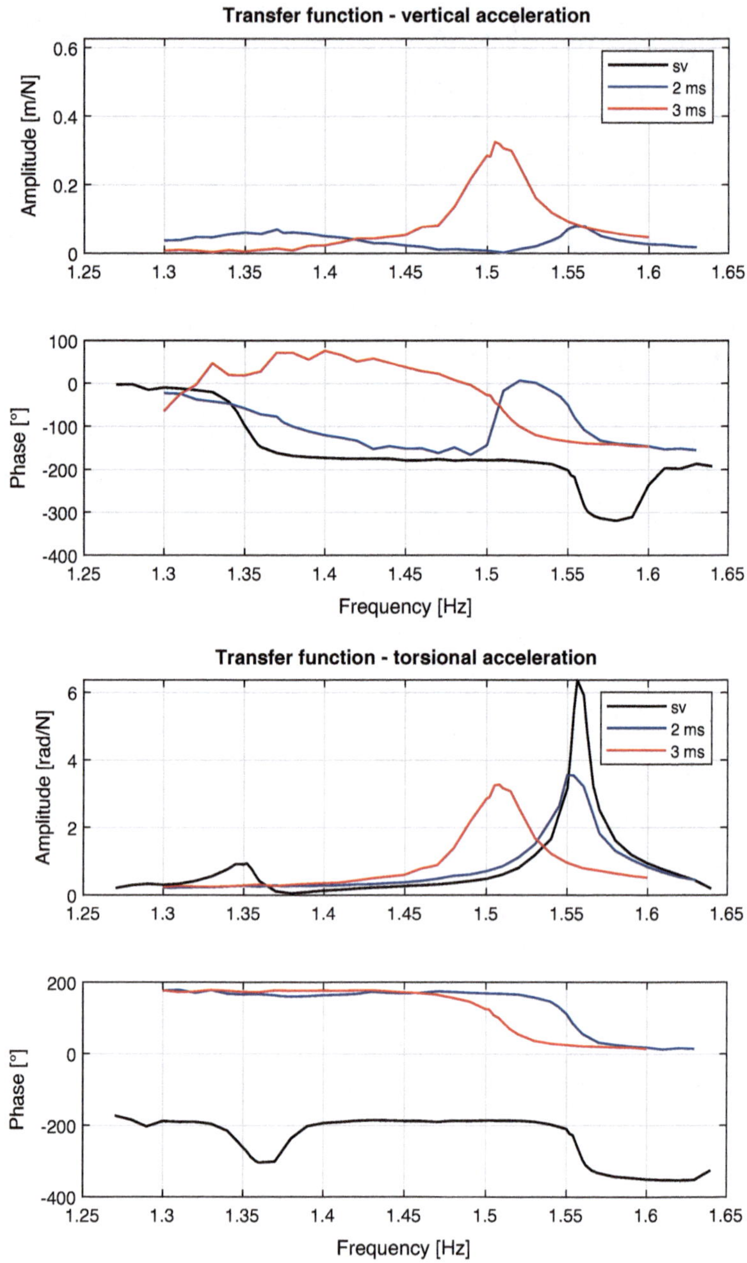

Fig. 5.53 Transfer function at different wind speed, measured at midspan at 0, 2, and 3 m/s mean wind speed, under smooth flow condition, for the aeroelastic model of the Messina Bridge (deck width $B = 0.2m$); **a** Vertical acceleration of the deck, **b** Torsional acceleration of the deck

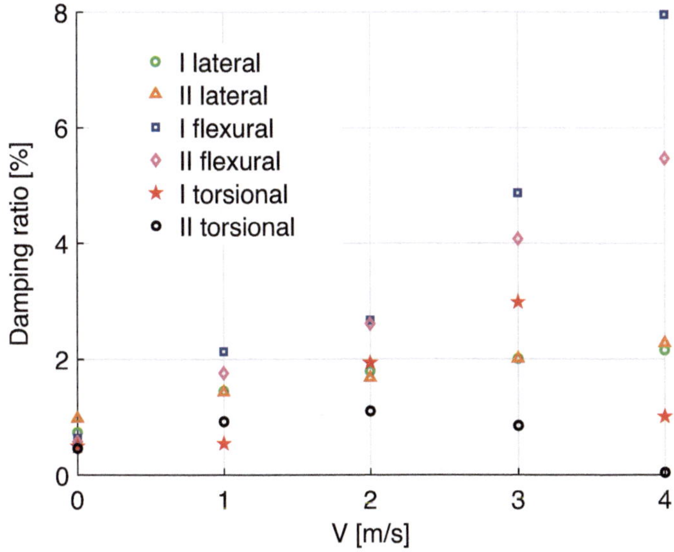

Fig. 5.54 Non dimensional damping of the first six modes varying the mean wind speed under smooth flow conditions (for the aeroelastic model of the Messina Bridge); $\lambda_L = 280$, $\lambda_U = 16.3$, $V = 67m/s$ Full-scale

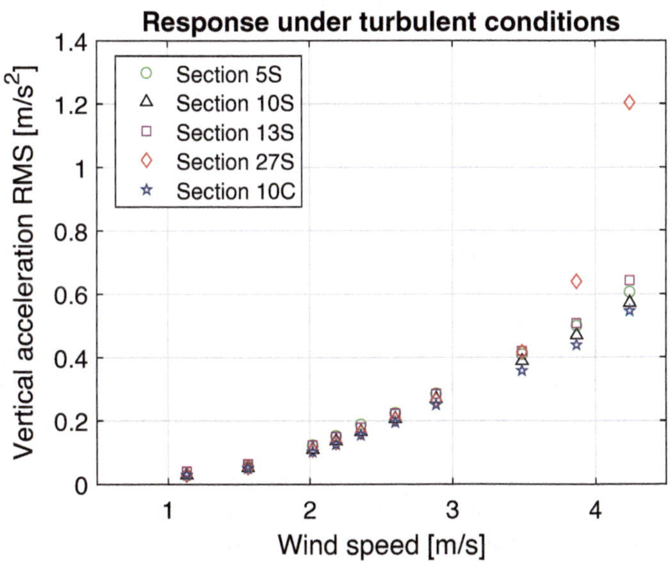

Fig. 5.55 RMS value of the vertical acceleration of different sections along the bridge axis varying the mean wind speed under turbulent conditions, for the aeroelastic model of the Messina Bridge

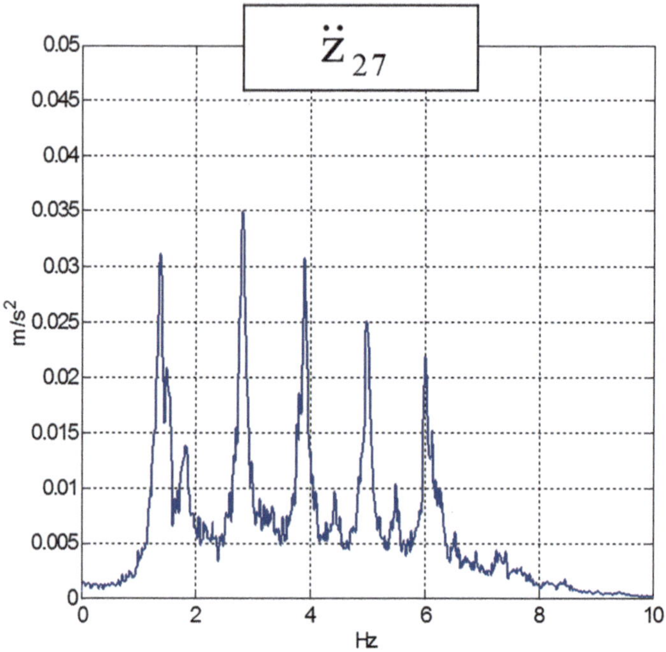

Fig. 5.56 Buffeting response: vertical acceleration measured at mid span

and increase the turbulence effects for some wind exposures. In this case, a scaled geometrical reproduction of the scenario close to the aeroelastic bridge model, which represents a construction stage, is visible.

5.5.2.2 Tower and Construction Stages

The main problem that must be addressed during the aerodynamic design of bridge towers is VIV, if we consider cable-stayed bridges that usually have single leg or reversed Y-shaped towers or suspension bridges that have H-shaped towers with two legs connected by one or more transverse beams. In both cases, specific wind tunnel tests are performed using sectional or aeroelastic models.

VIV of bridge towers can be investigated in the wind tunnel by measuring the dynamic response of sectional models that are elastically suspended in the flow in a similar manner to what already described for the deck. However, while the vortex shedding phenomenon is basically two-dimensional on the deck, on the contrary, for the tower, it is affected by the three-dimensional end effects at top and by the presence of the transverse beams along the height. Therefore, wind tunnel tests performed on aeroelastic models are aimed at representing the full 3D tower geometry and the related aerodynamic effects.

Fig. 5.57 Orography reproduction

Aeroelastic models are moreover used to investigate the different construction stages (Fig. 5.58) of the tower, since its dynamic characteristics may largely vary from the free standing condition to the bridge completion condition. If wind tunnel test results highlighted that VIV can reach dangerous values, suitable countermeasures must be implemented. The design must be driven by experimental results, which must consider the dynamic effects of all the scaffolding and cranes that are connected to the structure and participate to the dynamic response of the system.

In general, the optimization of the shape of the tower in order to avoid vortex shedding excitation is not an easy task. The final goal of the analysis is similar to what is done for the deck, where VIVs are reported as a function of Scruton number.

Going into detail, in Fig. 5.59 a typical tower section is reported together with the section of the tower legs. The cross sectional shape of the tower legs is generally a bluff body. The shape of the corner edge produces therefore flow detachment and then vortex shedding.

When the vortex shedding frequency becomes equal to one of the natural frequencies of the tower either in the stand-alone configuration during construction stage or in service with all the restraints applied by the cables and deck, then vibrations can be excited with vibration amplitudes that depend on the structural damping of the tower. The amplitudes are high if the damping is low, or, using a more representative figure for vortex shedding severity, if the Scruton number is low.

The Scruton number is, in this case, defined as

$$Sc = 2\pi \frac{m\,\xi}{\rho D^2}$$

Fig. 5.58 Free standing tower

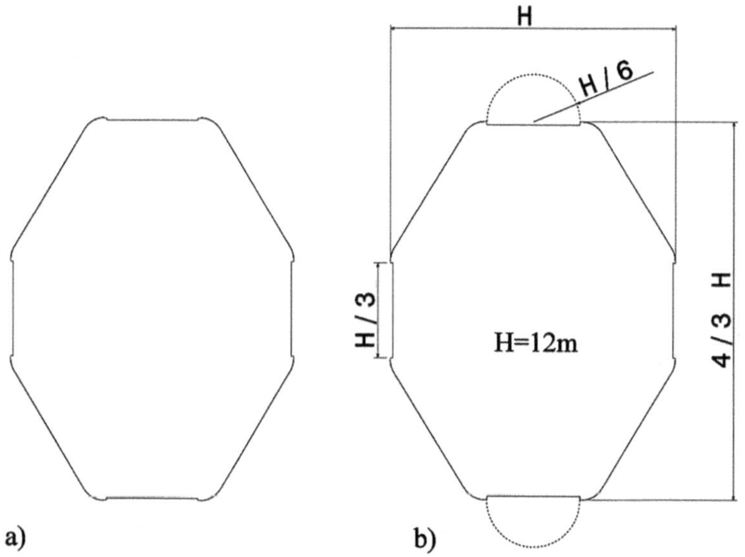

Fig. 5.59 Messina tower section: **a** Original configuration "Without Screens", **b** configuration "With Screens"

where m, is the tower linear mass, ξ the non-dimensional damping, ρ is the air density and D a reference sectional dimension of the tower leg.

In the stand-alone configuration during the construction stage it is relatively easy to prevent high tower vibrations by installing dampers or attaching temporary stays connected to dampers.

When the construction is complete and the tower reaches its final condition, the natural frequencies of the tower are generally higher than in the stand-alone configurations due to the restraint from the main cables. The wind speed at which severe vortex shedding vibrations may be excited increases, so the probability of occurrence decreases. Nevertheless, this problem must be fully investigated.

As an example, for the Messina bridge, the wind velocity at which the first vibration mode of the tower is excited is shown in Fig. 5.60 to be around 12 m/s for the free standing configuration, whereas it is much higher at around 57 m/s for the condition when the main cables are attached. This is due to the main cable restraint which raises the first mode frequency by a factor of four.

These tests have been performed on an aeroelastic model of the tower reproducing in the model scale the natural frequencies and modes of vibration of the real tower.

As it can be observed, the wind velocity at which vortex shedding may be excited can occur both in the stand-alone and in the final configuration, although with different probabilities.

Many attempts have been made to control this problem, common to the towers of all long span bridges. In 1992, for the design of the Messina towers, porous screens were placed at the edges of each leg section (Fig. 5.61) to mitigate vortex

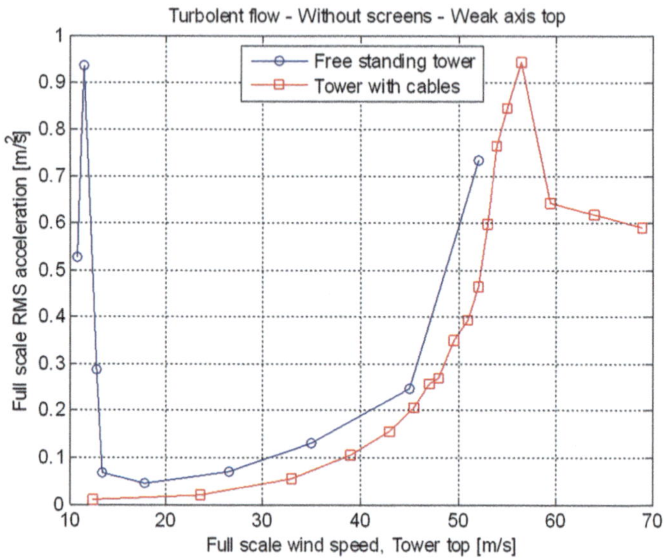

Fig. 5.60 Wind speeds at which are excited the first tower modes in stand alone

shedding excitation. Figure 5.61 shows that, due to the very effective role of the porous screens, no vortex shedding excitation is observed in either the free standing or final construction stage configuration.

This solution, however, has some disadvantages, such as the maintenance required by the screens and also the uncertainties introduced which are related to the difficulty of accurately representing Reynolds number in the model scale.

For the final tower design, it was decided to control vortex shedding excitation through the addition of large Tuned Mass Dampers (TMDs) in such a way as to increase the Scruton number and avoid dangerous vibrations of the tower. The same strategy was adopted on the Akashi Bridge towers, placing several TMDs in different locations within the tower legs.

5.6 Finite Element Models of the Full Bridge

5.6.1 Numerical Model for the Analysis of the Bridge Response to the Wind

It is common practice to develop finite element schematizations of the bridge (Fig. 5.62) to identify its response to the different actions of interest: wind, earthquake and traffic.

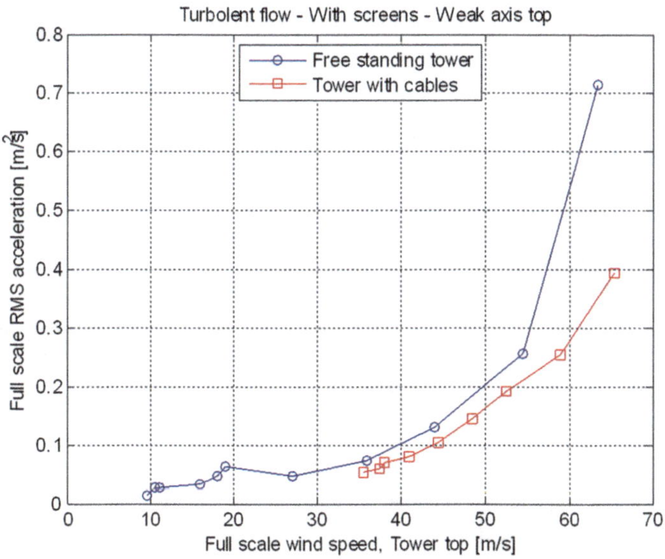

Fig. 5.61 Messina Tower section—with porous screens installed to mitigate vortex shedding excitation

For what concerns the wind action, generally beam elements are used for deck and tower, while taut string or tensioned beam elements are used for the cables. These models are generally used to compute the global behavior of the bridge to the different actions and they can be non linear or linear, as a function of the type of excitation.

A non linear approach is used to identify the static position of the bridge under the permanent load, or static load, like the traffic and the wind loads due to the average wind (Diana et al. 2005). For what concerns the dynamic load, a linear approach is generally used, being the linearization performed around the static equilibrium configuration (Diana et al. 2005; Chen et al. 2001; Jain et al. 1996; Katsuchi et al. 1999) as it will be better explained in the following.

To compute the stress induced by the dynamic load—or local effect—a more sophisticated schematization of the deck and tower is used, employing plate elements. In Fig. 5.64, as an example, the finite element schematization of the central box of the Messina deck is reported. This FEM has been used to assign the stresses on the boxes welding, in order to verify the fatigue problem.

The global bridge FEM (Fig. 5.62) is used to compute the bridge natural frequencies, which are fundamental for the analysis of the bridge response to the wind action.

As previously said, the ratio between the torsional and vertical frequencies of the same mode is very important for the flutter instability problem: the greater this ratio, the higher the flutter velocity. In order to reproduce the bridge response to the

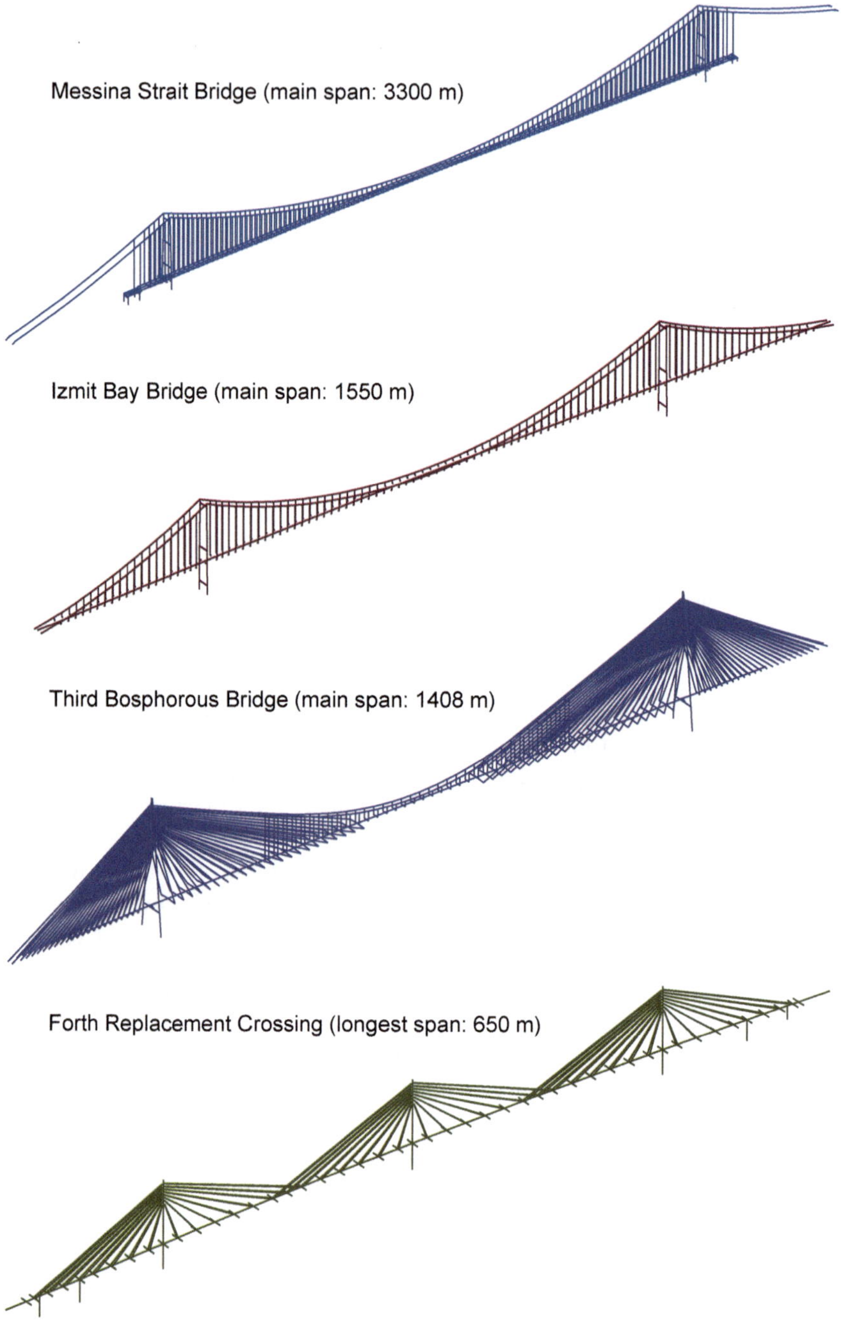

Fig. 5.62 Global FEM scheme of different bridges

turbulent wind, the dynamic load due to the incoming turbulence has to be applied to the FEM scheme.

Knowing the structure of the wind in the site surrounding the bridge, it is possible, by special tools (Chen and Kareem 2001a; Ding et al. 2006), to reproduce a space and time wind distribution—see Chap. 2. Figure 5.63 shows an example of a numerical simulated turbulent wind velocity w, for two different levels of index of turbulence I_w.

The forces acting on the bridge are a function of the incoming turbulence. On the other hand, as already observed, the bridge motion produces aerodynamic forces which are motion dependent. As a consequence, the aerodynamic forces acting on the bridge are functions both of the incoming turbulence and of the bridge motion.

In order to reproduce the over mentioned wind forces in the bridge FEM, the following equation has to be considered:

$$[M_s]\ddot{\underline{X}} + [R_s]\dot{\underline{x}} + [K_s]\underline{x} = \underline{F}_A\left(\underline{x}, \dot{\underline{x}}, U, \underline{b}\right) \tag{5.75}$$

where $[M_s]$, $[R_s]$, $[K_s]$ are mass, damping and stiffness matrices of the bridge structure linearized in the neighbourhood of the static equilibrium position, $\underline{x}$ is a vector containing the bridge degrees of freedom. The $\underline{x}$ vector defines the motion of the bridge components: deck, towers, cables. $\underline{F}_A$ is the vector defined by the aerodynamic forces acting on the different sections of the bridge deck, towers, cables. $\underline{F}_A$ is function, as previously said, of the bridge motion $\underline{x}$ and of the associated velocity, as well as of the incoming turbulence vector $\underline{b}(t)$.

The identification of this non linear function through wind tunnel tests is a complex matter and it is for sure one of the most advanced topics of the research in this area (Diana et al., 2008, 2010; Wu and Kareem, 2011). We have already explained the QSTC that is one the used approaches, with the limitation that it is valid for high reduced velocities, $V^* > 10 \div 15$.

The most used approach consists in linearizing the aerodynamic forces in the neighborhood of the static equilibrium position of the bridge $\underline{x}_0$, identified by a given average wind V_m. This leads to:

$$\begin{aligned}
&[M_s]\ddot{\tilde{\underline{X}}} + [R_s]\dot{\tilde{\underline{X}}} + [K_s](\tilde{\underline{X}} + \underline{x}_0) \\
&= \underline{F}_{A0}\left(\underline{X}_0, \underline{0}, V_m, \underline{0}\right) \\
&+ \left.\frac{\partial \underline{F}_A}{\partial \underline{x}}\right|_{\underline{X}_0, \underline{0}, V_m, \underline{0}} \tilde{\underline{X}} + \left.\frac{\partial \underline{F}_A}{\partial \dot{\underline{x}}}\right|_{\underline{X}_0, \underline{0}, V_m, \underline{0}} \dot{\tilde{\underline{X}}} + \left.\frac{\partial \underline{F}_A}{\partial \underline{b}}\right|_{\underline{X}_0, \underline{0}, V_m, \underline{0}} \underline{b}
\end{aligned} \tag{5.76}$$

where $\underline{F}_{A0}$ are the static aerodynamic forces due to the average wind V_m, applied to all the bridge members, function of the C_D, C_L and C_M coefficients of the different parts of the bridge—deck, towers, cables—and $\underline{b}(t)$ is a vector containing the incoming wind turbulent velocity in the different sections of the bridge, where $\underline{F}_A$ is applied—see Sect. 5.3.2, where the QST is developed.

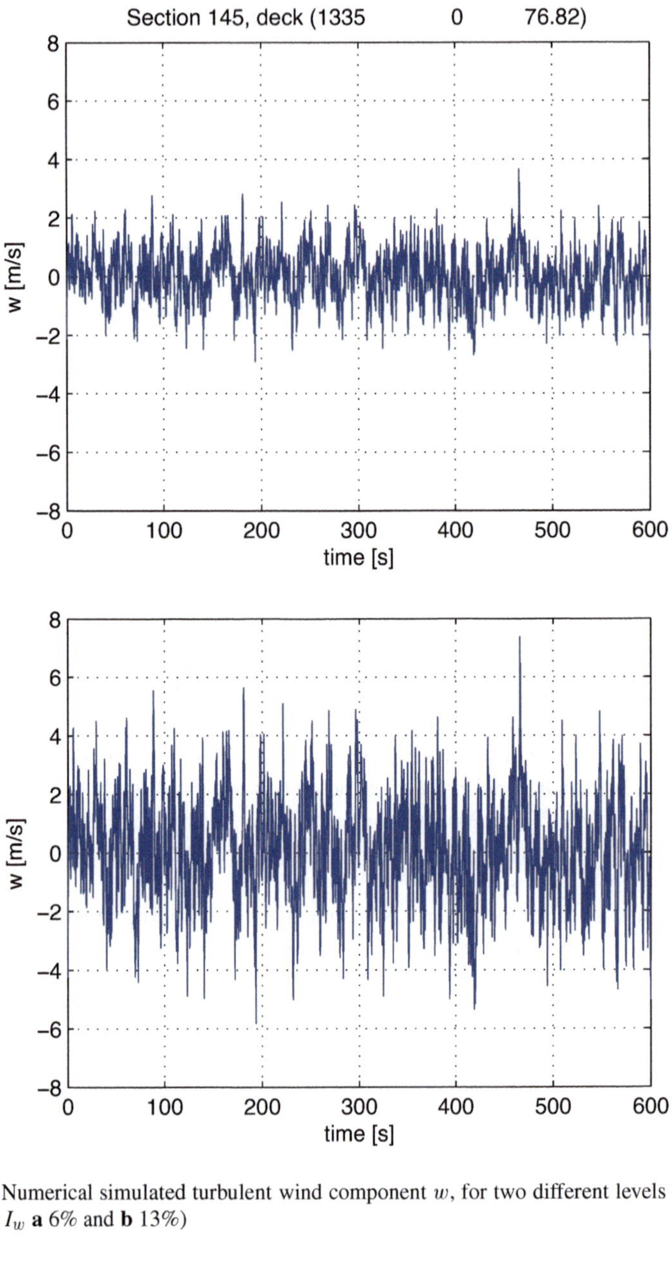

Fig. 5.63 Numerical simulated turbulent wind component w, for two different levels of index of turbulence I_w **a** 6% and **b** 13%)

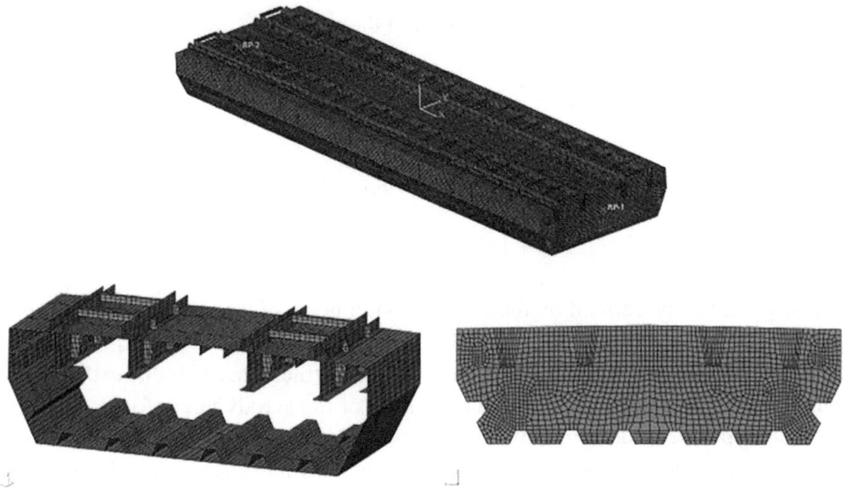

Fig. 5.64 FEM scheme (shell elements) of the Messina Bridge railway girder

The aerodynamic forces $\underline{F}_A$ and the associated derivatives $\left.\frac{\partial \underline{F}_A}{\partial \underline{x}}\right|_{\underline{X}_0,\underline{0},V_m,\underline{0}}$, $\left.\frac{\partial \underline{F}_A}{\partial \underline{\dot{x}}}\right|_{\underline{X}_0,\underline{0},V_m,\underline{0}}$, called flutter derivatives, must be identified through suitable wind tunnel tests on the deck, towers and cables, as will be better explained in the next paragraphs.

The aerodynamic forces playing the most important role in the bridge motion and stability are those acting on the deck.

The $\left.\frac{\partial \underline{F}_A}{\partial \underline{x}}\right|_{\underline{X}_0,\underline{0},V_m,\underline{0}}$ term gives rise to the equivalent elastic matrix $-[K_A]$, the $\left.\frac{\partial \underline{F}_A}{\partial \underline{\dot{X}}}\right|_{\underline{X}_0,\underline{0},V_m,\underline{0}}$ term gives rise to the equivalent damping matrix $-[R_A]$.

In the previous section the QSTC was described to define the aerodynamic forces $\underline{F}_A$ on the deck, together with the linearization procedure that allows to determine the above mentioned elastic and damping matrices $[K_A]$ and $[R_A]$.

The aerodynamic forces acting on a deck section are, as already said, function of the vertical (z), the horizontal (y) and the torsional (θ) deck motion, as well as of the incoming wind turbulent velocity.

As usual, the aerodynamic forces due to the incoming turbulence are defined by a lift component normal to the direction of the wind $V(t)$ and a drag component parallel to $V(t)$, both applied at the center of the section, plus a moment of the aerodynamic forces M_a.

Coming back to Eq. (5.75), the lift, drag and moment on each section of the deck represent a sub-vector of the vector containing all the lift, drag and moment forces applied to all the bridge sections while X is the vector of all the section displacements and of all the section velocities.

If we use a linearized approach, adopting the notation

$$\left.\frac{\partial \underline{F}_A}{\partial \underline{X}}\right|_{\underline{X}_0,\underline{0},V_m,\underline{0}} = -[K_A]; \quad \left.\frac{\partial \underline{F}_A}{\partial \underline{\dot{X}}}\right|_{\underline{X}_0,\underline{0},V_m,\underline{0}} = -[R_A]; \quad \left.\frac{\partial \underline{F}_A}{\partial \underline{b}}\right|_{\underline{X}_0,\underline{0},V_m,\underline{0}} = -[A_m];$$

equation (5.76) can be written in the following way:

$$[M_s]\ddot{\tilde{\underline{X}}} + ([R_s] + [R_a])\dot{\tilde{\underline{X}}} + ([K_s] + [K_a])\tilde{\underline{X}} = [A_m]\underline{b} \tag{5.77}$$

where $\underline{b}$ is a vector defined by the w, u, v components of the wind turbulence in the different points of the structure (see Sect. 5.3.2, where the QSTC is explained).

$[R_A]$ and $[K_A]$ are the equivalent damping and stiffness matrices due to the aerodynamic forces. In other words, the linearization of the aerodynamic forces produces, as already pointed out, an equivalent elastic and damping system that is added to the structural one, changing the natural frequencies and the overall damping. An aeroelastic system is produced. The $[K_A]$ matrix, as well as the $[R_A]$ matrix, are non symmetric and can give rise to flutter instability.

$[R_A]$, if non positive defined, can give rise to the so called one degree of freedom instability according to a torsional or vertical motion when the slope of the lift or moment coefficients is not positive. $[A_m]$ is the so called Admittance matrix.

All the parameters needed to define the $[K_A]$, $[R_A]$ and $[A_m]$ matrices must be identified through wind tunnel tests in order to reproduce the bridge response to the turbulent wind. Equation (5.77) can also be solved in the frequency domain Jain et al. 1996; Katsuchi et al. 1999; Minh et al. 2000: in this case the wind input is the spectrum of the incoming turbulence.

5.6.2 Comparison of Experimental and Numerical Results

Once all the above described aerodynamic tests have been completed, it is possible to compute the bridge response to the turbulent wind.

First of all the static equilibrium position of the bridge under the turbulent wind must be computed, as already explained.

For each section of a bridge a time history of the turbulent wind must be generated. The spectrum of the single time history can be identified and the Eq. (5.77) can be solved, for instance, in the frequency domain, taking into account that the flutter derivatives and the admittance function are frequency dependent.

Other methods are available to solve this problem in the time domain Caracoglia and Jones 2003, generally with modal approach Diana et al. 2005, and some of these methods, as already mentioned, take into account also the non linear effects of the aerodynamic forces Diana et al. 1995; Chen and Kareem 2001b.

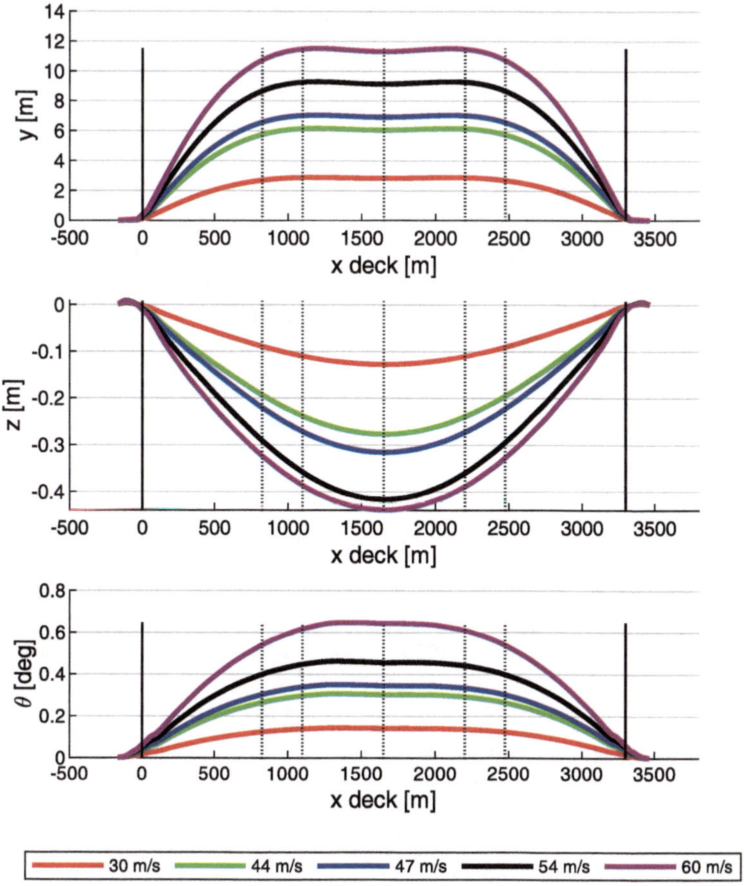

Fig. 5.65 Static deformation under different mean wind speeds

Some example of output result are reported in the following. Figure 5.65 reports the static deflection of the Messina bridge deck under the action of mean wind forces for different mean wind speeds. Figure 5.66 reports RMS values of the vertical acceleration of the different deck section along the bridge axis of the Messina bridge deck under the action of turbulent wind.

For the locations reported in Fig. 5.67, The stress in the road boxes for a wind speed 30 m/s and $I_u = 13.8\%$ are reported in Fig. 5.68 in terms of standard deviation of the normal stresses on the downwind road box ($y > 0$) at point A.

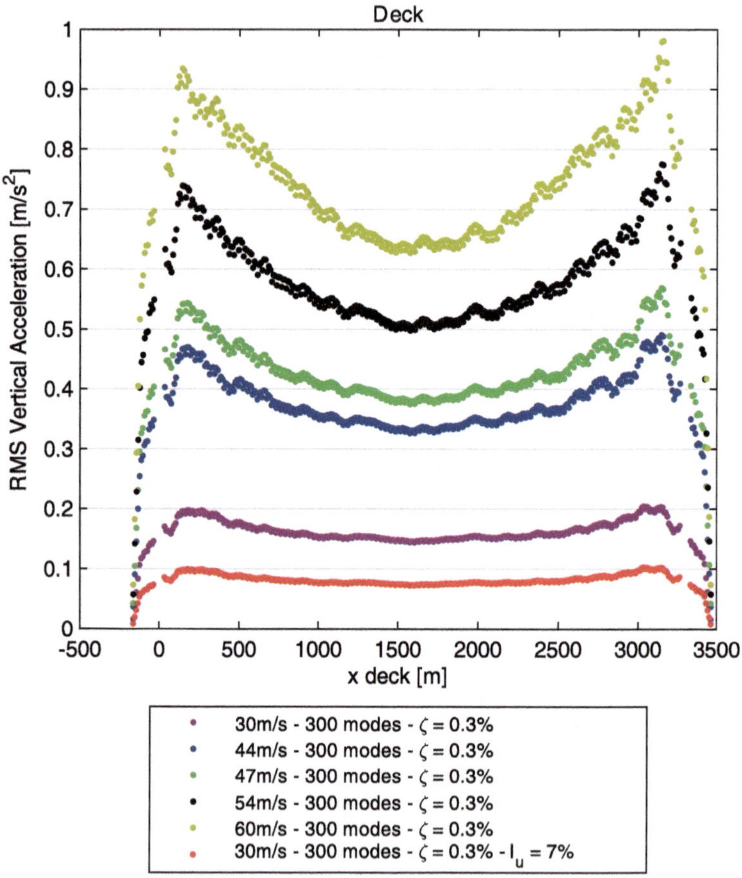

Fig. 5.66 RMS values of the vertical acceleration of the different deck section along the bridge axis

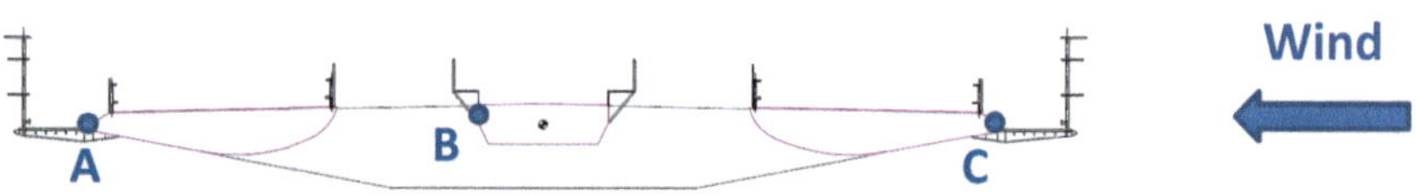

Fig. 5.67 Considered point on the road and railway boxes of the Messina bridge deck section

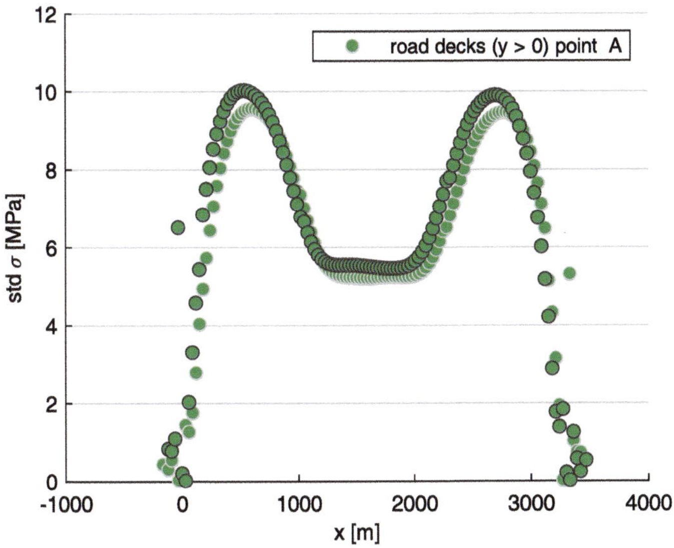

Fig. 5.68 Standard deviation of the stresses at the point A of the different deck sections along the bridge axis. Mean wind speed 30 m/s, index of turbulence 0.138

5.7 Appendix: from Modal Shapes to Section Models

5.7.1 Sectional Approach

The sectional approach in wind engineering, particularly when analyzing the aerodynamic behavior of bridges, refers to a method where the deck is modeled and studied based on its cross-sectional properties rather than as a full three dimensional structure. The flow around the section is treated as two-dimensional, simplifying the problem, although this assumption may overlook certain three-dimensional effects such as end effects or turbulence in real-world scenarios.

Typically, the deck is divided into several segments of length L_i, usually along the deck chord ($L_i = B$). As an example, refer to Fig. 5.69. Each section is considered independent of the others from an aerodynamic point of view. Thus, the aerodynamic coefficients remain the same for every deck section (assuming identical geometry), but the aeroelastic forces differ since each section moves differently. Additionally, buffeting forces vary because the wind conditions differ from one section to another.

A key assumption in this approach is that each section is rigid and the aerodynamic forces do not vary along it: therefore only three degrees of freedom are considered to define the section motion (horizontal y, vertical z, and torsional θ displacements) and the aerodynamic forces acting on the section are the forces per unit length multiplied by L_i.

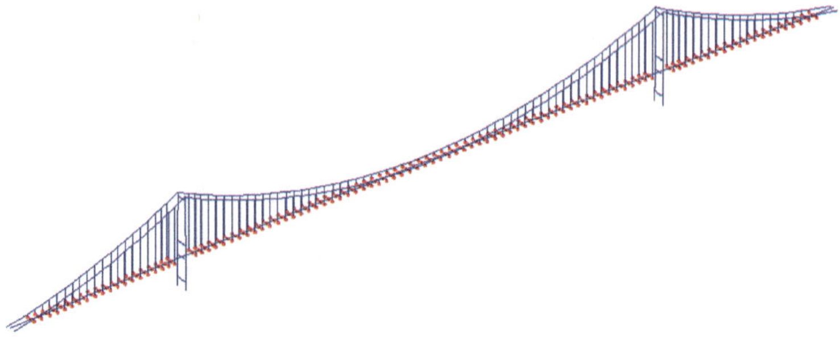

Fig. 5.69 Finite element model of a bridge using an equivalent beam scheme: deck sections are highlighted in red

5.7.2 Modal Approach

When modeling the bridge using a modal approach, it is clear that each section motion can be defined as:

$$\mathbf{x}_i = \begin{bmatrix} y_i \\ z_i \\ \theta_i \end{bmatrix} = \sum_{n=1}^{N} \begin{bmatrix} \phi_{y_i} \\ \phi_{z_i} \\ \phi_{\theta_i} \end{bmatrix}_n q_n = \boldsymbol{\Phi}_i \mathbf{q} \tag{5.78}$$

being $i = 1, \ldots, N_s$ the section number, q_n the nth modal coordinate ($n = 1, \ldots, N$), and $\boldsymbol{\phi}_{i,n}$ the nth mode shape for the ith deck section:

$$\boldsymbol{\phi}_{i,n} = \begin{bmatrix} \phi_{y_i} \\ \phi_{z_i} \\ \phi_{\theta_i} \end{bmatrix}_n \tag{5.79}$$

So recalling that the aeroelastic forces, for each section, can be defined as

$$\mathbf{F}_{A,i} = -\mathbf{K}_{A,i}\mathbf{x}_i - \mathbf{R}_{A,i}\dot{\mathbf{x}}_i \tag{5.80}$$

the Lagrangian component of the aeroelastic forces for the jth mode, Q_j, can be written as:

$$Q_j = \sum_{i=1}^{N_s} \boldsymbol{\phi}_{i,j}^T \mathbf{F}_{A,i} \tag{5.81}$$

$$= \sum_{i=1}^{N_s} \boldsymbol{\phi}_{i,j}^T \left(-\mathbf{K}_{A,i}\mathbf{x}_i - \mathbf{R}_{A,i}\dot{\mathbf{x}}_i\right) \tag{5.82}$$

$$= \sum_{i=1}^{N_s} \boldsymbol{\phi}_{i,j}^T \left(-\mathbf{K}_{A,i}\boldsymbol{\Phi}_i\mathbf{q} - \mathbf{R}_{A,i}\boldsymbol{\Phi}_i\dot{\mathbf{q}}\right) \tag{5.83}$$

$$= \sum_{i=1}^{N_s} \boldsymbol{\phi}_{i,j}^T \left(-\mathbf{K}_{A,i}\sum_{n=1}^{N}\boldsymbol{\phi}_{i,n}q_n - \mathbf{R}_{A,i}\sum_{n=1}^{N}\boldsymbol{\phi}_{i,n}\dot{q}_n\right) \tag{5.84}$$

Changing the order of the sum operators and manipulating the expressions, it is possible to notice that aeroelastic forces couples the modal coordinates:

$$Q_j = -\left(\sum_{i=1}^{N_s} \boldsymbol{\phi}_{i,j}^T \mathbf{K}_{A,i}\sum_{n=1}^{N}\boldsymbol{\phi}_{i,n}q_n + \sum_{i=1}^{N_s} \boldsymbol{\phi}_{i,j}^T \mathbf{R}_{A,i}\sum_{n=1}^{N}\boldsymbol{\phi}_{i,n}\dot{q}_n\right) \tag{5.85}$$

$$= -\sum_{n=1}^{N}\left(\sum_{i=1}^{N_s} \boldsymbol{\phi}_{i,j}^T \mathbf{K}_{A,i}\boldsymbol{\phi}_{i,n}\right) q_n - \sum_{n=1}^{N}\left(\sum_{i=1}^{N_s} \boldsymbol{\phi}_{i,j}^T \mathbf{R}_{A,i}\boldsymbol{\phi}_{i,n}\right) \dot{q}_n \tag{5.86}$$

$$= -\sum_{n=1}^{N}\left(k_{jn}^*\right) q_n - \sum_{n=1}^{N}\left(r_{jn}^*\right) \dot{q}_n \tag{5.87}$$

$$= -\left[k_{j1}^*\ k_{j2}^* \dots k_{jN}^*\right]\mathbf{q} - \left[r_{j1}^*\ r_{j2}^* \dots r_{jN}^*\right]\dot{\mathbf{q}} \tag{5.88}$$

In fact, collecting all the Q_j in a single vector $\mathbf{Q}$ it is possible to see that the aeroelastic forces are expressed as fully coupled modal stiffness and damping matrices:

$$\mathbf{Q} = -\begin{bmatrix} k_{11}^* & k_{12}^* & \dots & k_{1N}^* \\ k_{21}^* & k_{22}^* & \dots & k_{2N}^* \\ \vdots & \vdots & \ddots & \vdots \\ k_{N1}^* & k_{N2}^* & \dots & k_{NN}^* \end{bmatrix}\mathbf{q} - \begin{bmatrix} r_{11}^* & r_{12}^* & \dots & r_{1N}^* \\ r_{21}^* & r_{22}^* & \dots & r_{2N}^* \\ \vdots & \vdots & \ddots & \vdots \\ r_{N1}^* & r_{N2}^* & \dots & r_{NN}^* \end{bmatrix}\dot{\mathbf{q}} \tag{5.89}$$

$$= -\mathbf{K}_A^*\mathbf{q} - \mathbf{R}_A^*\dot{\mathbf{q}} \tag{5.90}$$

Therefore the coupled modal aeroelastic system is written in compact matrix form as:

$$\mathbf{M}^*\ddot{\mathbf{q}} + [\mathbf{R}^* + \mathbf{R}_A^*]\dot{\mathbf{q}} + [\mathbf{K}^* + \mathbf{K}_A^*]\mathbf{q} = \mathbf{0} \tag{5.91}$$

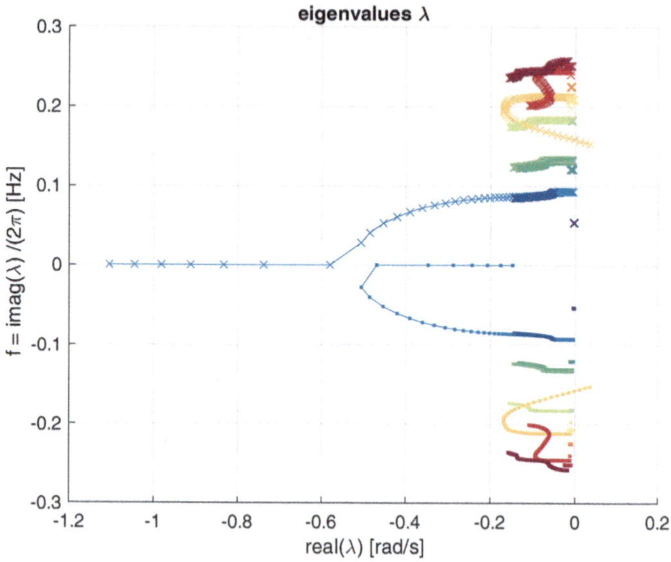

Fig. 5.70 Root locus

5.7.3 Stability Analysis

The stability analysis of Eq. 5.91 is performed with a classical generalized eigenvalue-eigenvector problem, where $\mathbf{q} = \boldsymbol{\psi} e^{\lambda t}$:

$$(\lambda^2 \mathbf{M}^* + \lambda[\mathbf{R}^* + \mathbf{R}_A^*] + [\mathbf{K}^* + \mathbf{K}_A^*])\boldsymbol{\psi} = \mathbf{0} \tag{5.92}$$

Either the quasi-steady or unsteady aeroelastic matrices are used, eigenvalue-eigenvector pairs $(\lambda, \boldsymbol{\psi})$ are computed and plotted as a function of the incoming wind speed U. A classical multi-modal root-locus as a function of U is shown in Fig. 5.70, where a suspension bridge stability is analyzed using 20 modes. It can be clearly seen that two complex-conjugate λ at a certain point reach a positive real part, i.e. instability.

The root locus is more clearly represented if we plot the modal frequencies and damping ratios as a function of U, as is is done in Figure 5.71. In this example it is clear that instability occurs at $U = 72$ m/s.

Looking at the eigenvector $\boldsymbol{\psi}$ associated to the unstable eigenvalue λ, it is possible to notice that mainly 2 modes are moving with the unstable eigenvalue (highlighted in red):

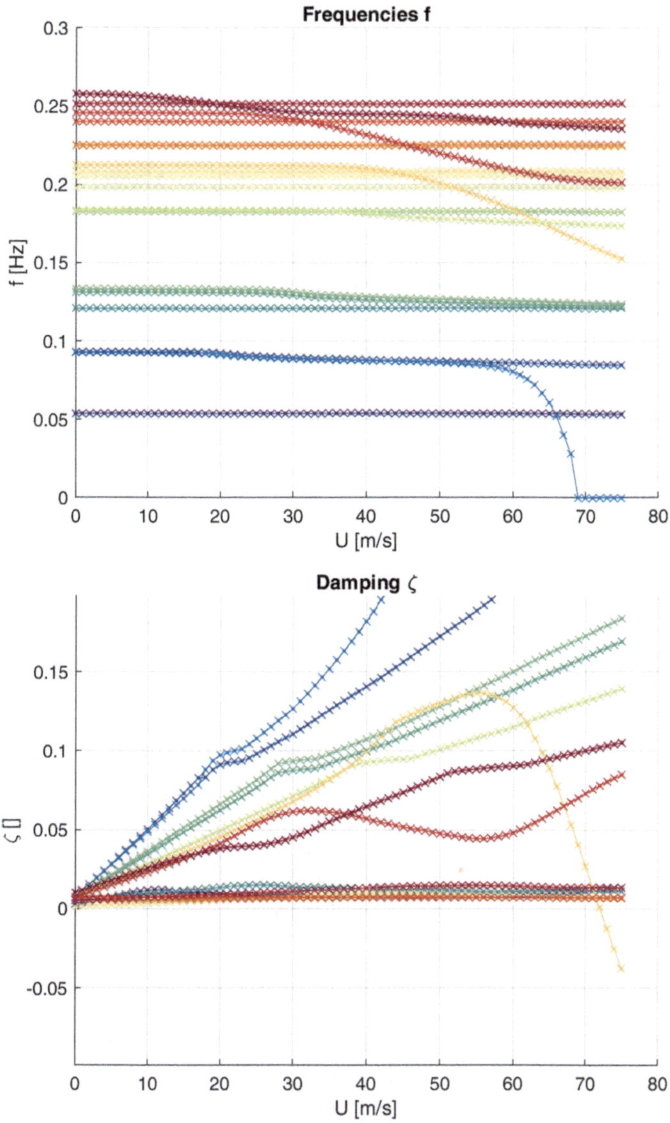

Fig. 5.71 Frequencies and damping as a function of U

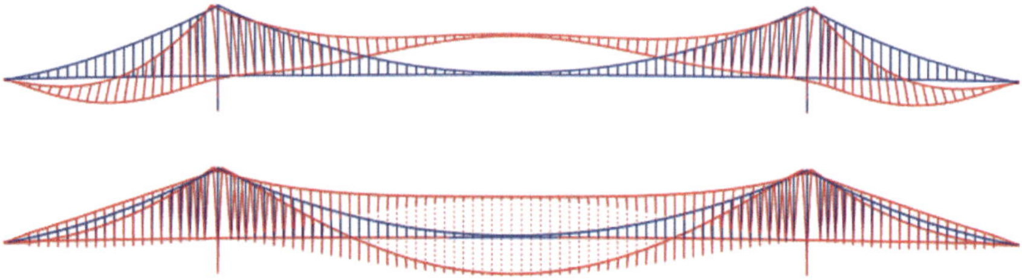

Fig. 5.72 First vertical and first torsional mode of a FE model of a suspension bridge

$$\|\boldsymbol{\psi}\| = \begin{bmatrix} 0.07 \\ 1.00 \\ 0.01 \\ 0.00 \\ 0.00 \\ 0.12 \\ 0.00 \\ 0.02 \\ 0.04 \\ 0.00 \\ 0.00 \\ 0.01 \\ 0.05 \\ 0.00 \\ 0.06 \\ 0.00 \\ 0.00 \\ 0.16 \\ 0.01 \\ 0.56 \end{bmatrix} \tag{5.93}$$

In this case the 2nd and the 20th elements of the eigenvector indicate that this two modes are coupled at instability (with relative amplitude 1 and 0.56). In this case the 2nd and the 20th modes are actually the first vertical and torsional modes, as shown in Fig. 5.72.

At the instability threshold, for the two complex conjugate $\lambda = i\omega$ and $\overline{\lambda} = -i\omega$, there are two complex conjugate eigenvectors $\boldsymbol{\psi}$ and $\overline{\boldsymbol{\psi}}$. So the solution related to the unstable eigenvalue is, in the time domain:

$$
\begin{aligned}
\mathbf{x}(t) &= \boldsymbol{\Phi}\mathbf{q}(t) && (5.94)\\
&= \boldsymbol{\Phi}(\alpha_1 \boldsymbol{\psi} e^{\lambda t} + \alpha_2 \overline{\boldsymbol{\psi}} e^{\overline{\lambda} t}) && (5.95)\\
&= \boldsymbol{\Phi}(\alpha_1 \|\boldsymbol{\psi}\| e^{i\boldsymbol{\varphi}} e^{\lambda t} + \alpha_2 \|\boldsymbol{\psi}\| e^{-i\boldsymbol{\varphi}} e^{\overline{\lambda} t}) && (5.96)\\
&= \boldsymbol{\Phi}\|\boldsymbol{\psi}\|(\alpha_1 e^{i\boldsymbol{\varphi}} e^{\lambda t} + \alpha_2 e^{-i\boldsymbol{\varphi}} e^{\overline{\lambda} t}) && (5.97)\\
&= \boldsymbol{\Phi}\|\boldsymbol{\psi}\|(\alpha_1 e^{i\boldsymbol{\varphi}} e^{i\omega t} + \alpha_2 e^{-i\boldsymbol{\varphi}} e^{-i\omega t}) && (5.98)\\
&= \boldsymbol{\Phi}\|\boldsymbol{\psi}\|(\alpha_1 e^{i(\omega t+\boldsymbol{\varphi})} + \alpha_2 e^{-i(\omega t+\boldsymbol{\varphi})}) && (5.99)\\
&= \boldsymbol{\Phi}\|\boldsymbol{\psi}\|(A_1 \cos(\omega t + \boldsymbol{\varphi}) + A_2 \sin(\omega t + \boldsymbol{\varphi})) && (5.100)\\
&= \boldsymbol{\Phi}\|\boldsymbol{\psi}\|(C \sin(\omega t + \beta + \boldsymbol{\varphi})) && (5.101)
\end{aligned}
$$

where the $\boldsymbol{\varphi}$ vector represents the phase shift between modes.

Using the previous example, naming $\boldsymbol{\phi}_z$ and $\boldsymbol{\phi}_\theta$ the 2nd and the 20th modes and assuming that the elements of $\boldsymbol{\psi}$ are null except for the 2nd and the 20th, the eigenvector is:

$$
\boldsymbol{\psi} = \begin{bmatrix} 0 \\ 1.00\, e^{i0} \\ 0 \\ 0 \\ 0 \\ 0 \\ 0 \\ 0 \\ 0 \\ 0 \\ 0 \\ 0 \\ 0 \\ 0 \\ 0 \\ 0 \\ 0 \\ 0 \\ 0 \\ 0.56\, e^{i2.8} \end{bmatrix} \qquad (5.102)
$$

thus the time solution is

$$
\begin{aligned}
\mathbf{x}(t) &= \boldsymbol{\Phi}\mathbf{q}(t) && (5.103)\\
&= \boldsymbol{\Phi}\|\boldsymbol{\psi}\|(C \sin(\omega t + \beta + \boldsymbol{\varphi})) && (5.104)\\
&= \boldsymbol{\phi}_z \cdot 1 \cdot (C \sin(\omega t + \beta + 0)) + \boldsymbol{\phi}_\theta \cdot 0.56 \cdot (C \sin(\omega t + \beta + 2.8)) && (5.105)
\end{aligned}
$$

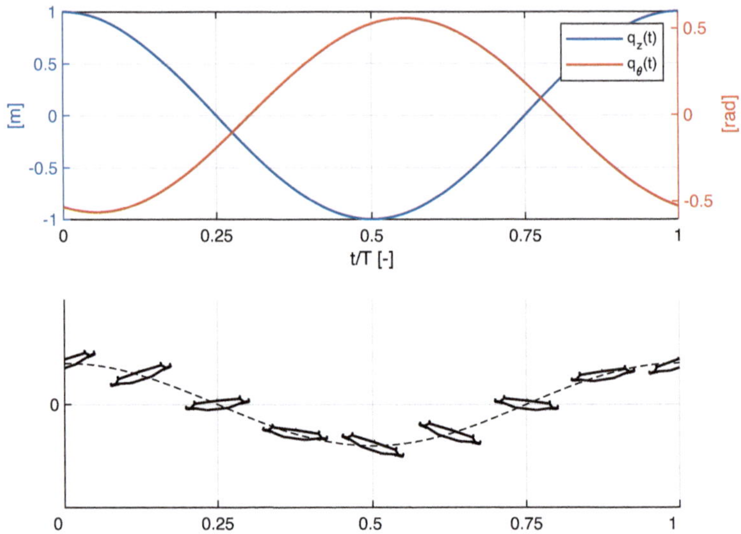

Fig. 5.73 Displacement of mid-span section at instability, with the typical flutter motion

which means that the two modes vertical-torsional modes are coupled and shifted approximately 160 deg (or 2.8 rad), nearly in opposite phase. Figure 5.73 represents graphically the time solution, for the mid-span deck section.

5.7.4 From Multi-modal Aeroelastic System to 2-DOF Sectional Model

If we consider only the first vertical and torsional modes (i.e. $N = 2$), flutter speed should be very well approximated, since we have seen from the previous analysis that the eigenvector at flutter consists mainly of these two modes.[5] This result is exploited in wind tunnel test to check stability using sectional models.

Let us consider pure modes, i.e. structurally uncoupled modes. This means that the vertical mode ($n = 1$) for each section will have only a vertical component:

$$\boldsymbol{\phi}_{i,1} = \begin{bmatrix} 0 \\ \phi_{z_i} \\ 0 \end{bmatrix}_1 \tag{5.106}$$

[5] This is not always the case: often multi-modal flutter occurs, i.e. more than two modes are aeroelastically coupled. Typically the torsional mode could be coupled to two or three vertical modes; considering only a bi-modal system could lead to inaccurate instability speed.

while the second mode ($n = 2$) for each section will have only a torsional component:

$$\boldsymbol{\phi}_{i,2} = \begin{bmatrix} 0 \\ 0 \\ \phi_{\theta_i} \end{bmatrix}_2 \tag{5.107}$$

$$Q_1 = -\sum_{n=1}^{2}\left(\sum_{i=1}^{N_s} \boldsymbol{\phi}_{i,1}^T \mathbf{K}_{A,i} \boldsymbol{\phi}_{i,n}\right) q_n - \sum_{n=1}^{2}\left(\sum_{i=1}^{N_s} \boldsymbol{\phi}_{i,1}^T \mathbf{R}_{A,i} \boldsymbol{\phi}_{i,n}\right) \dot{q}_n \tag{5.108}$$

$$= -\left(\sum_{i=1}^{N_s} \boldsymbol{\phi}_{i,1}^T \mathbf{K}_{A,i} \boldsymbol{\phi}_{i,1}\right) q_1 - \left(\sum_{i=1}^{N_s} \boldsymbol{\phi}_{i,1}^T \mathbf{K}_{A,i} \boldsymbol{\phi}_{i,2}\right) q_2 + \text{damping terms} \tag{5.109}$$

$$= -\left(\sum_{i=1}^{N_s} \begin{bmatrix} 0 \\ \phi_{z_i} \\ 0 \end{bmatrix}_1^T \mathbf{K}_{A,i} \begin{bmatrix} 0 \\ \phi_{z_i} \\ 0 \end{bmatrix}_1\right) q_1 - \left(\sum_{i=1}^{N_s} \begin{bmatrix} 0 \\ \phi_{z_i} \\ 0 \end{bmatrix}_1^T \mathbf{K}_{A,i} \begin{bmatrix} 0 \\ 0 \\ \phi_{\theta_i} \end{bmatrix}_2\right) q_2 + \text{damping terms} \tag{5.110}$$

$$= -\left(\sum_{i=1}^{N_s} \mathbf{K}_{A,i}^{(2,2)} \phi_{z_i,1}^2\right) q_1 - \left(\sum_{i=1}^{N_s} \mathbf{K}_{A,i}^{(2,3)} \phi_{z_i,1} \phi_{\theta_i,2}\right) q_2 + \text{damping terms} \tag{5.111}$$

where the notation $\mathbf{K}_A^{(r,c)}$ means the element in row r and column c of the 3×3 matrix $\mathbf{K}_A$.

Following analogous steps, for Q_2 we get:

$$Q_2 = -\sum_{n=1}^{2}\left(\sum_{i=1}^{N_s} \boldsymbol{\phi}_{i,2}^T \mathbf{K}_{A,i} \boldsymbol{\phi}_{i,n}\right) q_n - \sum_{n=1}^{2}\left(\sum_{i=1}^{N_s} \boldsymbol{\phi}_{i,2}^T \mathbf{R}_{A,i} \boldsymbol{\phi}_{i,n}\right) \dot{q}_n \tag{5.112}$$

$$= -\left(\sum_{i=1}^{N_s} \mathbf{K}_{A,i}^{(3,2)} \phi_{z_i,1} \phi_{\theta_i,2}\right) q_1 - \left(\sum_{i=1}^{N_s} \mathbf{K}_{A,i}^{(3,3)} \phi_{\theta_i,2}^2\right) q_2 + \text{damping terms} \tag{5.113}$$

Therefore the modal forcing $\mathbf{Q}$ is:

$$\mathbf{Q} = \begin{bmatrix} Q_1 \\ Q_2 \end{bmatrix} = -\begin{bmatrix} \sum_{i=1}^{N_s} \mathbf{K}_{A,i}^{(2,2)} \phi_{z_i,1}^2 & \sum_{i=1}^{N_s} \mathbf{K}_{A,i}^{(2,3)} \phi_{z_i,1} \phi_{\theta_i,2} \\ \sum_{i=1}^{N_s} \mathbf{K}_{A,i}^{(3,2)} \phi_{z_i,1} \phi_{\theta_i,2} & \sum_{i=1}^{N_s} \mathbf{K}_{A,i}^{(3,3)} \phi_{\theta_i,2}^2 \end{bmatrix} \begin{bmatrix} q_1 \\ q_2 \end{bmatrix} + \text{damping terms} \tag{5.114}$$

We introduce now another hypothesis: the aeroelastic matrices do not change among the different sections. This is in general true if: 1) the deck does not change geometry; 2) the mean static rotations of the sections (maximum at mid-span) do not influence much the values of the aeroelastic coefficients. This second condition should be verified accurately. Including this hypothesis leads to:

$$\mathbf{Q} = -\begin{bmatrix} \mathbf{K}_A^{(2,2)} \sum_i \phi_{z_i,1}^2 & \mathbf{K}_A^{(2,3)} \sum_i \phi_{z_i,1}\phi_{\theta_i,2} \\ \mathbf{K}_A^{(3,2)} \sum_i \phi_{z_i,1}\phi_{\theta_i,2} & \mathbf{K}_A^{(3,3)} \sum_i \phi_{\theta_i,2}^2 \end{bmatrix} \begin{bmatrix} q_1 \\ q_2 \end{bmatrix} + \text{damping terms} \tag{5.115}$$

The aeroelastic system in modal coordinates is therefore:

$$\begin{aligned} &\begin{bmatrix} m_1^* & \\ & m_2^* \end{bmatrix} \begin{bmatrix} \ddot{q}_1 \\ \ddot{q}_2 \end{bmatrix} + \begin{bmatrix} r_1^* & \\ & r_2^* \end{bmatrix} \begin{bmatrix} \dot{q}_1 \\ \dot{q}_2 \end{bmatrix} + \begin{bmatrix} k_1^* & \\ & k_2^* \end{bmatrix} \begin{bmatrix} q_1 \\ q_2 \end{bmatrix} \\ &= -\begin{bmatrix} \mathbf{K}_A^{(2,2)} \sum_i \phi_{z_i,1}^2 & \mathbf{K}_A^{(2,3)} \sum_i \phi_{z_i,1}\phi_{\theta_i,2} \\ \mathbf{K}_A^{(3,2)} \sum_i \phi_{z_i,1}\phi_{\theta_i,2} & \mathbf{K}_A^{(3,3)} \sum_i \phi_{\theta_i,2}^2 \end{bmatrix} \begin{bmatrix} q_1 \\ q_2 \end{bmatrix} \\ &\quad - \begin{bmatrix} \mathbf{R}_A^{(2,2)} \sum_i \phi_{z_i,1}^2 & \mathbf{R}_A^{(2,3)} \sum_i \phi_{z_i,1}\phi_{\theta_i,2} \\ \mathbf{R}_A^{(3,2)} \sum_i \phi_{z_i,1}\phi_{\theta_i,2} & \mathbf{R}_A^{(3,3)} \sum_i \phi_{\theta_i,2}^2 \end{bmatrix} \begin{bmatrix} \dot{q}_1 \\ \dot{q}_2 \end{bmatrix} \end{aligned} \tag{5.116}$$

where if we divide the first line by $\sum_i \phi_{z_i,1}^2$ and the second one by $\sum_i \phi_{\theta_i,1}^2$, we get the following result:

$$\begin{aligned} &\begin{bmatrix} \frac{m_1^*}{\sum_i \phi_{z_i,1}^2} & \\ & \frac{m_2^*}{\sum_i \phi_{\theta_i,2}^2} \end{bmatrix} \begin{bmatrix} \ddot{q}_1 \\ \ddot{q}_2 \end{bmatrix} + \begin{bmatrix} \frac{r_1^*}{\sum_i \phi_{z_i,1}^2} & \\ & \frac{r_2^*}{\sum_i \phi_{\theta_i,2}^2} \end{bmatrix} \begin{bmatrix} \dot{q}_1 \\ \dot{q}_2 \end{bmatrix} + \begin{bmatrix} \frac{k_1^*}{\sum_i \phi_{z_i,1}^2} & \\ & \frac{k_2^*}{\sum_i \phi_{\theta_i,2}^2} \end{bmatrix} \begin{bmatrix} q_1 \\ q_2 \end{bmatrix} \\ &= -\begin{bmatrix} \mathbf{K}_A^{(2,2)} & \mathbf{K}_A^{(2,3)} \sum_i \frac{\phi_{z_i,1}\phi_{\theta_i,2}}{\phi_{z_i,1}^2} \\ \mathbf{K}_A^{(3,2)} \sum_i \frac{\phi_{z_i,1}\phi_{\theta_i,2}}{\phi_{\theta_i,2}^2} & \mathbf{K}_A^{(3,3)} \end{bmatrix} \begin{bmatrix} q_1 \\ q_2 \end{bmatrix} \\ &\quad - \begin{bmatrix} \mathbf{R}_A^{(2,2)} & \mathbf{R}_A^{(2,3)} \sum_i \frac{\phi_{z_i,1}\phi_{\theta_i,2}}{\phi_{z_i,1}^2} \\ \mathbf{R}_A^{(3,2)} \sum_i \frac{\phi_{z_i,1}\phi_{\theta_i,2}}{\phi_{\theta_i,2}^2} & \mathbf{R}_A^{(3,3)} \end{bmatrix} \begin{bmatrix} \dot{q}_1 \\ \dot{q}_2 \end{bmatrix} \end{aligned} \tag{5.117}$$

Now, let us consider the aeroelastic equation of motion for a sectional model, long L, with mass per unit length m and mass moment of inertia per unit length J. The total mass will be $m_z = mL$, while the total polar moment of inertia $m_\theta = JL$. The equation of motion are:

$$
\begin{bmatrix} m_z & \\ & m_\theta \end{bmatrix}\begin{bmatrix} \ddot{z} \\ \ddot{\theta} \end{bmatrix} + \begin{bmatrix} r_z & \\ & r_\theta \end{bmatrix}\begin{bmatrix} \dot{z} \\ \dot{\theta} \end{bmatrix} + \begin{bmatrix} k_z & \\ & k_\theta \end{bmatrix}\begin{bmatrix} z \\ \theta \end{bmatrix}
$$
$$
= -\begin{bmatrix} \mathbf{K}_A^{(2,2)} & \mathbf{K}_A^{(2,3)} \\ \mathbf{K}_A^{(3,2)} & \mathbf{K}_A^{(3,3)} \end{bmatrix}\begin{bmatrix} z \\ \theta \end{bmatrix}
- \begin{bmatrix} \mathbf{R}_A^{(2,2)} & \mathbf{R}_A^{(2,3)} \\ \mathbf{R}_A^{(3,2)} & \mathbf{R}_A^{(3,3)} \end{bmatrix}\begin{bmatrix} \dot{z} \\ \dot{\theta} \end{bmatrix} \tag{5.118}
$$

In order to have Eq. 5.117 equal to Eq. 5.118, i.e. the sectional model is representative of the aeroelastic behavior of the bi-modal full bridge, the following relationships must hold. Let us start from the first line:

$$
m_z = \frac{m_1^*}{\sum_i \phi_{z_i,1}^2} \Longrightarrow m = \frac{m_1^*}{L\sum_i \phi_{z_i,1}^2} \tag{5.119}
$$

$$
k_z = \omega_z^2 m_z = \frac{k_1^*}{\sum_i \phi_{z_i,1}^2} = \frac{\omega_1^2 m_1^*}{\sum_i \phi_{z_i,1}^2} = \omega_1^2 m_z \Longrightarrow \omega_z = \omega_1 \tag{5.120}
$$

$$
r_z = 2\zeta_z\omega_z m_z = \frac{r_1^*}{\sum_i \phi_{z_i,1}^2} = \frac{2\zeta_1\omega_1 m_1^*}{\sum_i \phi_{z_i,1}^2} = 2\zeta_1\omega_z m_z \Longrightarrow \zeta_z = \zeta_1 \tag{5.121}
$$

$$
\sum_i \frac{\phi_{z_i,1}\phi_{\theta_i,2}}{\phi_{z_i,1}^2} = 1 \Longrightarrow \phi_{z_i,1} = \phi_{\theta_i,2} \tag{5.122}
$$

Analogously, for the second line:

$$
m_\theta = \frac{m_2^*}{\sum_i \phi_{\theta_i,2}^2} \Longrightarrow J = \frac{m_2^*}{L\sum_i \phi_{\theta_i,2}^2} \tag{5.123}
$$

$$
k_\theta = \omega_\theta^2 m_\theta = \frac{k_2^*}{\sum_i \phi_{\theta_i,2}^2} = \frac{\omega_2^2 m_2^*}{\sum_i \phi_{\theta_i,2}^2} = \omega_2^2 m_\theta \Longrightarrow \omega_\theta = \omega_2 \tag{5.124}
$$

$$
r_\theta = 2\zeta_\theta\omega_\theta m_\theta = \frac{r_2^*}{\sum_i \phi_{\theta_i,2}^2} = \frac{2\zeta_2\omega_2 m_2^*}{\sum_i \phi_{\theta_i,2}^2} = 2\zeta_2\omega_\theta m_\theta \Longrightarrow \zeta_\theta = \zeta_2 \tag{5.125}
$$

$$
\sum_i \frac{\phi_{z_i,1}\phi_{\theta_i,2}}{\phi_{\theta_i,2}^2} = 1 \Longrightarrow \phi_{z_i,1} = \phi_{\theta_i,2} \tag{5.126}
$$

Thus, if Eqs. 5.119–5.126 are satisfied, the sectional model is representative of the full bridge vertical-torsional modes. Obviously, the previous values are scaled in wind tunnel sectional models, according to the scaling law chosen for the tests.

5.7.4.1 Remarks

To better understand the previous results, we can model a simplified bridge as a simply uniform beam, long L and mass and inertia per unit length, $\tilde{m}$ and $\tilde{J}$. Boundary conditions are pinned-pinned for vertical displacement, while fixed-fixed for rotation.

In this case, the mode shapes are simple sine functions:

$$\phi_{z,1} = \sin\left(\frac{\pi}{L}x\right) \tag{5.127}$$

$$\int_0^L \phi_{z,1}^2 \mathrm{d}x = \int_0^L \sin^2\left(\frac{\pi}{L}x\right)\mathrm{d}x = L\int_0^1 \sin^2(\pi\xi)\, d\xi = L\sum_i \phi_{z_i,1}^2 = \frac{L}{2} \tag{5.128}$$

$$m_1^* = \int_0^L \tilde{m}\phi_{z,1}^2 \mathrm{d}x = \tilde{m}\int_0^L \phi_{z,1}^2 \mathrm{d}x = \frac{\tilde{m}L}{2} \tag{5.129}$$

$$\phi_{\theta,2} = \sin\left(\frac{\pi}{L}x\right) \tag{5.130}$$

$$\int_0^L \phi_{\theta,2}^2 \mathrm{d}x = \int_0^L \sin^2\left(\frac{\pi}{L}x\right)\mathrm{d}x = L\int_0^1 \sin^2(\pi\xi)\, d\xi = L\sum_i \phi_{\theta_i,2}^2 = \frac{L}{2} \tag{5.131}$$

$$m_2^* = \int_0^L \tilde{J}\phi_{\theta,2}^2 \mathrm{d}x = \tilde{J}\int_0^L \phi_{\theta,2}^2 \mathrm{d}x = \frac{\tilde{J}L}{2} \tag{5.132}$$

Moreover, the cross-integrals are unitary, i.e.:

$$\frac{\int_0^L \phi_{z,1}\phi_{\theta,2}\mathrm{d}x}{\int_0^L \phi_{z,1}^2 \mathrm{d}x} = \sum_i \frac{\phi_{z_i,1}\phi_{\theta_i,2}}{\phi_{z_i,1}^2} = 1 \tag{5.133}$$

$$\frac{\int_0^L \phi_{z,1}\phi_{\theta,2}\mathrm{d}x}{\int_0^L \phi_{\theta,2}^2 \mathrm{d}x} = \sum_i \frac{\phi_{z_i,1}\phi_{\theta_i,2}}{\phi_{\theta_i,1}^2} = 1 \tag{5.134}$$

Applying Eq. 5.119 we get the significant result

$$m = \frac{m_1^*}{L\sum_i \phi_{z_i,1}^2} = \frac{m_1^*}{L/2} = \tilde{m} \tag{5.135}$$

which means that the mass per unit length of the sectional model should be equal to that of the bridge deck. In actual bridges, the mass $m > \tilde{m}$ since in the modal mass m^* is composed not only by deck inertia, but also by cables and towers, as it can be seen in Fig. 5.72, where these elements clearly participate in the mode shape. Analogous results are obtained for the moment of inertia per unit length.

An additional remark is that if the mode shapes are orthogonal, there cannot be aeroelastic coupling. As an example, consider if we use the second vertical mode instead of the first one. The mode shape would be with half wavelength, namely:

$$\phi_{z,1} = \sin\left(\frac{2\pi}{L}x\right) \tag{5.136}$$

This would lead to a major difference:

$$\int_0^L \phi_{z,1}\phi_{\theta,2}\mathrm{d}x = \int_0^L \sin\left(\frac{2\pi}{L}x\right)\sin\left(\frac{\pi}{L}x\right)\mathrm{d}x = L\sum_i \phi_{z_i,1}\phi_{\theta_i,2} = 0 \tag{5.137}$$

This means that the extra-diagonal terms in the aeroelastic modal Eqs. 5.117 would be null, leading to two independent equation of motion for the two modes, where only direct aeroelastic terms can lead to 1-DOF instability.

References

Argentini T, Pagani A, Rocchi D, Zasso A (2014) Monte Carlo analysis of total damping and flutter speed of a long span bridge: effects of structural and aerodynamic uncertainties. J Wind Eng Ind Aerodyn 128:90–104. https://doi.org/10.1016/j.jweia.2014.02.010

Brancaleoni F, Diana G, Faccioli E, Fiammenghi G, Firth PT, Gimsing NJ et al (2009) The Messina Strait Bridge: a challenge and a dream. CRC Press, Boca Raton, FL, USA

Caracoglia L, Jones NP (2003) Time domain vs. frequency domain characterization of aeroelastic forces for bridge deck sections. J Wind Eng Ind Aerodyn 91(3), 371–402

Chen X, Kareem A (2001a) Aeroelastic analysis of bridges under multicorrelated winds: Integrated state-space approach. J Eng Mech 127(11):1124–1134

Chen X, Kareem A (2001b) Nonlinear response analysis of long-span bridges under turbulent winds. J Wind Eng Ind Aerodyn 89(14–15):1335–1350

Chen X, Kareem A, Matsumoto M (2001) Multimode coupled flutter and buffeting analysis of long span bridges. J Wind Eng Ind Aerodyn 89:649–664. https://doi.org/10.1016/S0167-6105(01)00064-2, http://www.sciencedirect.com/science/article/pii/S0167610501000642

Diana G, Cheli F (2010) Dinamica dei sistemi meccanici. Polipress

Diana G, Falco M, Bruni S, Cigada A, Larose G, Damsgaard A, Collina A (1995) Comparisons between wind tunnel tests on a full aeroelastic model of the proposed bridge over stretto di Messina and numerical results. J Wind Eng Ind Aerodyn 54–55:101–113

Diana G, Bruni S, Rocchi D (2005) A numerical and experimental investigation on aerodynamic non linearities in bridge response to turbulent wind. In: Proceedings of the EACWE4. Prague

Diana G, Resta F, Rocchi D (2008) A new numerical approach to reproduce bridge aerodynamic non-linearities in time domain. J Wind Eng Ind Aerodyn 96(10–11):1871–1884

Diana G, Rocchi D, Argentini T, Muggiasca S (2010) Aerodynamic instability of a bridge deck section model: linear and nonlinear approach to force modeling. J Wind Eng Ind Aerodyn 98(6–7):363–374

Ding Q, Zhu L, Xiang H (2006) Simulation of stationary gaussian stochastic wind velocity field. Wind Struct Int J 9(3):231–243

Frandsen JB (2001) Simultaneous pressures and accelerations measured full-scale on the Great Belt East suspension bridge. J Wind Eng Ind Aerodyn 89(1):95–129. https://doi.org/10.1016/S0167-6105(00)00059-3

Ge Y, Zhao L, Cao J (2022) Case study of vortex-induced vibration and mitigation mechanism for a long-span suspension bridge. J Wind Eng Ind Aerodyn 220:104866. https://doi.org/10.1016/j.jweia.2021.104866

Jain A, Jones NP, Scanlan RH (1996) Coupled aeroelastic and aerodynamic response analysis of long-span bridges. J Wind Eng Ind Aerodyn 60(1–3):69–80

Katsuchi H, Jones NP, Scanlan RH (1999) Multimode coupled flutter and buffeting analysis of the Akashi-Kaikyo bridge. J Struct Eng 125(1):60–69

Minh NN, Yamada H, Miyata T, Katsuchi H (2000) Aeroelastic complex mode analysis for coupled gust response of the Akashi Kaikyo bridge model. J Wind Eng Ind Aerodyn 88(2–3):307–324

Niu HW, Chen ZQ, Liu MG, Han Y, Hua XG (2011) Development of the 3-DOF forced vibration device to measure the aerodynamic forces on section models. In: Proceedings of the ICWE. Amsterdam

Simiu E, Scanlan RH (1996) Wind effects on structures fundamentals and applications to design. Wiley, New York

Wu T, Kareem A (2011) Modeling hysteretic nonlinear behavior of bridge aerodynamics via cellular automata nested neural network. J Wind Eng Ind Aerodyn 99(4):378–388

GPSR Compliance

The European Union's (EU) General Product Safety Regulation (GPSR) is a set of rules that requires consumer products to be safe and our obligations to ensure this.

If you have any concerns about our products, you can contact us on ProductSafety@springernature.com

In case Publisher is established outside the EU, the EU authorized representative is:

Springer Nature Customer Service Center GmbH
Europaplatz 3
69115 Heidelberg, Germany

Batch number: 10267894

Printed by Printforce, the Netherlands